Satellite Networking

Satellite Networking

Principles and Protocols

Zhili Sun

University of Surrey, UK

John Wiley & Sons, Ltd

Other Wiley Editorial Offices

John Wiley & Sons Inc., 111 River Street, Hoboken, NJ 07030, USA

Jossey-Bass, 989 Market Street, San Francisco, CA 94103-1741, USA

Wiley-VCH Verlag GmbH, Boschstr. 12, D-69469 Weinheim, Germany

John Wiley & Sons Australia Ltd, 42 McDougall Street, Milton, Queensland 4064, Australia

John Wiley & Sons (Asia) Pte Ltd, 2 Clementi Loop #02-01, Jin Xing Distripark, Singapore 129809

John Wiley & Sons Canada Ltd, 22 Worcester Road, Etobicoke, Ontario, Canada M9W 1L1

Wiley also publishes its books in a variety of electronic formats. Some content that appears
in print may not be available in electronic books.

Library of Congress Cataloging in Publication Data

Sun, Zhili.
   Satellite networking principles and protocols / Zhili Sun.
      p.   cm.
   Includes bibliographical references.
   ISBN-10: 0-470-87027-3
   ISBN-13: 978-0-470-87027-3
   1. Artificial satellites in telecommunication.   2. Computer network protocols.
   3. Internetworking (Telecommunication)   I. Title.
   TK5104.S78 2005
   621.382′5′028546—dc22

                                                2005012260

British Library Cataloguing in Publication Data

A catalogue record for this book is available from the British Library

ISBN-13 978-0-470-87027-3 (HB)
ISBN-10 0-470-87027-3 (HB)

Typeset in 10/12pt Times by Integra Software Services Pvt. Ltd, Pondicherry, India.
Printed and bound in Great Britain by Antony Rowe Ltd, Chippenham, Wiltshire.
This book is printed on acid-free paper responsibly manufactured from sustainable forestry
in which at least two trees are planted for each one used for paper production.

This book is dedicated to the memory of my grandparents

To my parents

To my wife

Contents

List of Tables

List of Figures

Preface

Satellite has played an important role in telephony communication and TV broadcasting services since the birth of telecommunication satellites. It is less known that satellite also plays an important role in broadband and Internet services and will continue to play an important role in the future generation networks. This is due to the satellite characteristics that make a niche position for satellites in the global network infrastructure (GNI).

Satellite networking is a special and important topic together with other networking technologies in recent years. Due to the nature of satellite links (long propagation delay, relative high bit error rate and limited bandwidth in comparison with terrestrial links, particularly optical links), some standard network protocols do not perform well and have to be adapted to support efficient connection over satellite. Satellite orbit directly affects the link characteristics and has a significant impact in satellite network design.

It is the ultimate goal of satellite networking to support the many different applications and services available in terrestrial networks. These applications and services generate different types of traffic having different requirements in terms of network resources and quality of service (QoS), particularly the recent development of integration of telecommunication, broadcast and computer networks and integration of telephone, TV, computer and global positioning system (GPS) terminals.

Satellite networking has evolved significantly since the first telecommunications satellite, from telephone and broadcast to broadband and Internet networks. It has adapted during the advancement for ISDN, ATM, Internet, digital broadcast, etc. The evolution has also been reflected in research and development, including the recent studies of onboard processing, onboard switching and onboard IP routing. There are also new developments and new issues in satellite networking such as resource management, security and quality of service, new services and applications including VoIP, multicast, video conference, DVB-S, DVB-RCS and IPv6 over satellite. There are always many practical constraints, such as cost, complexity, technologies and efficiency of space and ground segments in design, implementation and operation. Often trade-offs have to be made to achieve an optimal solution.

The technology development has stabilised and matured in satellite communication systems so that satellite networks can be addressed as an integral part of GNI rather than as a complicated system itself. Therefore, it is also a good time to publish a book to cover all these important and relevant developments.

This book is written based on my lecture notes and teaching experiences on the MSc in satellite communications, MSc in communications and software, BEng and MEng in electronic engineering, and industrial short courses in satellite communication, at the University of Surrey, and the MSc in computer and communications networks, at the Institute of National Telecommunications (INT), France. Therefore the book is intended to be written for MSc courses and undergraduate final stage in the areas related to satellite communications and networks.

The book also takes information from publications in international conference and journals produced by the research group and research community in general, from reports of a large number of research projects funded by the European Framework Programmes, UK Research Council and European Space Agency (ESA) and industries, and from PhD theses. Therefore, the book is also intended as a reference book for research students, professional engineers, satellite equipment manufacturers, satellite operators, telecom and network operators, network designers and Internet service providers.

This book covers satellite networking as a separate discipline, as well as an integrated part of the global network infrastructure. Unlike traditional satellite books, its emphasis is more on network aspects, network services and applications, and network principles and protocols, awareness of the characteristics of satellite networks and internetworking between the satellite and terrestrial networks. This book covers these topics with the following unique features by:

- Providing a balanced introduction of the principles and protocols of satellite communications networks, telecommunications networks, broadband networks and Internet networks to bridge the gaps between satellite and terrestrial networks.
- Following the time lines of technology development from analogue, to digital networks and to packet networks.
- Covering the developments of three major protocol reference models: ISO open systems interconnection (OSI), ITU-T asynchronous transfer mode (ATM) and IETF Internet protocol (IP) reference models.
- Focusing on satellite specific issues on networking QoS, security, performance and internetworking with terrestrial networks.
- Following the layering principle of network protocols and addressing the network issues from physical layer and link layer, to network and transport layer, and finally to application layers in the context of both satellite networks and terrestrial networks.
- Discussing the evolutionary development of PDH over satellite, SDH over satellite, N-ISDN over satellite, ATM and B-ISDN over satellite.
- Covering in-depth the developments of recent years on Internet protocol (IP) over satellite, IP multicast, TCP enhancement over satellite, VoIP and DVB over satellite (DVB-S and DVB-RCS) from different viewpoints including satellite centric, network centric and protocol centric.
- Providing insightful discussions on new services and applications, traffic modelling and traffic engineering, MPLS and QoS provisions.
- Introducing IPv6 and IPv6 over satellite using tunnelling and translation techniques, and important issues in the future development and convergence of satellite networking towards the next generation Internet (NGI).

The different views of the global networks reflect the logic behind this book. This will help with understanding the seamless integration between satellite and terrestrial networks and to achieve a common understanding of different network protocols and technologies, and the importance of pushing the complications to the network edges and services and applications into the end systems (client and server).

Any new book is an experiment, and this is certainly true here. Due to the limitation of my knowledge, continuous development of the technologies, the limited time and space available for the book, I may not be able to cover all the important topics in detail. The importance of fundamental concepts and principles for satellite networking and the role satellite plays in the GNI can never be overemphasised. Readers who wish to gain further details on some of the relevant topics from books written by other well-known authors are directed to the further reading sections at the end of each chapter.

As an extra resource for lecturers and instructors, this book has a companion website where a solutions manual and electronic versions of the figures are available. Please go to www.wiley.com/go/sun.

It is my first time of writing a complete book in a short period with full academic teaching and research duties; it is inevitable that I may have made mistakes of different types. I welcome feedback and comments from all the readers, but am especially keen to receive the following information: (1) any factual error in citation, attribution or interpretation; (2) recommendations concerning topics to include or delete; (3) how this book can best be used in academic and professional training courses as a text book or reference; and (4) information concerning tables, figures, equations, derivations, or ways of presentation and organisation would also be useful.

Acknowledgements

Writing this book has been a great challenge and also a learning experience. It has been impossible to complete the book without help and support from many people. Luckily, I have had the opportunities of working with many great scientists and researchers from research institutes, companies and universities throughout the world. They have contributed to my lecture notes and publications in journals, conferences, book chapters and international standards, and have hence enriched the contents of this book. All errors in the book are mine alone.

I would like to take this opportunity to thank all colleagues, friends and members of my family who helped me in many different ways to make this book possible. First, I would like to thank Professor B.G. Evans who has supported research in satellite networking since 1993 when I first joined the research team working in the CATALYST project led by Alcatel Space Industries France within the European Research in Advance Communications in Europe (RACE) programme to develop the first satellite ATM demonstrator to study the capability of satellite supporting broadband communications. He also provided valuable advice and comments on the book.

I would also like to thank the European Framework Programmes (FP) for providing over €3 million to my research at the University of Surrey for more than 12 different research projects over the decade. I would like to thank the project coordinators and managers who have invited me as a principal investigator leading a team representing the University of Surrey as a partner of consortia. In addition to many deliverables, the projects also produced a large number of publications in international journals, conferences and book chapters and contributions to international standards including ITU-T, ETSI and IETF.

These projects include: the European Advanced Communications and Technologies (ACTS) THESEUS project led by L3S in France together with the Paris, Brussels and Valence stock exchanges to study terminal at high speed for European stock exchanges demonstrations via advanced satellite links and terrestrial networks across Europe; the European ESPRIT COPARIS project led by Siemens in Germany to develop new chip and embedded system methods for ISDN high speed access interfaces; the European ESPRIT Broadband Integrated Satellite Network Traffic Evaluator (BISANTE) project led by Thales Group in France to study broadband traffic over satellite networks using simulation techniques and simulation models of satellite networks to evaluate multimedia traffic and its QoS; the European Trans-European Network (TEN) telecommunication programme (VIP-TEN) project led by Alenia Aerospzio in Italy to study QoS of IP telephony over satellite for trans-European networks using satellite links; the FP5 IST GEOCAST project led by

Alcatel Space Industries in France to investigate IP multicast over GEO satellite; the FP5 IST ICEBERGS project led by Telefonica in Spain to study IP conferencing with broadband multimedia over geostationary satellites; and the FP5 GROWTH programme ASP-NET project to study application service providers networks led by Archimedia in Greece. I would like to thank all members of the projects as colleagues at a professional level and as friends at a personal level.

I would also like to thank the continued support of the European Commission on the research in satellite networking. From 2004, the EU FP6 Specific Targeted Research Project (STRP) project 'SATLIFE – Satellite Access Technologies leading to improvements for Europe' led by Hispasat in Spain; the EU 6th Framework Network of Excellence (NoE) EuroNGI – European next generation Internet led by GET-Telecom in France; and FP6 NoE SatNex project on Satellite Communications Network of Excellence, led by DLR in Germany. I would also like to thank the European Space Agency (ESA) for the support of project 'Secure IP multicast over satellite' led by LogicaGMC in UK to study IP multicast security over satellite. Thanks to all members of the projects. I would also like to acknowledge the support from the EPSRC to the new joint project between the University of Aberdeen and University of Surrey to study secure reliable multicast protocols over satellite.

Particularly, I would like to thank some individual colleagues and friends including Professor M. Becker and Dr M. Marot of INT France, Dr R. Dhaou of ENSEE, Professor G. Maral and Professor M. Bousquet of ENST, Professor D. Kofman of GET-Telecom, Mr L. Claverotte, Mr M. Mazzella and Mr R. Mort of Alcatel Space Indutries, Dr R. Foka of Thales Group and Dr J. Robert of Franch Telecom SA in France, Professor E. Lutz of DLR, Professor P. Kuehn of University of Stuttgart and Professor K. Tutschku of University of Würzburg in Germany, Professor G. Corazza of University of Bologna in Italy, Professor G. Bi of National Technical University in Singapore, Professor Belén Carro of University of Valladolid, Dr A. Sánchez of Telefonica, Mr Juan Ramón López Caravantes of Hispasat and Mr R. Rey Gomez of Alcatel in Spain, Professor G. Haring and Dr Hlavacs of University of Vienna, and Professor G. Kotsis of Johannes Kepler University Linz in Austria, Dr G. Fairhurst of University of Aberdeen, Professor L. Cuthbert, Professor J. Pitts, Dr C. Phillips and Dr J. Shormmans of Queen Mary University of London, Professors D. Kouvatsos, Dr I. Awan, Professor R. Sheriff and Dr H. Fun of University of Bradford, Dr J. Wakeling and Dr M. Fitch of BT Satellite Systems, Dr T. Ors of Intelsat USA, Mr F. Zeppenfeldt and Mr R. Donadio of ESA, Mr P. Jauhiainen, Mr B. Barini and Mr P. De Sousa of the European Commission, and Mr C. Dvorak of AT&T Labs.

In the CCSR, I would like to thank all the members of the research team, in particular the current members: Dr H. Cruickshank, Dr M. Howarth, Dr D. He, Dr L. Fan, Dr K. Narenthiran, Dr V. Kueh, Mr S. Iyngar, Mr L. Liang, Mr B. Zhou, Mr Z. Luo and Mr W. Ng. I would also like to thank many former research fellows and PhDs and all members of the academic and support staff.

I would like to dedicate this book to my grandparents. I would like to thank my parents for their love and support. Finally, I would like to thank the rest of my family, in particular, my wife, for their love and support.

Zhili Sun

1

Introduction

This chapter aims to introduce the basic concepts of satellite networking including applications and services, circuit and packet switching, broadband networks, network protocols and reference models, characteristics of satellite networks, internetworking between satellite and terrestrial networks and convergence of network technologies and protocols. When you have completed this chapter, you should be able to:

- Understand the concepts of satellite networks and internetworking with terrestrial networks.
- Know the different satellite services, network services and quality of service (QoS).
- Appreciate the differences between satellite networking and terrestrial networking issues.
- Describe the functions of network user terminals and satellite user earth terminals and gateway earth terminals.
- Know the basic principles of protocols and the ISO reference model.
- Know the basic ATM reference model.
- Know the basic Internet TCP/IP protocol suite.
- Understand the basic concepts of multiplexing and multiple accessing.
- Understand the basic switching concepts including circuit switching, virtual circuit switching and routeing.
- Understand the evolution process and convergence of network technologies and protocols.

1.1 Applications and services of satellite networks

Satellites are manmade stars in the sky, and are often mistaken for real stars. To many people, they are full of mystery. Scientists and engineers love to give life to them by calling them birds – like birds, they fly where other creatures can only dream. They watch the earth from the sky, help us to find our way around the world, carry our telephone calls, emails

---

Satellite Networking: Principles and Protocols   Zhili Sun
© 2005 John Wiley & Sons, Ltd

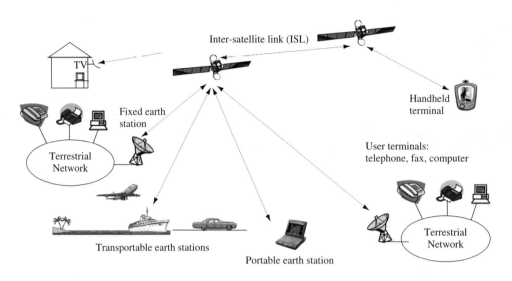

Figure 1.1   Typical applications and services of satellite networking

and web pages, and relay TV programmes across the sky. Actually the altitudes of satellites are far beyond the reach of any real bird. When satellites are used for networking, their high altitude enables them to play a unique role in the global network infrastructure (GNI).

Satellite networking is an expanding field, which has developed significantly since the birth of the first telecommunication satellite, from traditional telephony and TV broadcast services to modern broadband and Internet networks and digital satellite broadcasts. Many of the technological advances in networking areas are centred on satellite networking. With increasing bandwidth and mobility demands in the horizon, satellite is a logical option to provide greater bandwidth with global coverage beyond the reach of terrestrial networks, and shows great promise for the future. With the development of networking technologies, satellite networks are becoming more and more integrated into the GNI. Therefore, internet-working with terrestrial networks and protocols is an important part of satellite networking.

The ultimate goal of satellite networking is to provide services and applications. User terminals provide services and applications directly to users. The network provides trans-portation services to carry information between users for a certain distance. Figure 1.1 illustrates a typical satellite network configuration consisting of terrestrial networks, satellites with an inter-satellite link (ISL), fixed earth stations, transportable earth stations, portable and handheld terminals, and user terminals connecting to satellite links directly or through terrestrial networks.

1.1.1 Roles of satellite networks

In terrestrial networks, many links and nodes are needed to reach long distances and cover wide areas. They are organised to achieve economical maintenance and operation of the networks. The nature of satellites makes them fundamentally different from terrestrial networks in terms of distances, shared bandwidth resources, transmission technologies, design, development and operation, and costs and needs of users.

Functionally, satellite networks can provide direct connections among user terminals, connections for terminals to access terrestrial networks, and connections between terrestrial networks. The user terminals provide services and applications to people, which are often independent from satellite networks, i.e. the same terminal can be used to access satellite networks as well as terrestrial networks. The satellite terminals, also called earth stations, and are the earth segment of the satellite networks, providing access points to the satellite networks for user terminals via the user earth station (UES) and for terrestrial networks via the gateway earth station (GES). The satellite is the core of satellite networks and also the centre of the networks in terms of both functions and physical connections. Figure 1.2 illustrates the relationship between user terminal, terrestrial network and satellite network.

Typically, satellite networks consist of satellites interconnecting a few large GES and many small UES. The small GES are used for direct access by user terminals and the large UES for connecting terrestrial networks. The satellite UES and GES define the boundary of the satellite network. Like other types of networks, users access satellite networks through the boundary. For mobile and transportable terminals, the functions of user terminal and satellite UES are integrated into a single unit, but for transportable terminals their antennas are distinguishably visible.

The most important roles of satellite networks are to provide access by user terminals and to internetwork with terrestrial networks so that the applications and services provided by terrestrial networks such as telephony, television, broadband access and Internet connections can be extended to places where cable and terrestrial radio cannot economically be installed and maintained. In addition, satellite networks can also bring these services and applications to ships, aircraft, vehicles, space and places beyond the reach of terrestrial networks. Satellites also play important roles in military, meteorology, global positioning systems (GPS), observation of environments, private data and communication services, and future development of new services and applications for immediate global coverage such as

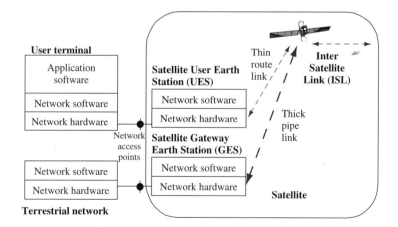

Figure 1.2 Functional relationships of user terminal, terrestrial network and satellite network

broadband network, and new generations of mobile networks and digital broadcast services worldwide.

1.1.2 Network software and hardware

In terms of implementation, the user terminal consists of network hardware and software and application software. The network software and hardware provide functions and mechanisms to send information in correct formats and to use the correct protocols at an appropriate network access point. They also receive information from the access point.

Network hardware provides signal transmission making efficient and cost-effective use of bandwidth resources and transmission technologies. Naturally, a radio link is used to ease mobility of the user terminals associated with access links; and high-capacity optical fibre is used for backbone connections.

With the advance of digital signal processing (DSP), traditional hardware implementations are being replaced more and more by software to increase the flexibility of reconfiguration, hence reducing costs. Therefore the proportion of implementation becomes more and more in software and less and less in hardware. Many hardware implementations are first implemented and emulated in software, though hardware is the foundation of any system implementation.

For example, traditional telephone networks are mainly in hardware; and modern telephone networks, computer and data networks and the Internet are mainly in software.

1.1.3 Satellite network interfaces

Typically, satellite networks have two types of external interfaces: one is between the satellite UES and user terminals; and the other is between the satellite GES and terrestrial networks. Internally, there are three types of interfaces: between the UES and satellite communication payload system; between the GES and satellite communication payload system; and the inter-satellite link (ISL) between satellites. All use radio links, except that the ISL may also use optical links.

Like physical cables, radio bandwidth is one of the most important and scarce resources for information delivery over satellite networks. Unlike cables, bandwidth cannot be manufactured, it can only be shared and its use maximised. The other important resource is transmission power. In particular, power is limited for user terminals requiring mobility or for those installed in remote places that rely on battery supply of power, and also for communication systems on board satellites that rely on battery and solar energy. The bandwidth and transmission power together within the transmission conditions and environment determine the capacity of the satellite networks.

Satellite networking shares many basic concepts with general networking. In terms of topology, it can be configured into star or mesh topologies. In terms of transmission technology, it can be set up for point-to-point, point-to-multipoint and multipoint-to-multipoint connections. In terms of interface, we can easily map the satellite network in general network terms such as user network interface (UNI) and network nodes interface (NNI).

When two networks need to be connected together, a network-to-network interface is needed, which is the interface of a network node in one network with a network node in another network. They have similar functions as NNI. Therefore, NNI may also be used to denote a network-to-network interface.

1.1.4 Network services

The UES and GES provide network services. In traditional networks, such services are classified into two categories: teleservices and bearer services. The teleservices are high-level services that can be used by users directly such as telephone, fax service, video and data services. Quality of service (QoS) at this level is user centric, i.e. the QoS indicates users' perceived quality, such as mean objective score (MOS). The bearer services are lower level services provided by the networks to support the teleservices. QoS at this level is network centric, i.e. transmission delay, delay jitter, transmission errors and transmission speed.

There are methods to map between these two levels of services. The network needs to allocate resources to meet the QoS requirement and to optimise the network performance. Network QoS and user QoS have contradicting objectives adjustable by traffic loads, i.e. we can increase QoS by reducing traffic load on the network or by increasing network resources, however, this may decrease the network utilisation for network operators. Network operators can also increase network utilisation by increasing traffic load, but this may affect user QoS. It is the art of traffic engineering to optimise network utilisation with a given network load under the condition of meeting user QoS requirements.

1.1.5 Applications

Applications are combinations of one or more network services. For example, tele-education and telemedicine applications are based on combinations of voice, video and data services. Combinations of voice, video and data are also called multimedia services. Some applications can be used with the network services to create new applications.

Services are basic components provided by the network. Applications are built from these basic components. Often the terms application and service are used interchangeably in the literature. Sometimes it is useful to distinguish them.

1.2 ITU-R definitions of satellite services

Satellite applications are based on the basic satellite services. Due to the nature of radio communications, the satellite services are limited by the available radio frequency bands. Various satellite services have been defined, including fixed satellite service (FSS), mobile satellite service (MSS) and broadcasting satellite service (BSS) by the ITU Radiocommunication Standardisation Sector (ITU-R) for the purpose of bandwidth allocation, planning and management.

1.2.1 Fixed satellite service (FSS)

The FSS is defined as a radio communication service between a given position on the earth's surface when one or more satellites are used. These stations at the earth surface are called earth stations of FSS. Stations located on board satellites, mainly consisting of the satellite transponders and associated antennas, are called space stations of the FSS. Of course, new-generation satellites have onboard sophisticated communication systems including onboard switching. Communications between earth stations are through one satellite or more satellites interconnected through ISL. It is also possible to have two satellites interconnected through a common earth station without an ISL. FSS also includes feeder links such as the link between a fixed earth station and satellite for broadcasting satellite service (BSS) and mobile satellite service (MSS). The FSS supports all types of telecommunication and data network services such as telephony, fax, data, video, TV, Internet and radio.

1.2.2 Mobile satellite service (MSS)

The MSS is defined as a radio communication service between mobile earth stations and one or more satellites. This includes maritime, aeronautical and land MSS. Due to mobility requirements, mobile earth terminals are often small, and some are even handheld terminals.

1.2.3 Broadcasting satellite service (BSS)

The BSS is a radio communication service in which signals transmitted or retransmitted by satellites are intended for direct reception by the general public using a TV receiving only antenna (TVRO). The satellites implemented for the BSS are often called direct broadcast satellites (DBS). The direct receptions include individual direct to home (DTH) and community antenna television (CATV). The new generation of BSS may also have a return link via satellite.

1.2.4 Other satellite services

Some other satellite services are designed for specific applications such as military, radio determination, navigation, meteorology, earth surveys and space exploration. A set of space stations and earth stations working together to provide radio communication is called a satellite system. For convenience, sometimes the satellite system or a part of it is called a satellite network. We will see in the context of network protocols that the satellite system may not need to support all the layers of functions of the protocol stack (physical layer, link layer or network layer).

1.3 ITU-T definitions of network services

During the process of developing broadband communication network standards, the ITU Telecommunication Standardisation Sector (ITU-T) has defined telecommunication services provided to users by networks. There are two main classes of services: interactive and distribution services, which are further divided into subclasses.

1.3.1 Interactive services

Interactive services offer one user the possibility to interact with another user in real-time conversation and messages or to interact with information servers in computers. It can be seen that different services may have different QoS and bandwidth requirements from the network to support these services. The subclasses of the interactive services are defined as the following:

- *Conversational services*: conversational services in general provide the means for bidirectional communication with real-time (no store-and-forward) end-to-end information transfer from user to user or between user and host (e.g. for data processing). The flow of the user information may be bidirectional symmetric, bidirectional asymmetric and in some specific cases (e.g. such as video surveillance), the flow of information may be unidirectional. The information is generated by the sending user or users, and is dedicated to one or more of the communication partners at the receiving site. Examples of broadband conversational services are telephony, videotelephony, and videoconference.
- *Messaging services*: messaging services offer user-to-user communication between individual users via storage units with store-and-forward, mailbox and/or message handling (e.g. information editing, processing and conversion) functions. Examples of broadband messaging services are message-handling services and mail services for moving pictures (films), high-resolution images and audio information.
- *Retrieval services*: the user of retrieval services can retrieve information stored in information centres provided for public use. This information will be sent to the user by demand only. The information can be retrieved on an individual basis. Moreover, the time at which an information sequence starts is under the control of the user. Examples are broadband retrieval services for film, high-resolution images, audio information and archival information.

1.3.2 Distribution services

This is modelled on traditional broadcast services and video on demand to distribute information to a large number of users. The requirement of bandwidth and QoS are quite different from interactive services. The distribution services are further divided into the following subclasses:

- *Distribution services without user individual presentation control*: these services include broadcast services. They provide a continuous flow of information, which is distributed from a central source to an unlimited number of authorised receivers connected to the network. The user can access this flow of information without the ability to determine at which instant the distribution of a string of information will be started. The user cannot control the start and order of the presentation of the broadcasted information. Depending on the point of time of the user's access, the information will not be presented from the beginning. Examples are broadcast services for television and radio programmes.
- *Distribution services with user individual presentation control*: services of this class also distribute information from a central source to a large number of users. However, the information is provided as a sequence of information entities (e.g. frames) with cyclical repetition. So, the user has the ability of individual access to the cyclical distributed information and can control the start and order of presentation. Due to

the cyclical repetition, the information entities selected by the user will always be presented from the beginning. One example of such a service is video on demand.

1.4 Internet services and applications

Like computers, in recent years the Internet has been developed significantly and the use of it has been extended from research institutes, universities and large organisations into ordinary family homes and small businesses.

The Internet was originally designed to interconnect different types of networks including LANs, MANs and WANs. These networks connect different types of computers together to share resources such as memory, processor power, graphic devices and printers. They can also be used to exchange data and for users to access data in any of the computers across the Internet.

Today the Internet is not only capable of supporting data, but also image, voice and video on which different network services and applications can be built such as IP telephony, videoconferencing, tele-education and telemedicine.

The requirements of new services and applications clearly changed the original objectives of the Internet. Therefore the Internet is evolving towards a new generation to support not only the traditional computer network services but also real-time user services including telephony. Eventually, this will lead to a convergence of the Internet and telecommunication networks towards the future global network infrastructures of which satellite will play an important part.

1.4.1 World wide web (WWW)

The WWW enables a wide range of Internet services and applications including e-commerce, e-business and e-government. It also enables virtual meetings with a new style of work, communication, leisure and lives. The WWW is an application built on top of the Internet, but is not the Internet itself. It can be seen that the basic principle of the Internet hasn't change much in the last 40 years, but applications of the Internet have changed significantly, particularly the user terminals, user software, services and applications, and human–computer interface (HCI).

The WWW is a distributed, hypermedia-based Internet information system including browsers for users to request information, servers to provide information and the Internet to transport users' requests from users to servers and information from servers to users.

The hypertext transfer protocol (HTTP) was created in 1990, at CERN, the European particle physics laboratory in Geneva, Switzerland, as a means for sharing scientific data internationally, instantly and inexpensively. With hypertext a word or phrase can contain a link to other text. To achieve this, the hypertext mark up language (HTML), a subset of general mark up language (GML), is used to enable a link within a web page to point to other pages or files in any server connected to the network. This non-linear, non-hierarchical method of accessing information was a breakthrough in information sharing. It quickly became the major source of traffic on the Internet. There are a wide variety of types of information (text, graphics, sounds, movies, etc.). It is possible to use the web to access information from almost every server connected to the Internet in world. The basic elements for access to the WWW are:

- HTTP: the protocol used for the WWW to transport web pages.
- URL (uniform resource locator): defines a format to address the unique location of the web page identified by the IP address of a computer, port number within the computer system and location of the page in the file system.
- HTML: the programming 'tags' added to text documents that turn them into hypertext documents.

In the original WWW, the URL identified a static file. Now it can be a dynamic web page created according to information provided by users; and it can also be an active web page, which is a piece of program code to be downloaded and run on the user's browser computer when clicked.

1.4.2 File transfer protocol (FTP)

FTP is an application layer protocol providing a service for transferring files between a local computer and a remote computer. FTP is a specific method used to connect to another Internet site to receive and send files. FTP was developed in the early days of the Internet to copy files from computer to computer using a command line. With the advent of WWW browser software, we no longer need to know FTP commands to copy to and from other computers, as web browsers have integrated the commands into their browser functions.

1.4.3 Telnet

This is one of the earliest Internet services providing text-based access to a remote computer. We can use telnet in a local computer to login to a remote computer over the Internet. Normally, an account is needed in the remote host so that the user can enter the system. After a connection is set up between the local computer and remote computer, it allows users to access the remote computer as if it were a local computer. Such a feature is called location transparency, i.e., the user cannot tell the difference between the responses from the local machine or remote machine. It is called time transparency if the response is so fast that user cannot tell the difference between local machine and remote machine by response time. Transparency is an important feature in distributed information systems.

1.4.4 Electronic mail (email)

The email is like our postal system but much quicker and cheaper, transmitting only information without papers or other materials, i.e. you can order a pizza through the Internet but cannot receive any delivery from it. The early email allowed only text messages to be sent from one user to another via the Internet. Email can also be sent automatically to a number of addresses. Electronic mail has grown over the past 20 years, from a technical tool used by research scientists, to a business tool as common as faxes and letters. Everyday, millions and millions of emails are sent through intranet systems and the Internet. We can also use mailing lists to send an email to groups of people. When an email is sent to a mailing list, the email system distributes the email to the listed group of users. It is also possible to send very large files, audio and video clips.

The success of email systems also causes problems for the Internet, e.g. viruses and junk mail are spread through email, threatening the Internet and the many computers linked to it.

1.4.5 Multicast and content distribution

Multicast is a generalised case of broadcast and unicast. It allows distribution of information to multiple receivers via the Internet or intranets. Example applications are content distributions including news services, information on stocks, sports, business, entertainment, technology, weather and more. It also allows real-time video and voice broadcast over Internet. This is an extension to the original design of the Internet.

1.4.6 Voice over internet protocol (VoIP)

VoIP is one of the important services under significant development. This type of service is real time and is more suitable for traditional telecommunication networks. It is different in many ways from the original Internet service. It has quite different traffic characteristics, QoS requirements and bandwidth and network resources.

Digitised streams of voices are segmented into voice 'frames'. These frames are encapsulated into a voice packet using a real-time transport protocol (RTP) that allows additional information for real-time service including time stamps to be included. The real-time transport control protocol (RTCP) is designed to carry control and signalling information used for VoIP services.

The RTP packets are put into the user datagram protocol (UDP), which is carried through the Internet by IP packets. The QoS of VoIP depends on network conditions in terms of congestion, transmission errors, jitter and delay. It also depends on the quality and available bandwidth of the network such as the bit error rate and transmission speed.

Though the RTP and RTCP were originally designed to support telephony and voice services, they are not limited to these, as they can also support real-time multimedia services including video services. By making use of the time-stamp information generated at source by the sender, the receiver is able to synchronise different media streams to reproduce the real-time information.

1.4.7 Domain name system (DNS)

The DNS is an example of application layer services. It is not normally used by users, but is a service used by the other Internet applications. It is an Internet service that translates domain names into IP addresses. Because domain names are alphabetical, they are easier to remember. The Internet, however, is really based on IP addresses. Every time you use a domain name, therefore, a DNS service must translate the name into the corresponding IP address. For example, the domain name www.surrey.ac.uk will translate to IP address: 131.227.102.18. The IP address can also be used directly.

The DNS is, in fact, a distributed system in the Internet. If one DNS server does not know how to translate a particular domain name, it asks another one, and so on, until the correct IP address is returned.

The DNS is organised as a hierarchical distributed database that contains mapping of domain names to various types of information including IP addresses. Therefore, the DNS can also be used to discover other information stored in the database.

1.5 Circuit-switching network

The concept of circuit-switching networks comes from the early analogue telephony networks. The network can be of different topologies including star, hierarchical and mesh at different levels to achieve coverage and scalability. Figure 1.3 shows typical topologies of networks.

An example of telephone networks is shown in Figure 1.4. At local exchange (LEX) level, many telephones connect to the exchange forming a star topology (a complete mesh topology is not scalable). Each trunk exchange (TEX) connects several local exchanges to

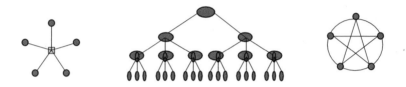

Figure 1.3   Typical topologies of networks: star, hierarchy and mesh

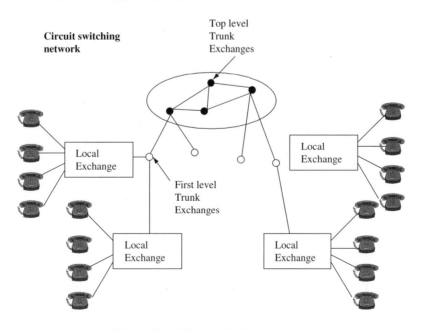

Figure 1.4   Circuit switching networks

form the first level of the hierarchy. Depending on the scale of the network, there may be several levels in the hierarchy. At the top level, the number of exchanges is small, therefore a mesh topology is used by adding redundancy to make efficient use of network circuits.

All the telephones have a dedicated link to the local exchange. A circuit is set up when requested by a user dialling the telephone number, which signals the network for a connection.

1.5.1 Connection set up

To set up a connection, a set of circuits has to be connected, joining two telephone sets together. If two telephones are connected to the same LEX, the LEX can set up a circuit directly. Otherwise, additional steps are taken at a higher level TEX to set up a circuit across the switching network to connect to the remote LEX then to the destination telephone.

Each TEX follows routing and signalling procedures. Each telephone is given a unique number or address to identify which LEX it is connected to. The network knows which TEX the LEX is connected to. The off-hook signal and dialled telephone number provide signalling information for the network to find an optimum route to set up a group of circuits to connect the two telephones identified by the calling telephone number and called telephone number.

If the connection is successful, communication can take place, and the connection is closed down after communication has ended. If the connection fails or is blocked due to lack of circuits in the network, we have to try again.

At this point, you may imagine that due to the wide coverage of satellite systems, it is possible to have satellites acting as a LEX to connect the telephones directly, or to act as a link to connect LEX to TEX, or connect TEX together. The roles of the satellite in the network have a significant impact on the complexity and cost of the satellite systems, as the different links require different transmission capacities. Satellites can be used for direct connection without strict hierarchy for the scalability needed in terrestrial networks.

1.5.2 Signalling

Early generation of switches could only deal with very simple signalling. Signalling information was kept to the minimum and the signal used the same channel as the voice channel.

Modern switches are capable of dealing with a large amount of channels, hence the signalling. The switches themselves have the same processing power as computers, are very flexible and are capable of dealing with data signals. This leads to separation of signal and user traffic, and to the development of common channel signalling (CCS). In CCS schemes, signals are carried by the same channel over a data network, separated from the voice traffic.

Combination of the flexible computerised switch and CCS enables a better control and management of the telephone network and facilitates new services such as call forwarding, call back and call waiting.

Signalling between network devices can be very fast, but responses from people are still the same. The processing power of devices can be improved significantly but not people's ability to react. People used to cause stress to network technologies, but now they are often stressed by technologies.

1.5.3 Transmission multiplexing hierarchy based on FDM

Frequency division multiplexing (FDM) is a technique to share bandwidth between different connections in the frequency domain. All transmission systems are design to transmit signals within a bandwidth limit measured in hertz (Hz). The system may allocate a fraction of the bandwidth-called channel to a connection to support a network service such as telephony rather than allocate a physical cable to the connection. This effectively increases the capacity.

When the bandwidth is divided into channels, each channel can support a connection. Therefore, connections from many physical links can be multiplexed into a single physical link with many channels. Similarly, multiplexed connections in one physical connection can be de-multiplexed into many physical connections. Figure 1.5 illustrates the concept of multiplexing in the frequency domain.

The given channel can be used to transmit digital as well as analogue signals. However, analogue transmission is more convenient to process in the frequency domain. A traditional telephone channel transmits audio frequency at a bandwidth of 3.1 kHz (from 0.3 to 3.4 kHz). It is transmitted in the form of a single-sideband (SSB) signal with suppressed carriers at 4 kHz spacing. Through multiplexing, 12 or 16 single channels can form a group. Five groups can form a super-group, super-group to master-group or hyper-group, and to super-group and master-group. Figure 1.6 shows the analogue transmission hierarchy.

1.5.4 Transmission multiplexing hierarchy based on TDM

Digital signals can be processed conveniently in the time domain. Time division multiplexing (TDM) is a technique to share bandwidth resources in the time domain. A period of time called a frame can be divided into time slots. Each time slot can be allocated to a connection. The frame can support the same number of connections as the number of slots. For example, the basic digital connection for telephony is 64 kbit/s. Each byte will take 125 microseconds to transmit. If the transmission speed is very fast, each byte can be transmitted in a fraction

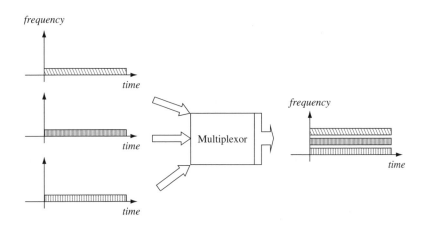

Figure 1.5 Concept of multiplexing in the frequency domain

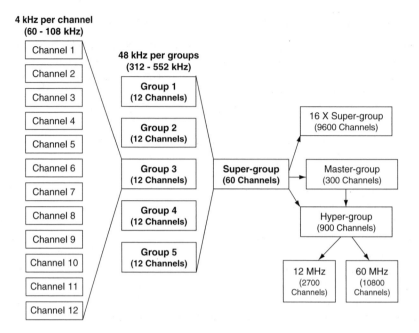

Figure 1.6   Analogue transmission multiplexing hierarchy

of the 125 microseconds, and then a time frame of 125 microseconds can be divided into more time slots to support one connection for each slot. Several slow bit streams can be multiplexed into one high-speed bit stream. Figure 1.7 illustrates the concept of multiplexing in the time domain.

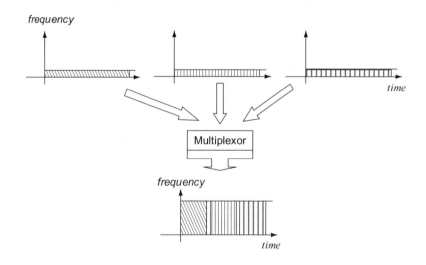

Figure 1.7   Concept of multiplexing in the time domain

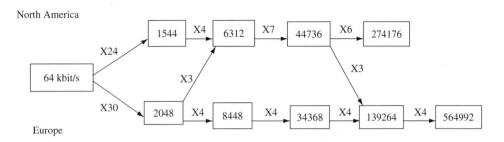

Figure 1.8   Digital transmission hierarchies

The digital streams in the trunk and access links are organised into the standard digital signal (DS) hierarchy in North America: DS1, DS2, DS3, DS4 and higher levels starting from 1.544 Mbit/s; in Europe, they are organised into E1, E2, E3, E4 and higher levels starting from 2.048 Mbit/s. The two hierarchies can only internetwork at certain levels, however, the basic rate is the same 64 kbit/s needed to accommodate one telephone circuit. Additional bits or bytes are added to the multiplexed bit stream for signalling and synchronisation purposes, which are also different between North America and European systems. Figure 1.8 shows the transmission multiplexing hierarchies.

1.5.5 Space switching and time switching

In telephony networks and broadcasting networks, the usage of each channel normally is in the order of minutes or hours. The requirements for bandwidth resources are also well defined. For example, channels for telephony services and broadcast services are all well defined.

If a switch cannot buffer any information, space in terms of bandwidth or time slots has to be reserved to allow information to flow and switched across the switch as shown in Figure 1.9. This means that the switch can only perform space switching.

If a switch can buffer a frame of time slots, the output of slot contents in the frame can be switched as shown in Figure 1.10. This means that the switch can perform time switching.

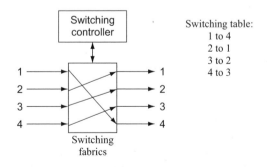

Figure 1.9   Space switching concept

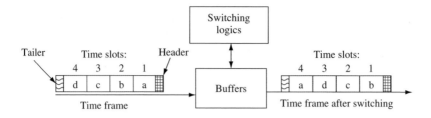

Figure 1.10 Time switching concept

Switch designs can use either/or a combination of space switching and time switching, such as space-time-space or time-space-time combinations.

1.5.6 Coding gain of forward error correction (FEC)

In satellite networking, the transmission from satellite to the earth station is normally power limited. To make it worse, there may be propagation loss and increased noise power. Therefore, it is important to introduce an error correction coding, i.e., to add additional information to the data so that some errors can be corrected by the receiver. This is called forward error correction (FEC), because the additional information and processing take place before any error occurs.

Depending on modulation schemes, bit error probability (BEP) is expressed as a function of E_b/N_0 which is related to E_c/N_0 by expression:

$$E_b/N_0 = E_c/N_0 - 10 \log \rho \tag{1.1}$$

where E_b is the energy per bit without coding, E_c is the energy per bit with coding, N_0 is the noise spectral density (W/Hz) and $\rho = n/(n+r)$ is the code rate (where r is the number of bits added for n information bits). It can be seen that we can use less power to improve the BEP at the cost of additional bits (hence bandwidth). The value ($10 \log \rho$) is called the coding gain. There is also a trade-off between power and bandwidth for a given BEP.

Using $C = E_c R_c$, we calculate:

$$E_c/N_0 = (C/R_c)/N_0 = (C/N_0)/R_c \tag{1.2}$$

where C is carrier power, and R_c is the channel bit rate.

1.6 Packet-switching networks

The packet switching concept was developed for computer networks, because streams of bits or bytes do not make much sense to computers. The computer needs to know the start and end of the data transmission.

In a data network, it is important to be able to identify where transmission of data starts and where transmission ends. The data, together with identifiers of the start and end of the data, is called a frame. In addition, addresses, frame checks and other information are added so that the sending computer can tell the receiving computer what to do based on a protocol sent when the frame is received. If the frame is exchanged on a link between two computers, it is defined by the link layer protocol. The frame is special packet on links. Therefore, the frame is related to link layer functions.

Information can also be added to the frame to create a packet so that the computer can make use of it to route the packet from the source to the destination across the network. Therefore, the packet is related to network layer functions.

The initial packet network was design for transmission of messages or data. The start and end of the data, correctness of transmission and mechanisms to detect and recover errors are all important. If the communication channel is perfect, a complete message can be handled efficiently as a whole, however, in the real world, this assumption cannot be met easily. Therefore, it is practical to break down the message into smaller segments using packets for transmission. If there is any error in the message, only the error packet needs to be dealt with rather than the whole message.

With packets, we don't need to divide bandwidth resources into narrow channels or small time slots to meet service requirement. We can use the complete bandwidth resources to transmit packets at high speed. If we need more bandwidth, we can simply use more or larger packets to send our data. If we use less bandwidth, we use fewer and smaller packets. Packets provide flexibility for bandwidth resource allocations, particularly when we don't know the requirement of bandwidth resources from some new multimedia services.

The meaning of broadband has been defined by the ITU-T as a system or transmission capable of dealing with data rates higher than the primary rates, which are 1.544 Mbit/s in North America and 2.048 Mbit/s in Europe.

There are two approaches for the packet-switching network. One is used in traditional telephony networks and the other is used in the computer and data networks.

1.6.1 Connection-oriented approach

In a packet-switching network, each physical connection has a much wider bandwidth, which is capable of supporting high-speed data transmissions. To divide this bandwidth for more connections, the concept of a virtual channel is used. The packet header carries an identification number to identify different logical connections within the same physical connection.

On receiving the packet, the packet switch can forward the packet to the next switch using another virtual channel until the packet reaches its destination. For switching, the network needs to be set up before the packet is transmitted. That is, a switching table needs to be set up in the switch to connect the incoming virtual channels to the outgoing virtual channels. If connection requirements are known, the network can reserve resources for the virtual connections in terms of packets and their payload.

This approach is called the virtual channel approach. Like telephony networks, the virtual channel based approach is connection oriented, i.e., a connection needs to be set up before communication. All packets follow the same connection from source to destination. The connection is called virtual connection.

In circuit switching, physical paths are set up to switch from input channels to output channels. In virtual channel switching, channels are identified by logic numbers; hence changing the logic number identifier virtually switches the packets to a different logical channel. Virtual channel switching is also called virtual circuit switching. Figure 1.11 illustrates the concept of virtual channel switching.

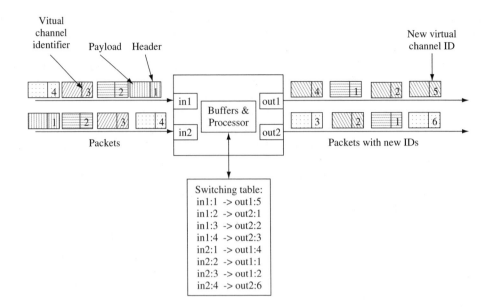

Figure 1.11   Virtual channel switching concept

The network node is called a packet switch, and functions like traditional circuit switching, but it gives flexibility of allocating different amounts of resources to each virtual connection. Therefore it is a useful concept for a broadband network, and is used in the asynchronous transfer mode (ATM) network. The virtual connection identifiers are only significant to each switch for identifying logical channels.

This kind of network is quite similar to our telephony and railway networks. Resources can be reserved to guarantee QoS during the connection set-up stage. The network blocks the connection request if there are not enough resources to accommodate the additional connection.

1.6.2 Connectionless approach

In computer and data networks, transmission of information often takes a very short period of time compared to telephone connections. It becomes inefficient to set up a connection for the computer and data networks for each packet transmission.

To overcome the problem with the virtual channel approach, the connectionless approach is used to transmit packets from sources to destinations without pre-setting connections. Such a packet is called the datagram approach because it consists of source and destination addresses rather than connection identifiers to allow the network node (also called the router) to route the packet from source to destination. Figure 1.12 illustrates the concept of connectionless approach.

In a connectionless network, the packet header needs to carry the destination address so that the network can use it to route the packet from source to destination, and also the source address for response by the destination computer. The network packet switch is called a router to distinguish it from the connection-oriented switch or traditional channel-based

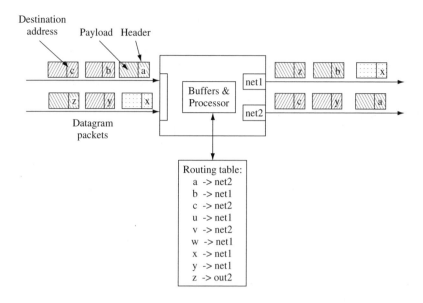

Figure 1.12   Datagram routing concept

switch. The router has a routing table containing information about destination and the next node leading to the destination with minimum costs.

The connectionless approach has flexibility for individual packets to change to different routes if there is congestion or failure in the route to destination. This kind of network is quite similar to postal delivery and motorway networks in the UK. There is no way to make a reservation, hence there is no guarantee of QoS. When traffic conditions are good, one car journey can give a good estimate of travel time. Otherwise, it may take much more time to reach the destination and sometimes it can be too late to be useful. However, there is flexibility to change its route after starting the journey to avoid any congestion or closure in the route. The Internet is an example of this kind of network, hence the information highway is a good description of the information infrastructure widely used today.

1.6.3 Relationship between circuit switching and packet switching

Circuit switching relates more closely to transmission technologies than packet switching. It provides physical transmission of signals carrying information in the networks. The signals can be analogue and digital. For analogue signals it provides bandwidth resources in term of Hz, kHz or MHz, treated in the frequency domain such as FDM; and for digital signals it provides bandwidth resources in term of bit/s, kbit/s or Mbit/s, treated in the time domain such as TDM. It is also possible to take into account both time and frequency domains such as CDMA. At this level, switches deal with streams of bits and bytes of digital signals to flow along the circuits or analogue signals with defined bandwidth. There is no structure in the signal.

Packets provide a level of abstraction above the bit or byte level, by providing structure to bit streams. Each packet consists of a header and payload. The header carries information to be used by the network for processing, signalling, switching and controlling purposes. The payload carries information to be received and processed by user terminals.

On top of a circuit it is possible to transmit packets. With packets it is possible to emulate the circuit by continuous streams of packets. These allow internetworking between circuit networks and packet networks. The emulated circuit is called a virtual circuit. It can be seen that virtual circuit, frame and packet are different levels of abstract from physical transmissions to network layer functions.

1.6.4 Impacts of packet on network designs

A packet is a layer of functions introduced to the networks. It separates the user services and applications from transmission technologies. A packet provides flexibility for carrying voice, video and data without involving transmission technologies and media. The network only deals with packets rather than different services and applications. The packets can be carried by any network technology including satellite.

Introducing packets into networks brings tremendous benefit for developing new services and applications and for exploring new network technologies, and also brings a great challenge to network designers.

What size should the packet be? There should be a trade-off between requirements from applications and services and the capabilities of transmission technologies. If is too small, it may not be capable of meeting the requirements, but if it is too big it may not be fully utilised and may also cause problems in transmission. Large packets are more likely to get bit errors than small ones, as transmission channels are never perfect in real life. For large packets it takes a long time to transmit and process and they also need large memory space to buffer them. Real-time services may not be able to tolerant long delays, hence there is a preference for small packets.

1.6.5 Packet header and payload

How many bits should be used for the packet header and how many for payload? With a large header, it is possible to carry more control and signal information. It also allows more bits to be used for addresses for end systems, but it can be very inefficient if services need only a very small payload. There are also special cases for large headers, for example, a large header may be needed for secure transmission of credit card transactions.

1.6.6 Complexity and heterogeneous networks

The complexity is due to a large range of services and applications and different transmission technologies. Many different networks have been developed to support a wide range of services and applications and to better utilise bandwidth resources based on packet-switching technologies. Systems may not work together if they are developed with different specifications of packets. Therefore such issues have to be dealt with in a much wider community in order for systems to interwork globally. This is often achieved by developing common international standards.

1.6.7 Performance of packet transmissions

At bit or byte level, transmission errors are overcome by increasing transmission power and/or bandwidth using better channel coding and modulation techniques. In real systems, it is impossible to eliminate bit errors completely. The errors at bit level will propagate to packet levels. Retransmission mechanisms are used to recover the error/lost packets, thus controlling the error at packet levels. Therefore, packet transmission can be made reliable even if bit transmissions are unreliable. However, this additional error recovery capability is at the cost of additional transmission time and buffer space. It also relies on efficient error detection schemes and acknowledgement packets to confirm a successful transmission. For the retransmission scheme, the efficiency of channel utilisation can be calculated as:

$$\eta = t_t/(t_t + 2t_p + t_r) \tag{1.3}$$

where t_t is the time for transmission of a packet onto the channels, t_p is the time for propagation of the packet along the channel to the receiver, and t_r is the processing time of the acknowledgement packet by the receiver. It can be seen that large packet transmission times or small propagation times and packet processing times are good for packet transmission performance.

1.6.8 Impact of bit level errors on packet level

We may quickly realise that a large packet can also lead to a high probability of packet error. If P_b is the probability of a bit error, the probability of packet error P_p of n bits can be calculated as:

$$P_p = 1 - (1 - P_b)^n \tag{1.4}$$

Figure 1.13 shows the packet error probabilities for given bit error probabilities and packet sizes.

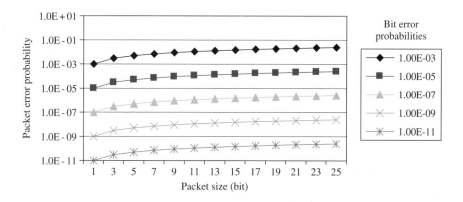

Figure 1.13 Packet error probabilities for given bit error probabilities and packet sizes

1.7 OSI/ISO reference model

Protocols are important for communications between entities. There are many options available to set protocols. For global communications, protocols are important to be internationally acceptable. Obviously, the International Standards Organisation (ISO) has played a very important role in setting and standardising a reference model so that any implementations following the reference model will be able to internetwork and communicate with each other.

Like any international protocol, it is easy to agree in principle how to define the reference model but always difficult to agree about details such as how many layers the model should have, how many bytes a packet should have, how many headers a packet should have to accommodate more functionalities but minimise overheads, whether to provide best-effort or guaranteed services, whether to provide connection-oriented services or connectionless services, etc. There are endless possible options and trade-offs with many technological selections and political considerations.

1.7.1 Protocol terminology

A protocol is the rules and conventions used in conversation by agreement between the communicating parties. A reference model provides all the roles so that all parties will be able to communicate with each other if they follow the roles defined in the reference model in their implementation.

To reduce design complexity, the whole functions of systems and protocols are divided into layers, and each layer is designed to offer certain services to higher layers, shielding those layers from the details of how the services are actually implemented.

Each layer has an interface with the primitive operations, which can be used to access the offered services. Network protocol architecture is a set of layers and protocols.

A protocol stack is a list of protocols (one protocol per layer). An entity is the active element in each layer, such as user terminals, switches and routers. Peer entities are the entities in the same layer capable of communication with the same protocols.

Basic protocol functions include segmentation and reassembly, encapsulation, connection control, ordered delivery, flow control, error control, and routing and multiplexing.

Protocols are needed to enable communicating parties to understand each other and make sense of received information. International standards are important to achieve a global acceptance. Protocols described in the standards are often in the context of reference models, as many different standards have been developed.

1.7.2 Layering principle

The layering principle is an important concept for network protocols and reference models. In the 1980s, the ISO derived the seven-layer reference model shown in Figure 1.14 called the open systems interconnection (OSI) reference model, which is based on clear and simple principles.

It is the first complete reference model developed as an international standard. The principles that were applied to arrive at the seven layers can be summarised as:

- A layer defines a level of abstraction which should be a different from any other layer.
- Each layer performs a well-defined function.

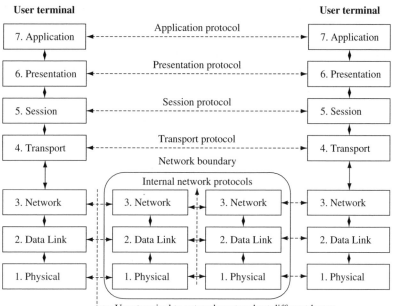

Figure 1.14   OSI/ISO seven-layer reference model

- The function of each layer should be chosen to lead to internationally standardised protocols.
- The layer boundaries should be chosen to minimise information flow across the interface.
- The number of layers should be large enough but not too large.

1.7.3 Functions of the seven layers

The following are brief descriptions of the functions of each layer.

- Layer 1 – the physical layer (bit stream) specifies mechanical, electrical and procedure interfaces and the physical transmission medium. In satellite networks, radio links are the physical transmission media; modulation and channel coding enable the bit stream to be transmitted in defined signals and allocated frequency bands.
- Layer 2 – the data link layer provides a line that appears free of undetected transmission errors to the network layer. Broadcasting media have additional issues in data link layer, i.e., how to control access to the shared medium. A special sublayer called the medium access control (MAC) schemes, such as Polling, Aloha, FDMA, TDMA, CDMA, DAMA, deals with this problem.
- Layer 3 – the network layer routes packets from source to destination. The functions include network addressing, congestion control, accounting, disassembling and reassembling, coping with heterogeneous network protocols and technologies. In broadcast networks, the routing problem is simple: the routing protocol is often thin or even non-existent.

- Layer 4 – the transport layer provides a reliable data delivery service for high layer users. It is the highest layer of the services associated with the provider of communication services. The higher layers are user data services. It has functions of ordered delivery, error control, flow control and congestion control.
- Layer 5 – the session layer provides the means of cooperating presentation entities to organise and synchronise their dialogue and to manage the data exchange.
- Layer 6 – the presentation layers are concerned with data transformation, data formatting and data syntax.
- Layer 7 – the application layer is the highest layer of the ISO architecture. It provides services to application processes.

1.7.4 Fading of the OSI/ISO reference model

Today we can see the development of many types of new applications, services, networks and transmission media. No one expected such a fast development of the Internet and new services and applications. New technologies and new service and application developments have changed the conditions of the optimisation points of the layering functions as one of the reasons leading to the fading of the international standards.

There are also many other reasons, including technical, political and economical reasons, or too complicated to be used in a practical world. The reference model is not much used in today's networks. However, the principles of layering protocol are still widely used in network protocol design and implementation. It is the classical and true reference model that all modern protocols always try to use as a reference to discuss and describe the functions of their protocols and evaluate their performance by analysis, simulation and experiment.

1.8 The ATM protocol reference model

The asynchronous transfer model (ATM) is based on fast packet switching techniques for the integration of telecommunications and computer networks. Historically, telephone networks and data networks were developed independently. Development of integrated services digital network (ISDN) standards by the ITU-T was the first attempt to integrate telephony and data networks.

1.8.1 Narrowband ISDN (N-ISDN)

N-ISDN provides two 64 kbit/s digital channels, which replace the analogue telephone services plus a 16 kbit/s data channel for signalling and data services from homes to local exchanges. The ISDN follows the concept of circuit networks very closely, as the envisaged main services, telephony and high-speed data transfer, need no more than 64 kbit/s. The primary rates are 1.5 Mbit/s for North America and 2 Mbit/s for Europe.

1.8.2 Broadband ISDN (B-ISDN)

ATM is a further effort by ITU-T to develop a broadband integrated services digital network (B-ISDN) following the development of ISDN, which is called narrowband ISDN (N-ISDN) to distinguish it from B-ISDN.

As soon as standardisation of the N-ISDN was complete, it was realised that the N-ISDN based on circuit networks could not meet the increasing demand by new services and applications and data networks.

The standardisation processes of B-ISDN led to the development of ATM based on the concept of packet switching. It provides flexibility of allocating bandwidth to user services and applications from tens of kbit/s used for telephony services to hundreds of Mbit/s for high-speed data and high definition TV.

The ITU-T recommended that the ATM is the target solution for broadband ISDN. It is the first time in its history that standards were set up before development.

1.8.3 ATM technology

The basic ATM technology is very simple. It is based on a fixed packet size of 53 bytes of which 5 bytes are for the header and 48 for payload. The ATM packet is called a cell, due to the small and fixed size.

It is based on the virtual channel switching approach providing a connection-oriented service and allowing negotiation of bandwidth resources and QoS for different applications. It also provides control and management functions to manage the systems, traffic and services for generating revenue from the network operations.

1.8.4 The reference model

The reference model covers three plans: user, control and management. All transportation aspects are in the form of ATM, as shown in Figure 1.15 including the:

- physical layer provides physical media-related transmissions such as optical, electrical and microwave;
- ATM layer defines ATM cells and related ATM functions; and
- ATM adaptation layer adapts high-layer protocols including the services and applications and divides data into small segments so that they can be suitable for transportation in the ATM cells.

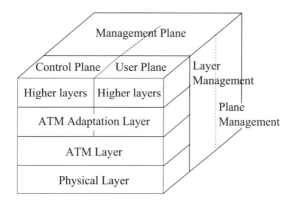

Figure 1.15 B-ISDN ATM reference model

1.8.5 Problems: lack of available services and applications

The ATM has been influenced by the development of optical fibre, which provides very large bandwidths and very low transmission errors. However, such transmission conditions are hardly possible in satellite transmission systems.

Services and applications are considered as parts of functions in user terminals rather than as parts of the network. The networks are designed to be able to meet all the requirements of services and applications. However, the higher layers were never defined and so few services and applications were developed on the ATM network. ATM has tried to internetwork with all different sorts of networks including some legacy networks together with the management and control functions making ATM very complicated and expensive to implement.

1.9 Internet protocols reference model

Originally, the Internet protocols were not developed by any international standardisation organisation. They were developed by the Department of Defense (DoD) research project to connect a number of different networks designed by different vendors into a network of networks (the 'Internet'). It was initially successful because it delivered a few basic services that everyone needed (file transfer, electronic mail, telnet for remote logon) across a very large number of different systems.

The main part of the Internet protocol reference model is the suite of transmission control protocol (TCP) and Internet protocol (IP) known as the TCP/IP protocols. Several computers in a small department can use TCP/IP (along with other protocols) on a single LAN or a few interconnected LANs. The Internet protocols allow the construction of very large networks with less central management.

As all other communications protocol, TCP/IP is composed of different layers but is much simpler than the ATM. Figure 1.16 shows the Internet reference model.

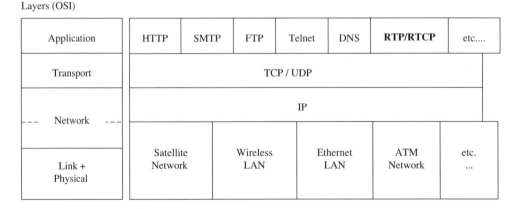

Figure 1.16   The Internet reference model

1.9.1 Network layer: IP protocol

The network layer is the Internet protocol (IP) based on the datagram approach, proving only best effort service without any guarantee of quality of service. IP is responsible for moving packets of data from node to node. IP forwards each packet based on a four-byte destination address (the IP address). The Internet authorities assign ranges of numbers to different organisations. The organisations assign groups of their numbers to departments.

1.9.2 Network technologies

The network technologies, including satellite networks, LANs, ATM, etc., are not part of the protocols. They transport IP packets from one edge of the network to the other edge. The source host sends IP packets and the destination host receives the packets. The network nodes route the IP packets to the next routers or gateways until they can route the packets directly to the destination hosts.

1.9.3 Transport layer: TCP and UDP

The transmission control protocol (TCP) and user datagram protocol (UDP) are transport layer protocols of the Internet protocol reference model. They provide ports or sockets for services and applications at user terminals to send and receive data across the Internet.

The TCP is responsible for verifying the correct delivery of data between client and server. Data can be lost in the intermediate network. TCP adds support to detect errors or lost data and to trigger retransmission until the data is correctly and completely received. Therefore TCP provides a reliable service though the network underneath may be unreliable, i.e., operation of Internet protocols do not require reliable transmission of packets, but reliable transmission can reduce the number of retransmissions and hence increase performance.

UDP provides the best-effort service without trying to recover any error or loss. Therefore, it is also a protocol providing unreliable transport of user data. However, this is very useful for real-time application, as retransmission of any packet may cause more problems than the lost packets.

1.9.4 Application layer

The application layer protocols are designed as functions of the user terminals or server. The classical Internet application layer protocols include HTTP for WWW, FTP for file transfer, SMTP for email, telnet for remote login, DNS for domain name service and more including real-time protocol (RTP) and real-time control protocol (RTCP) for real-time services and others for dynamic and active web services. All these should be independent from the networks.

1.9.5 Problems: no QoS and no control on resources

Most functions of the Internet define the high layer protocols. Current Internet protocol version 4 (IPv4) provides only best-effort services, hence it does not support any control functions and cannot provide any quality of services. The problems are addressed in the next generation of the Internet protocol version 6 (IPv6).

1.10 Satellite network

There are two types of transmission technologies: broadcast and point-to-point transmissions. Satellite networks can support both broadcast and point-to-point connections. Satellite networks are most useful where the properties of broadcast and wide coverage are important. Satellite networking plays an important role in providing global coverage. There are three types of roles that satellites can play in communication networks: access network, transit network and broadcast network.

1.10.1 Access network

The access network provides access for user terminals or private networks. Historically in telephony networks, it provided connections from telephone or private branch exchanges (PBX) to the telephony networks. The user terminals link to the satellite earth terminals to access satellite links directly. Today, in addition to the telephony access network, the access networks can also be the ISDN access, B-ISDN access and Internet access.

1.10.2 Transit network

The transit network provides connection between networks or network switches. It often has a large capacity to support a large number of connections for network traffic. Users do not have direct access to it. Therefore they are often transparent to users, though they may notice some differences due to propagation delay or quality of the link via a satellite network. Examples of satellite as transit networks include interconnect international telephony networks, ISDN, B-SDN and Internet backbone networks. Bandwidth sharing is often pre-planned using fixed assignment multiple access (FAMA).

1.10.3 Broadcast network

Satellite supports both telecommunication service and broadcast service. Satellite can provide very efficient broadcasting services including digital audio and video broadcast (DVB-S) and DVB with return channels via satellite (DVB-RCS).

1.10.4 Space segment

The main components of a communication satellite system consist of the space segment: satellites, and the ground segment: earth stations. The design of satellite networks is concerned with service requirements, orbit and coverage and frequency band selection (see Figure 1.17).

The satellite is the core of the satellite network consisting of a communication subsystem and platform. The platform, also called a bus, provides the structure support and power supply of the communication subsystems, and also includes altitude control, orbit control, thermal control, tracking, telemetry and telecommand (TT&T) to maintain normal operations of the satellite system.

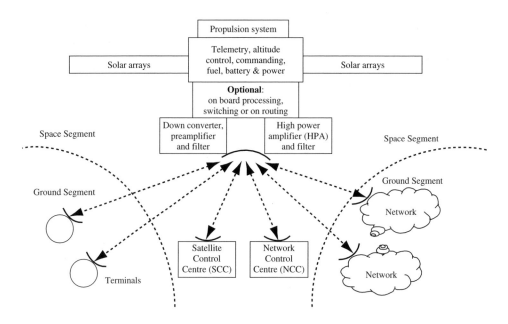

Figure 1.17   Illustration of the space segment and ground segment

The telecommunication subsystems consist of transponders and antenna. The antennas associated with the transponders are specially designed to provide coverage for the satellite network. Modern satellites may also have onboard processing (OBP) and onboard switching (OBS). There are different types of transponders:

- Transparent transponders provide the function of relaying radio signals. They receive transmissions from the earth station and retransmit them to the earth station after amplification and frequency translation. Satellites with transparent transponders are called transparent satellites.
- OBP transponders provide addition functions including digital signal processing (DSP), regeneration and base band signal processing before retransmitting the signal from satellite to the earth station. Satellites with OBP transponders are called OBP satellites.
- OBS transponders have additional functions than OBP transponders, providing switching functions. Similarly, satellites with OBS transponders are called OBS satellites. With the rapid development of the Internet, experiments are also in progress to fly onboard routers.

In addition, the satellite control centre (SCC) and network control centre (NCC) or network management centre (NMC), are parts of the space segment, though they are located at ground level:

- Satellite control centre (SCC): it is the on-ground system responsible for the operation of the satellite. It monitors the status of the different satellite subsystems through telemetry links, controls the satellite on its nominal orbit through telecommand links. It communicates with the satellite using dedicated links, which are different from the communication links. It normally consists of typically one earth station and GEO or non-GEO satellite systems, receiving telemetry from the satellites and sending telecommands to the satellites. Sometimes, a backup centre is built at a different location to improve reliability and availability.

- Network control centre (NCC) or network management centre (NMC): this has different functions from the SCC. Its main functions are to manage the network traffic and associated resources on board the satellite and on ground to achieve efficient use of the satellite network for communications.

1.10.5 Ground segment

The earth station is part of the satellite network. It provides functions of transmitting and receiving traffic signals to and from satellites. It also provides interfaces to terrestrial networks or to user terminals directly. The earth station may consist of the following parts:

- The transmitting and receiving antenna are the most visible parts of the earth station. There are different sizes typically ranging from below 0.5 metres to 16 metres and above.
- Low noise amplifier of the receiver system with noise temperature ranging from about 30 K to a few hundred K.
- High performance amplifier (HPA) of the transmitter with power from a few watts to a few thousands kilowatts depending on capacity.
- Modulation, demodulation and frequency translation.
- Signal processing.
- Interfaces to terrestrial networks or user terminals.

1.10.6 Satellite orbits

Orbits are one of the importance resources for satellite in space, as satellites need to be in a right orbit to provide coverage to the service areas. There are different ways to classify satellite orbits (see Figure 1.18).

According to the altitude of satellites, satellite orbits can be classified as the following types:

- Low earth orbit (LEO) has an altitude range of less than 5000 km. Satellites in this type of orbit are called LEO satellites. The period of the satellite is about 2–4 hours.
- Media earth orbit (MEO) has an altitude range between 5000 to 20 000 km. Satellites in this type of orbit are called MEO satellites. The period of the satellite is about 4–12 hours.
- Highly elliptical earth orbit (HEO) has an altitude range of more than 20 000 km. Satellites in this type of orbit are called HEO satellites. The period of the satellite is more than 12 hours.

Please note that the space surrounding the earth is not as empty as it looks. There are mainly two kinds of space environment constraints to be considered when choosing orbit altitude.

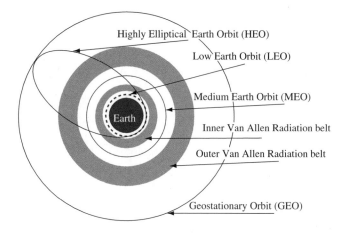

Figure 1.18   Satellite orbits

- The Van Allen radiation belts where energetic particles such as protons and electrons are confined by the earth's magnetic field. They can cause damage to the electronic and electrical components of the satellite.
- Space debris belts where spacecraft are abandoned at end of their lifetime. They are becoming of increasing concern to the international community as they can also cause damage to satellite networks particularly satellite constellations and to space missions in the future.

1.10.7 Satellite transmission frequency bands

Frequency bandwidth is another important resource of satellite networking and also a scarce resource. The radio frequency spectrum extends from about 3 kHz to 300 GHz, communications above 60 GHz are generally not practical because of the high power needed and equipment costs. Parts of this bandwidth are used for terrestrial microwave communication links historically, and for terrestrial mobile communications such as GSM and 3G networks and wireless LANs today.

In addition, the propagation environment between the satellite and earth station due to rain, snow, gas and other factors and limited satellite power from solar and battery limits further suitable bandwidth for satellite communications. Figure 1.19 shows attenuations of different frequency bands due to rain, fog and gas.

Link capacity is limited by the bandwidth and transmission power used for transmission. Frequency bandwidths are allocated by the ITU. There are several bands allocated for satellite communications. Table 1.1 shows the different available bandwidths for satellite communications.

Historically, bandwidths around 6 GHz for uplink and 4 GHz for downlink have been commonly paired in the C band. Many FSS still use these bands. Military and governmental systems use bands around 8/7 GHz in the X band. There are also some systems that operate around 14/12 GHz in the Ku band. New-generation satellites try to use the Ka band to explore wide bandwidth due to saturation of the Ku band. Table 1.2 gives examples of uses of frequency bands.

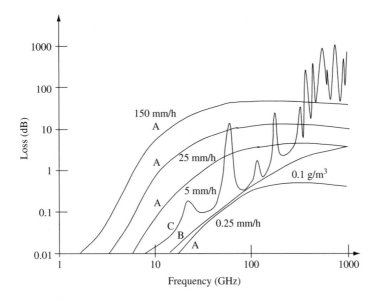

Figure 1.19   Attenuations of different frequency bands due to A: rain, B: fog and C: gas

Table 1.1   Typical frequency bands
of satellite communications

Denomination	Frequency bands (GHz)
UHF	0.3–1.12
L band	1.12–2.6
S band	2.6–3.95
C band	3.95–8.2
X band	8.2–12.4
Ku band	12.4–18
K band	18.0–26.5
Ka band	26.5–40

1.11 Characteristics of satellite networks

Most of the presently employed communication satellites are radio frequency (RF) repeaters or 'bent pipe' satellites. A processing satellite, as a minimum, regenerates the received digital signal. It may decode and recode a digital bit stream. It also may have some bulk switching capability and inter satellite links (ISL).

Radio link (microwave LOS) provides real transmission of the bits and bytes at the physical layer of the layered reference model. There are three basic technical problems in the satellite radio link due to the satellite being located at great distances from the terminal earth stations.

Table 1.2 Example usages of frequency bands for GEO

Denomination	Uplink (bandwidth)	Downlink (bandwidth)	Typical utilisation in FSS for GEO
6/4 C band	5.850–6.425 (575 MHz)	3.625–4.2 (575 MHz)	International and domestic satellites: Intelsat, USA, Canada, China, France, Japan, Indonesia
8/7 X band	7.925–8.425 (500 MHz)	7.25–7.75 (500 MHz)	Governmental and military satellites
		10.95–11.2	International and domestic satellites in Region 1 and 3
		11.45–11.7	
		12.5–12.75 (1000 MHz)	Intelsat, Eutelsat, France, German, Spain, Russia
13–14/11–12 Ku band	13.75–14.5 (750 MHz)		
		10.95–11.2	International and domestic satellites in Region 2
		11.45–11.7	
		12.5–12.75 (700 MHz)	Intelsat, USA, Canada, Spain
18/12	17.3–18.1 (800 MHz)	BSS bands	Feeder link for BSS
30/20 Ka band	27.5–30.0 (2500 MHz)	17.7–20.2 (2500 MHz)	International and domestic satellites Europe, USA, Japan
40/20 Ka band	42.5–45.5 (3000 MHz)	18.2, 21.2 (3000 MHz)	Governmental and military satellites

1.11.1 Propagation delay

The first problem to deal with is very long distances. For GEO satellites, the time required to traverse these distances – namely, earth station to satellite to another earth station – is in the order of 250 ms. Round-trip delay will be of 2×250 or 500 ms. These propagation times are much greater than those encountered in conventional terrestrial systems. One of the major problems is propagation time and resulting echo on telephone circuits. It delays the reply of certain data circuits for block or packet transmission systems and requires careful selection of telephone signalling systems, or call set-up time may become excessive.

1.11.2 Propagation loss and power limited

The second problem is that there are far greater losses. For LOS microwave we encounter free-space losses possibly as high as 145 dB. In the case of a satellite with a range of 22 300 miles operating on 4.2 GHz, the free-space loss is 196 dB and at 6 GHz, 199 dB. At 14 GHz the loss is about 207 dB. This presents no insurmountable problem from earth to satellite, where comparatively high-power transmitters and very high-gain antennas may be used. From satellite to earth the link is power-limited for two reasons:

1. In bands shared with terrestrial services, such as the popular 4-GHz band, to ensure non-interference with those services; and
2. In the satellite itself, which can derive power only from solar cells. It takes a great number of solar cells to produce the RF power necessary; thus the downlink, from satellite to earth, is critical, and received signal levels will be much lower than on comparative radio links, as low as $-150\,\mathrm{dBW}$.

1.11.3 Orbit space and bandwidth limited for coverage

The third problem is crowding. The equatorial orbit is filling up with geostationary satellites. Radio-frequency interference from one satellite system to another is increasing. This is particularly true for systems employing smaller antennas at earth stations with their inherent wider beam widths. It all boils down to a frequency congestion of emitters.

1.11.4 Operational complexity for LEO

In addition to the GEO satellite, we also see several new low earth orbit satellite systems in operation, which can explore the potential of satellite capabilities. These satellites typically have much lower altitude orbits above the earth. This may reduce the problems of delay and loss, but introduce more complexity in maintaining communication links between earth terminals and satellites due to the fast movement of LEO constellation satellites.

1.12 Channel capacity of digital transmissions

In the frequency domain, greater bandwidth can support more communication channels. In the time domain, the digital transmission capacity is also directly proportional to the bandwidth.

1.12.1 The Nyquist formula for noiseless channels

For a noiseless channel, the Nyquist formula is used to determine the channel capacity:

$$C = 2B\log_2 M \tag{1.5}$$

where C is the maximum channel capacity for data transfer rate in bit/s, B is bandwidth in hertz and M is the number of levels per signalling element.

1.12.2 The Shannon theorem for noise channels

The Shannon and Hartley capacity theorem is used to determine the maximum bit rate C over a band-limited channel giving a specific signal-to-noise ratio (S/N). The theorem is:

$$C = B\log_2(1 + S/N) \tag{1.6}$$

where C is the maximum capacity in bit/s, B is bandwidth of the channel, S is signal power and N is noise power.

As $S = (RE_b)$ and $N = (N_0 B)$ the formula can be rewritten in a different form as the following:

$$C = B\log_2[1 + RE_b/(N_0 B)]$$
$$= B\log_2[1 + (R/B)(E_b/N_0)] \tag{1.7}$$

where E_b is energy per bit, R is transmission bit rate and $N = N_0 B$ where N_0 is noise power spectral density.

1.12.3 Channel capacity boundary

Let $R = C$ in Equation (1.7), we get the capacity boundary function between bandwidth efficiency C/B and given E_b/N_0:

$$C/B = \log_2[1 + (C/B)(E_b/N_0)] \tag{1.8}$$

Then:

$$E_b/N_0 = (2^{C/B} - 1)/(C/B) \tag{1.9}$$

Figure 1.20 shows the relationship of the capacity boundary of the communication channel with E_b/N_0. If the transmission data rate is within the capacity limit, i.e., if $R < C$, we may be able to achieve transmission rate with properly designed modulation and coding mechanisms, and if $R > C$, it is impossible to achieve error free transmission.

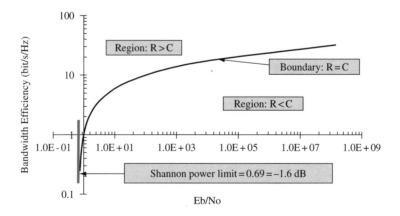

Figure 1.20   Capacity boundary of communication channel

1.12.4 The Shannon power limit (−1.6 dB)

We can increase the bandwidth to reduce transmission power as a trade-off. If we let the transmission bit rate R achieve the maximum, then we can get from Equation (1.8) the following:

$$(E_b/N_0)^{-1} = \log_2[1 + (C/B)(E_b/N_0)]^{(B/C)/(Eb/N0)} \tag{1.10}$$

As $(1 + 1/x)^x \to e$ when $x \to \infty$, let $B \to \infty$ we can get the Shannon power limit:

$$E_b/N_0 = \log_2(1/e) = \log_e 2 \approx 0.69 = -1.6 \, \text{dB} \tag{1.11}$$

This tell us, no matter how much bandwidth we have, the transmission power in terms of E_b/N_0 should be larger than the Shannon limit, though there is a trade-off between bandwidth and power.

1.12.5 Shannon bandwidth efficiency for large E_b/N_0

Similarly we can derive the formula of Shannon bandwidth efficiency from Equation (1.8) for large E_b/N_0, as the following:

$$\log_2[(C/B)(E_b/N_0)] \le C/B \le 1 + \log_2[(C/B)(E_b/N_0)]$$

Hence, $(C/B) \approx \log_2(E_b/N_0)$, when $(E_b/N_0) \to \infty$

Figure 1.21 shows the convergence between (C/B) and $\log_2(E_b/N_0)$. It also shows that when transmission power is low, increasing the power by a small amount will have a large impact on the bandwidth efficiency; and when transmission power is high reducing bandwidth efficiency by a small amount will have a large saving on transmission power.

Therefore engineers can trade between transmission bandwidth and transmission power, but should not go too far to benefit from such a trade off.

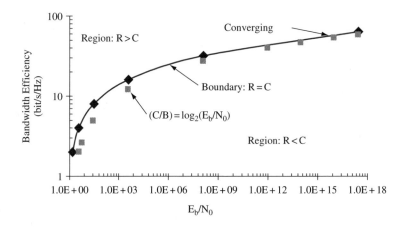

Figure 1.21   The Shannon bandwidth efficiency for large E_b/N_0

1.13 Internetworking with terrestrial networks

Internetworking techniques have been well developed in terrestrial networks. When we have different types of networks we face problems at different layers of the protocol stacks, such as different transmission media, different transmission speeds, different data formats and different protocols. Since networking only involves the lower three layers of the protocols, satellite networking with other types of networks could involve any of the three layers.

1.13.1 Repeaters at the physical layer

At the physical layer, internetworking is at bit level. The internetworking repeater needs to have a function to deal with the digital signal. It is relatively easy to internetwork between the terrestrial network and satellite networks, as the physical layer protocol functions are very simple. The main problem is dealing with data transmission rate mismatch, as terrestrial networks may have much higher data transmission rates.

The main disadvantage of this solution is that it is inflexible due to the nature of implementation at the physical layer. One may have quickly noticed that the communication payload of transparent satellites, relay satellites or bent-pipe satellites deals with bit streams as functions of a repeater.

1.13.2 Bridges at link layer

A bridge is a store and forward device and is normally used in the context of LANs, interconnecting one or more LANs at the link layer. In satellite networking, we borrow the term to refer to the internetworking unit between the satellite network and terrestrial networks. As it works at the link layer, it also relies on the physical layer transmission, i.e., the bridge deals with the functions of two layers: physical and link layers.

A frame arriving from the satellite network will be checked to decide if the frame should be forwarded to the terrestrial networks according to its routing table and the destination address. If yes, the bridge forwards the frame to the terrestrial networks, otherwise it discards it. Before forwarding, the frame is formatted based on the protocol of the terrestrial networks.

Similar procedures are also carried out when frames flow from the terrestrial networks to the satellite network. The main disadvantage is that the satellite has to deal with a large number of different types of networks and protocol translations. It has more complicated functions than repeaters.

The main advantage is that the satellite network will be able to make use of the link layer functions such as error detection, flow control and frame retransmission. The satellite payload can also implement the bridge functions. Otherwise the link layer functions have to be carried out on the other side of the satellite networks.

1.13.3 Switches at the physical, link and network layers

Switches can work at any layer of the three layers depending on the nature of the networks. Switching networks can set up end-to-end connections to transport bit streams, frames and even network layer packets.

The main advantage is that switching networks can reserve network resources when setting up connections.

The disadvantages are that they are not very efficient when dealing with short data transmission and supporting connectionless network protocols such as the Internet, and that it is difficult to deal with heterogeneous networks.

1.13.4 Routers for interconnecting heterogeneous networks

Router here refers to an Internet router or an IP router. It deals with only Internet protocol (IP) packets. Figure 1.22 shows how routers can be used to internetwork with heterogeneous terrestrial networks. Here it requires that all user terminals use the IP protocol.

1.13.5 Protocol translation, stacking and tunnelling

It can be seen that there are three basic techniques for interconnecting heterogeneous networks:

1. Protocol translation: this technique is normally used at the physical layer dependent sublayers of the link layer. Protocol translations are carried out between the different sublayers.
2. Protocol stacking: this technique is normally used for different layers. One layer is stacked on top of the other network.
3. Protocol tunnelling: this technique is similar to protocol staking, but with two of the same type of networks communicating through a tunnelling of other networks.

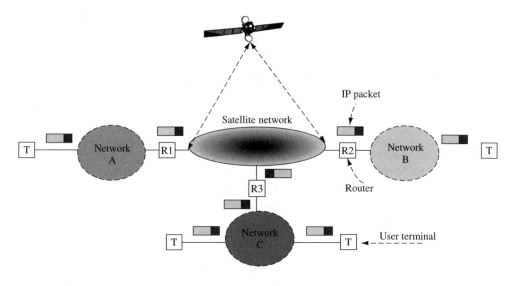

Figure 1.22 Using routers to internetwork with heterogeneous terrestrial networks

1.13.6 Quality of service (QoS)

The term quality of service (QoS) is extensively used today. It is not only used in analogue and digital transmission in telephony networks but also in broadband networks, wireless networks, multimedia services and even the Internet. Networks and systems are gradually being designed with consideration of the end-to-end performance required by user applications. Most traditional Internet applications such as email and ftp are sensitive to packet loss but can tolerate delays. For multimedia applications (voice and video) this is generally the opposite. They can tolerate some packet loss but are sensitive to delay and variation of the delay.

Therefore, networks should have mechanisms for allocating bandwidth resources to guarantee a specific QoS for real-time applications. QoS can be described as a set of parameters that describes the quality of a specific stream of data.

1.13.7 End-user QoS class and requirements

Based on the end-user application requirements, ITU-T recommendation G.1010 defines classification of performance requirements into end-user QoS categories.

Based on the target performance requirements, the various applications can be mapped onto axes of packet loss and one-way delay as shown in Figure 1.23. The size and shape of the boxes provide a general indication of the limit of delay and information loss tolerable for each application class.

It can be seen that there are eight distinct groups, which encompass the range of applications identified. Within these eight groupings there is a primary segregation between applications that can tolerate some information loss and those that cannot tolerate any information loss at all, and four general areas of delay tolerance.

This mapping is summarised in Figure 1.24, which provides a recommended model for end-user QoS categories, where the four areas of delay are given names chosen to illustrate the type of user interaction involved.

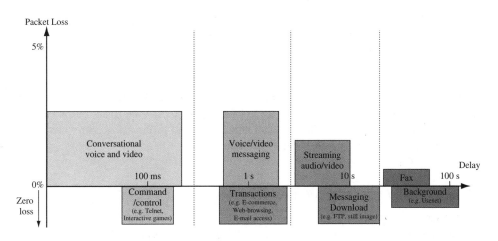

Figure 1.23 Mapping of user-centric QoS requirements into network performance (ITUT-G1010) (Reproduced with the kind permission of ITU.)

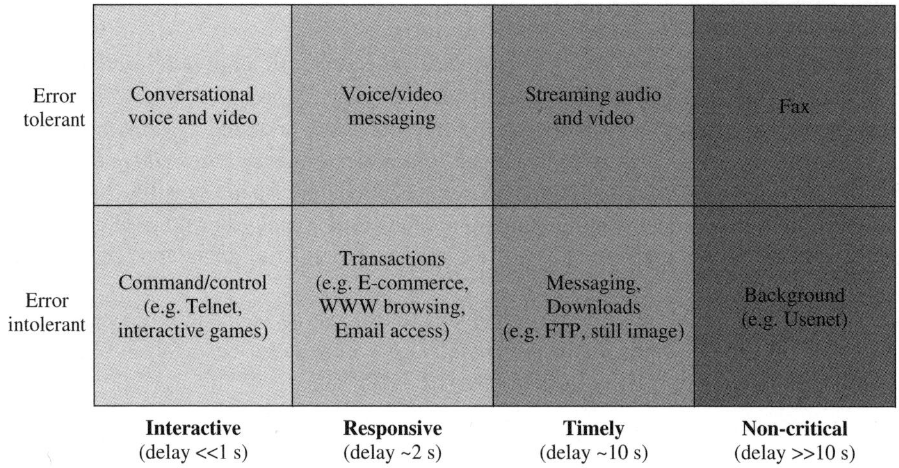

	Interactive	Responsive	Timely	Non-critical
Error tolerant	Conversational voice and video	Voice/video messaging	Streaming audio and video	Fax
Error intolerant	Command/control (e.g. Telnet, interactive games)	Transactions (e.g. E-commerce, WWW browsing, Email access)	Messaging, Downloads (e.g. FTP, still image)	Background (e.g. Usenet)
	Interactive (delay <<1 s)	**Responsive** (delay ~2 s)	**Timely** (delay ~10 s)	**Non-critical** (delay >>10 s)

Figure 1.24 Model for user-centric QoS categories (ITU-T-G1010) (Reproduced with the kind permission of ITU.)

1.13.8 Network performance

Network performance (NP) contributes towards QoS as experienced by the user/customer. Network performance may or may not be on an end-to-end basis. For example, access performance is usually separated from the core network performance in the operations of a single IP network, while Internet performance often reflects the combined NP of several autonomous networks.

There are four viewpoints of QoS defined by the ITU-T G.1000 recommendation, corresponding with different perspectives, as shown in Figure 1.25:

- customer QoS requirements;
- service provider offerings of QoS (or planned/targeted QoS);
- QoS achieved or delivered;
- customer survey ratings of QoS.

Among these four viewpoints, the customer's QoS requirements may be considered as the logical starting point. A set of customer's QoS requirements may be treated in isolation as far as its capture is concerned. This requirement is an input to the service provider for the determination of the QoS to be offered or planned.

1.13.9 QoS and NP for satellite networking

The definitions of QoS given by the ITU-T are based on a user-centric approach, but these may not reflect well on the QoS and NP related to networking. Therefore it is useful to employ the layering approach to define QoS and NP parameters related to networks (see Figure 1.26).

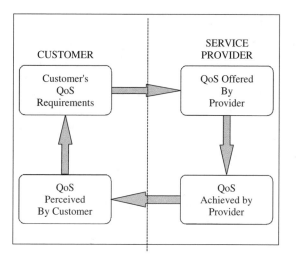

Figure 1.25 The four viewpoints of QoS (ITU-T-G1000) (Reproduced with the kind permission of ITU.)

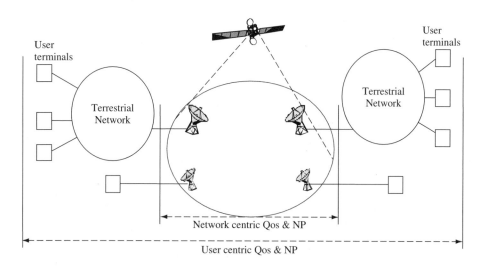

Figure 1.26 User- and network-centric views of QoS and NP concepts

The network centric approach enables us to quantify the QoS and NP parameters without the uncertainty of terminal performance, higher layer protocol functions and user factors. Typical parameters are:

- at analogue transmission level: signal to noise power ratio (S/N);
- at digital transmission level: bit error ratio (BER), propagation delay and delay variation; and
- at packet level: packet propagation delay and packet delay variation, packet error ratio, packet loss ratio and network throughput.

1.14 Digital video broadcasting (DVB)

Digital video broadcasting (DVB) technology allows broadcasting of 'data containers', in which all kinds of digital data can be transmitted. It simply delivers compressed images, sound or data to the receiver within these 'containers'. No restrictions exist as to the kind of information in the data containers. The DVB 'service information' acts like a header to the container, ensuring that the receiver knows what it needs to decode.

A key difference of the DVB approach compared to other data broadcasting systems is that the different data elements within the container can carry independent timing information. This allows, for example, audio information to be synchronised with video information in the receiver, even if the video and audio information does not arrive at the receiver at exactly the same time.

This facility is, of course, essential for the transmission of conventional television programmes. The DVB approach provides a good deal of flexibility. For example, a 38 Mbit/s data container could hold eight standard definition television (SDTV) programmes, four enhanced definition television (EDTV) programmes or one high definition television (HDTV) programme, all with associated multi-channel audio and ancillary data services.

Alternatively, a mix of SDTV and EDTV programmes could be provided or even multimedia data containing little or no video information. The content of the container can be modified to reflect changes in the service offer over time (e.g. migration to a widescreen presentation format).

At present, the majority of DVB satellite transmissions convey multiple SDTV programmes and associated audio and data. DVB is also useful for data broadcasting services (e.g. access to the World Wide Web).

1.14.1 The DVB standards

Digital video broadcasting (DVB) is a term that is generally used to describe digital television and data broadcasting services that comply with the DVB 'standard'.

In fact, there is no single DVB standard, but rather a collection of standards, technical recommendations and guidelines. These were developed by the Project on Digital Video Broadcasting, usually referred to as the 'DVB Project'.

The DVB Project was initiated in 1993 in liaison with the European Broadcasting Union (EBU), the European Telecommunications Standards Institute (ETSI) and the European Committee for Electrotechnical Standardisation (CENELEC). The DVB Project is a consortium of some 300 member organisations. As opposed to traditional governmental agency standards activities round the world, the DVB Project is market-driven and consequently works on commercial terms, to tight deadlines and realistic requirements, always with an eye toward promoting its technologies through achieving economies of scale. Though based in Europe, the DVB Project is international, and its members are in 57 countries round the globe. DVB specifications concern:

- source coding of audio, data and video signals;
- channel coding;
- transmitting DVB signals over terrestrial and satellite communications paths;
- scrambling and conditional access;

- the general aspects of digital broadcasting;
- software platforms in user terminals;
- user interfaces supporting access to DVB services;
- the return channel, as from a user back to an information or programme source to support interactive services.

The DVB specifications are interrelated with other recognised specifications. DVB source coding of audio-visual information as well as multiplexing is based on the standards evolved by the Moving Picture Experts Group (MPEG), a joint effort of the International Organisation for Standards (ISO) and the International Electrotechnical Commission (IEC). The principal advantage of MPEG compared to other audio and audio coding formats is that the sophisticated compression techniques used make MPEG files far smaller for the same quality. For instance, the first standard, MPEG1, was introduced in 1991 and supports 52:1 compression, while the more recent MPEG2 supports compression of up to 200:1.

The DVB Project is run on a voluntary basis and brings together experts from more than 300 companies and organisations, representing the interests of manufacturing industries, broadcasters and services providers, network and satellite operators and regulatory bodies. Its main intent is to reap the benefits of technical standardisation, while at the same time satisfying the commercial requirements of the project members. Although a large part of the standardisation work is now complete, work is still ongoing on issues such as the Multimedia Home Platform. Much of the output of the DVB Project has been formalised by ETSI.

1.14.2 DVB-S satellite delivery

One of the earliest standards developed by the DVB Project and formulated by ETSI was for digital video broadcasting via satellite (usually referred to as the 'DVB-S standard'). Specifications also exist for the retransmission of DVB signals via cable networks and satellite master antenna television (SMATV) distribution networks.

The techniques used for DVB via satellite are classical in the sense that they have been used for many years to provide point-to-point and point-to-multipoint satellite data links in 'professional' applications. The key contribution of the DVB Project in this respect has been the development of highly integrated and low-cost chip sets that adapt the DVB baseband signal to the satellite channel. Data transmissions via satellite are very robust, offering a maximum bit error rate in the order of 10^{-11}.

In satellite applications, the maximum data rate for a data container is typically about 38 Mbit/s. This container can be accommodated in a single 33 MHz satellite transponder. It provides sufficient capacity to deliver, for example, four to eight standard television programmes, 150 radio channels, 550 ISDN channels, or any combination of these services. This represents a significant improvement over conventional analogue satellite transmission, where the same transponder is typically used to accommodate a single television programme with far less operational flexibility.

A single modern high-power broadcasting satellite typically provides at least twenty 33 MHz transponders, allowing delivery of about 760 Mbit/s of data to large numbers of users equipped with small (around 60 cm) satellite dishes.

A simple generic model of a digital satellite transmission channel comprises several basic building blocks, which include baseband processing and channel adaptation in the transmitter and the complementary functions in the receiver. Central to the model is, of course, the satellite transmission channel. Channel adaptation would most likely be done at the transmit satellite earth station, while the baseband processing would be performed at a point close to the programme source.

1.14.3 MPEG-2 baseband processing

MPEG is a group of experts drawn from industry who contribute to the development of common standards through an ITU-T and ISO/IEC joint committee. The established MPEG-2 standard was adopted in DVB for the source coding of audio and video information and for multiplexing a number of source data streams and ancillary information into a single data stream suitable for transmission. Therefore, many of the parameters, fields and syntax used in DVB baseband processing are specified in the relevant MPEG-2 standards. The MPEG-2 standards are generic and very wide in scope. Some of the parameters and fields of MPEG-2 are not used in DVB.

The processing function deals with a number of programme sources. Each programme source comprises any mixture of raw data and uncompressed video and audio, where the data can be, for example, teletext and/or subtitling information and graphical information such as logos.

Each of the video, audio and programme-related data is called an elementary stream (ES). It is encoded and formatted into a packetised elementary stream (PES). Thus each PES is a digitally encoded component of a programme.

The simplest type of service is a radio programme, which would consist of a single audio elementary stream. A traditional television broadcast would comprise three elementary streams: one carrying coded video, one carrying coded stereo audio and one carrying teletext.

1.14.4 Transport stream (TS)

Following packetisation, the various elementary streams of a programme are multiplexed with packetised elementary streams from other programmes to form a transport stream (TS). Each of the packetised elementary streams can carry timing information, or 'time stamps', to ensure that related elementary streams, for example, video and audio, are replayed in synchronism in the decoder. Programmes can each have a different reference clock, or can share a common clock. Samples of each 'programme clock', called programme clock references (PCRs), are inserted into the transport stream to enable the decoder to synchronise its clock to that in the multiplexer. Once synchronised, the decoder can correctly interpret the time stamps and can determine the appropriate time to decode and present the associated information to the user.

Additional data is inserted into the transport stream, which includes programme specific information (PSI), service information (SI), conditional access (CA) data and private data. Private data is a data stream whose content is not specified by MPEG.

The transport stream is a single data stream that is suitable for transmission or storage. It may be of fixed or variable data rate and may contain fixed or variable data rate elementary streams. There is no form of error protection within the multiplex. Error protection is implemented within the satellite channel adaptor.

1.14.5 Service objectives

The DVB-S system is designed to provide so-called 'quasi error free' (QEF) quality. This means less than one uncorrected error event per transmission hour, corresponding to a bit error rate (BER) of between 10^{-10} and 10^{-11} at the input of the MPEG-2 demultiplexer (i.e. after all error correction decoding). This quality is necessary to ensure that the MPEG-2 decoders can reliably reconstruct the video and audio information.

This quality target translates to a minimum carrier-to-noise ratio (C/N) requirement for the satellite link, which in turn determines the requirements for the transmit earth station and the user's satellite reception equipment for a given satellite broadcasting network. The requirement is actually expressed in E_b/N_0 (energy per bit to noise density ratio), rather than C/N, so that it is independent of the transmission rate.

The DVB-S standard specifies the E_b/N_0 values at which QEF quality must be achieved when the output of the modulator is directly connected to the input of the demodulator (i.e. in an 'IF loop'). An allowance is made for practical implementation of the modulator and demodulator functions and for the small degradation introduced by the satellite channel. The values range from 4.5 dB for rate 1/2 convolutional coding to 6.4 dB for rate 7/8 convolutional coding.

The inner code rate can be varied to increase or decrease the degree of error protection for the satellite link at the expense of capacity. The reduction or increase in capacity associated with a change in the code rate and the related increase or reduction in the E_b/N_0 requirement. The latter is also expressed as an equivalent increase or reduction in the diameter of the receive antenna (the size of user's satellite dish), all other link parameters remaining unchanged.

1.14.6 Satellite channel adaptation

The DVB-S standard is intended for direct-to-home (DTH) services to consumer integrated receiver decoders (IRD), as well as for reception via collective antenna systems (satellite master antenna television (SMATV)) and at cable television head-end stations. It can support the use of different satellite transponder bandwidths, although a bandwidth of 33 MHz is commonly used. All service components ('programmes') are time division multiplexed (TDM) into a single MPEG-2 transport stream, which is then transmitted on a single digital carrier.

The modulation is classical quadrature phase shift keying (QPSK). A concatenated error protection strategy is employed based on a convolutional 'inner' code and a shortened Reed–Solomon (RS) 'outer' code. Flexibility is provided so that transmission capacity can be traded off against increased error protection by varying the rate of the convolutional code. Satellite links can therefore be made more robust, at the expense of reduced throughput per satellite transponder (i.e. fewer DVB services).

The standard specifies the characteristics of the digitally modulated signal to ensure compatibility between equipment developed by different manufacturers. The processing at the receiver is, to a certain extent, left open to allow manufacturers to develop their own proprietary solutions. It also defines service quality targets and identifies the global performance requirements and features of the system that are necessary to meet these targets.

1.14.7 DVB return channel over satellite (DVB-RCS)

The principal elements of a DVB return channel over satellite (DVB-RCS) system are the hub station and user satellite terminals. The hub station controls the terminals over the forward (also called outbound link), and the terminals share the return (also called inbound link). The hub station continuously transmits the forward link in time division multiplex (TDM). The terminals transmit as needed, sharing the return channel resources using multi-frequency time division multiple access (MF-TDMA). The DVB-RCS system supports communications on channels in two directions:

- Forward channel, from the hub station to many terminals.
- Return channels, from the terminals to the hub station.

The forward channel is said to provide 'point-to-multipoint' service, because it is sent by a station at a single point to stations at many different points. It is identical to a DVB-S broadcast channel and has a single carrier, which may take up the entire bandwidth of a transponder (bandwidth-limited) or use the available transponder power (power limited). Communications to the terminals share the channel by using different slots in the TDM carrier.

The terminals share the return channel capacity of one or more satellite transponders by transmitting in bursts, using MF-TDMA. In a system, this means that there is a set of return channel carrier frequencies, each of which is divided into time slots which can be assigned to terminals, which permits many terminals to transmit simultaneously to the hub. The return channel can serve many purposes and consequently offers choices of some channel parameters. A terminal can change frequency, bit rate, FEC rate, burst length, or all of these parameters, from burst to burst. Slots in the return channel are dynamically allocated.

The uplink and downlink transmission times between the hub and the satellite are very nearly fixed. However, the terminals are at different points, so the signal transit times between them and the satellite vary. On the forward channel, this variation is unimportant. Just as satellite TV sets successfully receive signals whenever they arrive, the terminals receive downlink signals without regard to small differences in their times of arrival.

However, on the uplink, in the return direction from the terminals to the hub, small differences in transit time can disrupt transmission. This is because the terminals transmit in bursts that share a common return channel by being spaced from each other in time. For instance, a burst from one terminal might be late because it takes longer to reach the satellite than a burst sent by another terminal. A burst that is earlier or later than it should be can collide with the bursts sent by the terminals using the neighbouring TDMA slots.

The difference in transmission times to terminals throughout the footprint of a satellite might be compensated for by using time slots that are considerably longer than the bursts transmitted by the terminals, so both before and after a burst there is a guard time sufficiently long to prevent collisions with the bursts in neighbouring slots in the TDMA frame. The one-way delay time between a hub and a terminal varies from 250 to 290 ms, depending on the geographical location of the terminal with respect to the hub. So the time differential, T, might be as large as 40 ms. So most TDMA satellite systems minimise guard time by incorporating various means of timing adjustment to compensate for satellite path differences.

DVB-RCS has two built-in methods of pre-compensating the burst transmission time of each terminal:

- Each terminal 'knows' its local GPS coordinates and therefore can calculate its own burst transmission time.
- The hub monitors the arrival times of bursts, and can send correction data to terminals if need be.

1.14.8 TCP/IP over DVB

DVB-RCS uses the MPEG-2 digital wrappers, in which 'protocol-independent' client traffic is enclosed within the payloads of a stream of 188-byte packets. The MPEG-2 digital wrapper offers a 182-byte payload and has a 6-byte header. The sequence for transmission of Internet TCP/IP traffic includes:

- The TCP/IP message arrives and is subjected to TCP optimisation.
- The IP packets are divided into smaller pieces and put into data sections with 96-bit digital storage medium – command and control (DSM-CC) headers.
- The DSM-CC data sections are further divided into 188-byte MPEG2-TS packets in the baseband processing.
- The MPEG2-TS packets then are subjected to channel coding for satellite transmissions.

1.15 Historical development of computer and data networks

Telecommunication systems and broadcasting systems have been developing for over 100 years. The basic principles and services have changed little since their beginnings and we can still recognise the earliest telephony systems and televisions. However, computers and the Internet have changed greatly in the last 40 years. Today's systems and terminals are completely different from those used 40 or even 10 years ago. The following gives a quick review of these developments to show the pace of technology progress.

1.15.1 The dawn of the computer and data communications age

The first electronic digital computer was developed during 1943–6. Early computer interfaces used punched tapes and cards. Later terminals were developed and the first communication between terminals and computer over long distances was in 1950, which used voice-grade telephone links at low transmission speeds of 300 to 1200 kbit/s. Automatic repeat requests (ARQ) for error correction were mainly used for data transmission.

1.15.2 Development of local area networks (LANs)

From 1950 to 1970 research carried out on computer networks led to the development of different types of network technologies – local area networks (LANs), metropolitan area networks (MANs) and wide area networks (WANs).

A collection of standards, known as IEEE 802, was developed in the 1980s including the Ethernet as IEEE802.3, token bus as IEEE802.4, token ring as IEEE802.5, DQDB as IEEE802.6 and others. The initial aim was to share file systems and expensive peripheral devices such as high-quality printers and graphical plot machines at fast data rates.

1.15.3 Development of WANs and ISO/OSI

The ISO developed the Open System Interconnection (OSI) reference model with seven layers for use in wide area networks in the 1980s. The goal of the reference model was to provide an open standard so that different terminals and computer systems could be connected together if they conformed to the standard. The terminals considered in the reference model were connected to a mainframe computer over a WAN in text mode and at slow speed.

1.15.4 The birth of the Internet

Many different network technologies were developed during the 1970s and 1980s and many of them did not fully conform to international standards. Internetworking between different types of networks used protocol translators and interworking units, and became more and more complicated as the protocol translators and interworking units became more technology dependent.

In the 1970s, the Advanced Research Project Agency Network (ARPARNET) sponsored by the US Department of Defense developed a new protocol, which was independent of network technologies, to interconnect different types of networks. The ARPARNET was renamed as the Internet in 1985. The main application layer protocols included remote telnet for terminal access, FTP for file transfer and email for sending mail through computer networks.

1.15.5 Integration of telephony and data networks

In the 1970s, the ITU-T started to develop standards called integrated services digital networks with end-to-end digital connectivity to support a wide range of services, including voice and non-voice services. User access to the ISDN was through a limited set of standard multipurpose customer interfaces. Before ISDN, access networks, also called local loops, to the telecommunication networks were analogue, although the trunk networks, also called transit networks, were digital. This was the first attempt to integrate telephony and data networks and integration of services over a single type of network. It still followed the fundamental concepts of channel- and circuit-based networks used in traditional telecommunication networks.

1.15.6 Development of broadband integrated networks

As soon as the ISDN was completed in the 1980s, the ITU-T started to develop broadband ISDN. In addition to broadband integrated services, ATM technology was developed to support the services based on fast packet-switching technologies. New concepts of virtual channels and circuits were developed. The network is connection oriented, which allows negotiation of bandwidth resources and applications. It was expected to unify the telephony networks and data networks and also unify LANs, MANs and WANs.

From the LAN aspect, ATM faced fierce competition from fast Ethernet. From application aspects, it faced competition from the Internet.

1.15.7 The killer application WWW and Internet evolutions

In 1990, Tim Berners-Lee developed a new application called the World Wide Web (WWW) based on hypertext over the Internet. This significantly changed the direction of network research and development. A large number of issues needed to be addressed to cope with the requirements of new services and applications, including real-time services and their quality of service (QoS), which were not considered in traditional Internet applications.

1.16 Historical development of satellite communications

Satellite has been associated with telecommunications and television from its beginning, but few people have noticed this. Today, satellites broadcast television programmes directly to our homes and allow us to transmit messages and surf the Internet. The following gives a quick review of satellite history.

1.16.1 Start of satellite and space eras

Satellite technology has advanced significantly since the launch of the first artificial satellite *Sputnik* by the USSR on 4 October 1957 and the first experiment of an active relaying communications satellite *Courier-1B* by the USA in August 1960.

The first international cooperation to explore satellite for television and multiplexed telephony services was marked by the experimental pre-operation transatlantic communications between the USA, France, Germany and the UK in 1962.

1.16.2 Early satellite communications: TV and telephony

Establishment of the Intelsat organisation started with 19 national administration and initial signatories in August 1964. The launch of the REARLY BIRD (Intelsat-1) marked the first commercial geostationary communication satellite. It provided 240 telephone circuits and one TV channel between the USA, France, Germany and the UK in April 1965. In 1967, Intelsat-II satellites provided the same service over the Atlantic and Pacific Ocean regions. From 1968 to 1970, Intelsat-III achieved worldwide operation with 1500 telephone circuits and four TV channels. The first Intelsat-IV satellite provided 4000 telephone circuits and two TV channels in January 1971 and Intelsat-IVa provided 20 transponders of 6000 circuits and two TV channels, which used beam separation for frequency reuse.

1.16.3 Development of satellite digital transmission

In 1981, the first Intelsat-V satellite achieved capacity of 12 000 circuits with FDMA and TDMA operations, 6/4 GHz and 14/11 GHz wideband transponders, and frequency reuse by beam separation and dual polarisation. In 1989, the Intelsat-VI satellite provided onboard satellite-switched TDMA of up to 120 000 circuits. In 1998, Intelsat VII, VIIa and Intelsat-VIII satellites were launched. In 2000, the Intelsat-IX satellite achieved 160 000 circuits.

1.16.4 Development of direct-to-home (DTH) broadcast

In 1999, the first K-TV satellite provided 30 14/11-12 GHz transponders for 210 TV programmes with possible direct-to-home (DTH) broadcast and VSAT services.

1.16.5 Development of satellite maritime communications

In June 1979, the International Maritime Satellite (Inmarsat) organisation was established to provide global maritime satellite communication with 26 initial signatories. It explored the mobility feature of satellite communications.

1.16.6 Satellite communications in regions and countries

At a regional level, the European Telecommunication Satellite (Eutelsat) organisation was established with 17 administrations as initial signatories in June 1977. Many countries also developed their own domestic satellite communications systems, including the USA, the USSR, Canada, France, Germany, the UK, Japan, China and other nations.

1.16.7 Satellite broadband networks and mobile networks

Since the 1990s, significant development had been carried out on broadband networks including onboard-switching satellite technologies. Various non-geostationary satellites have been developed for mobile satellite services (MSSs) and broadband fixed satellite services (FSSs).

1.16.8 Internet over satellite networks

Since the late 1990s and the start of the twenty-first century, we have seen a dramatic increase in Internet traffic over the communication networks. Satellite networks have been used to transport Internet traffic in addition to telephony and television traffic for access and transit networks. This brings great opportunities as well as challenges to the satellite industry. On one hand, it needs to develop internetworking with many different types of legacy networks; and on the other hand, it needs to develop new technologies to internetwork with future networks. We have also see the convergence of different types of networks including network technologies, network protocols and new services and applications.

1.17 Convergence of network technologies and protocols

The convergence is the natural progression of technologies pushing and user demands pulling and the development of business cases. Obviously, satellite networking closely follows the development of terrestrial networks, but is capable of overcoming geographical barriers and the difficulty of wide coverage faced by terrestrial networks. Figure 1.27 illustrates the vision of a future satellite network in the context of the global information infrastructure.

1.17.1 Convergence of services and applications in user terminals

In the early days, user terminals were designed for particular types of services and had very limited functions. For example, we had telephone handsets for voice services, computer terminals for data services, and television for receiving television services. Different networks were developed to support these different types of terminals.

As the technology developed, additional terminals and services were introduced into the existing networks. For example, fax and computer dialup services were added to telephone networks. However, the transmission speeds of fax and dialup links were limited by the capacity of the telephone channel supported by the telephony networks.

Computer terminals have become more and more sophisticated and are now capable of dealing with voice and video services in real time. Naturally, in addition to data services, there are increasing demands to support real time voice and video over data networks.

Multimedia services, a combination of voice, video and data, were developed. These complicate the QoS requirements requiring complicated user terminal and network design, implementation and operation.

To support such services over satellite networks for applications such as aeronautics, shipping, transport and emergency services brings even more challenges. We are starting to see the convergence of different user terminals for different types of services into a single user terminal for all types of services.

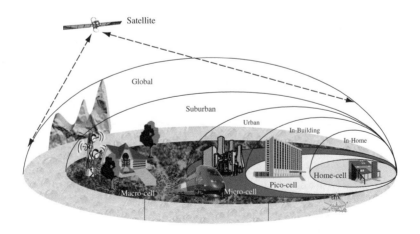

Figure 1.27 Satellite in the global information infrastructure

1.17.2 Convergence of network technologies

Obviously, network services are closely related to the physical networks. To support a new generation of services we need a new generation of networks. However, the design of new services and networks needs significant amounts of investment and a long period of time for research and development. To get users to accept new services and applications is also a great challenge.

How about building new services on the existing networking infrastructure? Yes, this approach has been tried as far as possible, as mentioned previously, fax and computer dialup were added to telephony networks, and voice and video services to data networks. This approach does not ease the task of developing new services and networks, as the original designs of the networks were optimised for original services. Therefore, new networks have to be developed for new services and applications.

Luckily we do not need to start from scratch. The telephone networks and services on the existing networks have been developing over the past 100 years; during that time we have accumulated a huge amount of knowledge and experience.

1.17.3 Convergence of network protocols

Following the concepts of telecommunication networking principles, attempts were made to develop new services and networks. Examples of these are the integrated services digital networks (ISDN), synchronous transfer mode (STM) networks, broadband ISDN (B-ISDN) and asynchronous transfer mode (ATM) networks. As telephony services are historically the major services in the telecommunication networks, the new networks are biased towards these services, and are perhaps emphasising too much on real time and QoS. The results are not completely satisfactory.

Computer and data networks have been developing for about 50 years, during this time we have also accumulated a significant amount of knowledge and experience in the design of computers and data networks. All the computer and data network technologies have converged to the Internet technologies. In LAN, Ethernet is the dominating technology; other LANs, such as token ring and token bus networks are disappearing. Of course, wireless LANs are becoming popular. The Internet protocols are now the protocol for computer and data networking. One of the most important successful factors is backward compatibility, i.e., new network technology should be capable of supporting the existing services and applications and internetworking with the existing user terminals and network without any modifications.

Following the success of the Internet, significant research and development have been carried out to support telephony services and other real-time services. As the original design of the Internet was for data services, it was optimised for reliable data services without much thought given to real-time services and QoS. Therefore, IP telephony cannot be achieved easily with the level of QoS provided by telecommunication networks.

Convergence of network design is inevitable, however, we have to learn from the telecommunication networks for QoS and reliable transmission of data from the Internet.

The principles of networking are still the same: to improve reliability; increase capacity; support integrated services and applications; to reach anywhere and anytime; and particularly important for satellite networking to fully utilise limited resources and reduce costs.

1.17.4 Satellite network evolution

It can be seen that satellite communication started from telephony and TV broadcast terrestrial networks. It went on to increase capacity, extend coverage to the oceans for mobile service, and extend services to data and multimedia services.

Satellites have become more sophisticated, and have progressed from single transparent satellites to onboard processing and onboard switching satellites, and further to non-geostationary satellite constellations with inter-satellite links (ISL).

Basic satellites have a repeater to relay signals from one side to the other. Satellites with this type of payload are called transparent satellites. They also called pent-pipe satellites as they simply provide links between terminals without processing.

Some satellites have onboard processing (OBP) as part of the communication subsystems to provide error detection and error correction to improve the quality of the communication links, and some have onboard switching (OBS) to form a network node in the sky to explore efficient use of radio resources. Experiments have also been also carried out to fly IP router onboard satellites due to the recent rapid development of Internet.

Satellites have played an important role in telecommunications networks supporting telephony, video, broadcast, data, broadband and Internet services and have become an important integrated part of the global information infrastructure providing the next generation of integrated broadband and Internet network.

Further reading

[1] Brady, M. and M. Rogers, *Digital Video Broadcasting Return Channel via Satellite (DVB-RCS) Background Book*, Nera Broadband Satellite AS (NBS), 2002.
[2] Eutelsat, *Overview of DVB, Annex B to Technical Guide*, June 1999.
[3] Haykin, S., *Communication Systems*, 4th edition, John Wiley & Sons, Inc., 2001.
[4] ITU, *Handbook on Satellite Communications*, 3rd edition, John Wiley & Sons, Inc., 2002.
[5] Joel, A., Retrospective: telecommunications and the IEEE communications society, *IEEE Communications*, May 2002.
[6] Khader, M. and W.E. Barnes, *Telecommunications Systems and Technology*, Prentice-Hall, 2000.

Exercises

1. Explain the meaning of broadband, using the definition given in the ITU-T recommendations.
2. Explain the basic concepts of satellite networking and internetworking with terrestrial networks.
3. Explain the terms satellite services, network services and quality of service (QoS).
4. Discuss the differences between satellite networking and terrestrial networking issues.
5. Explain the functions of network user terminals and satellite terminals.
6. Derive the Shannon power limit and the Shannon bandwidth capacity for large E_b/N_0.
7. Explain the basic principles of protocols and the ISO reference model.

Exercises (*continued*)

8. Explain the basic ATM reference model.
9. Explain the Internet protocol TCP/IP suite.
10. Explain the basic concepts of multiplexing and multiple accessing.
11. Explain the basic switching concepts including circuit switching, virtual circuit switching and routeing.
12. Explain the evolution process and convergence of network technologies and protocols.

2

Satellite Orbits and Networking Concepts

This chapter aims to provide an introduction to the physical layer of satellite networking concepts including principles of satellite orbits, satellite link characteristics, transmission techniques, multiple access, bandwidth allocation, and satellite availability and diversity. The physical layer is the lowest layer of network protocol stacks, providing real signal transmission. In the context of the protocol reference model, it provides service to the link layer above, however, there are no layers below it. It is the layer directly related to transmission media technologies and performs many of the functions of radio communication systems. In this chapter we will focus on satellite networking related issues. When you have completed this chapter, you should be able to:

- Review the laws of physics including Kepler's laws and Newton's laws.
- Make use of the laws to explain the characteristics of satellite orbits and calculate satellite orbit parameters.
- Make use of the laws to design orbits for a single satellite or a constellation of satellites for different requirements of satellite networking coverage.
- Appreciate the characteristics of satellite links and calculate the values of the link parameters.
- Understand different types of modulation techniques and why the phase shift modulation technique is more suitable for satellite transmission.
- Know the important error correcting coding schemes.
- Know different bandwidth resource allocation schemes and their applications.
- Describe the satellite networking design issues.
- Understand the concept of quality of service (QoS) at the physical layer.
- Know the quality of a satellite system in terms of availability and the techniques to improve satellite availability.

Satellite Networking: Principles and Protocols   Zhili Sun
© 2005 John Wiley & Sons, Ltd

2.1 Laws of physics

Like terrestrial mobile base stations, satellite communications systems have to be installed on a platform or bus. The laws of physics determine how and where we can put the base station in the sky to form an integrated part of our network.

2.1.1 Kepler's three laws

The German astronomer Johannes Kepler (1571–1630) formulated three laws of planetary motion that also apply to the motion of satellites around earth. Kepler's three laws are:

1. The orbit of any smaller body about a large body is always an ellipse, with the centre of mass of the large body as one of the two foci.
2. The orbit of the smaller body sweeps out equal areas in equal time.
3. The square of the period of revolution of the smaller body about a large body equals a constant multiplied by the third power of the semi major axis of the orbital ellipse.

2.1.2 Newton's three laws of motion and the universal law of gravity

In 1687, the British astronomer, mathematician, physicist and scientist Issac Newton discovered the three laws of motion as the following:

1. A body stays motionless, or continues moving in a straight line, unless a force acts on it.
2. Any change in movement of a body is always proportional to the force that acts on it, and is made in the direction of the straight line in which the force acts. It can be described mathematically as the sum of all vector forces $\vec{F}$ acting on a body with a mass of m equals the product of the mass and acceleration of vector $\vec{r}$ of the body:

$$\vec{F} = m\frac{d^2\vec{r}}{dt^2} \tag{2.1}$$

3. Every action has an opposite and equal reaction. In addition to the three laws, Newton also discovered the force of gravity (the force that made the apple fall to the ground).

More importantly, he provided a mathematical proof of the force of gravity forming the universal law of gravity as the 'two-body problem':

$$\vec{F} = Gm_1m_2\frac{1}{r^2}\frac{\vec{r}}{r} \tag{2.2}$$

where $\vec{F}$ is the vector force of mass m_1 on the mass m_2 in the direction from m_1 to m_2, $G = 6.672 \times 10^{-11}\,\mathrm{m^3/kg/s^2}$ is the universal gravity constant, r is the distance between the two bodies, and $\frac{\vec{r}}{r}$ is the unit vector showing the direction from m_1 to m_2. Clearly this can be used to describe the force between the sun and earth by letting m_1 be the mass of the sun and m_2 be the mass of the earth.

2.1.3 Kepler's first law: satellite orbits

Newton derived Kepler's laws mathematically. Mathematics is the most important tool in systems design and analysis. Here, we should make use of our analytical skills to look into the fundamental and theoretical aspects of satellite orbits. By taking the following steps, we can approve Kepler's first law for the satellite case mathematically.

First, apply Newton's third law to get:

$$\vec{F} = GMm \frac{1}{r^2} \frac{\vec{r}}{r} = \mu m \frac{1}{r^2} \frac{\vec{r}}{r} \tag{2.3}$$

where the mass of the earth $M = 5.974 \times 10^{24}$ kg, Kepler's constant $\mu = GM = 3.986 \times 10^{14}\,\text{m}^3/\text{s}^2$, and satellite mass is m kg.

Second, applying Newton's second law of motion, force = mass × acceleration, we get:

$$\frac{d^2 \vec{r}}{dt^2} + \mu \frac{\vec{r}}{r^3} = 0 \tag{2.4}$$

Then, referring to Figure 2.1 and letting $\vec{x} = \vec{r}/r$ and $\vec{y}$ is the unit vector orthogonal to $\vec{x}$, we get:

$$\frac{d\vec{r}}{dt} = \frac{dr}{dt}\vec{x}(t) + r\frac{d\theta}{dt}\vec{y}(t) \tag{2.5}$$

$$\frac{d^2 \vec{r}}{dt^2} = \left[\frac{d^2 r}{dt^2} - r\left(\frac{d\theta}{dt}\right)^2\right]\vec{x}(t) + \left[r\frac{d^2\theta}{dt^2} + 2\frac{dr}{dt}\frac{d\theta}{dt}\right]\vec{y}(t) \tag{2.6}$$

Substituting Equation (2.6) into Equation (2.4) gives:

$$\left\{\left[\frac{d^2 r}{dt^2} - r\left(\frac{d\theta}{dt}\right)^2\right] + \frac{u}{r^2}\right\}\vec{x}(t) + \left[r\frac{d^2\theta}{dt^2} + 2\frac{dr}{dt}\frac{d\theta}{dt}\right]\vec{y}(t) = 0 \tag{2.7}$$

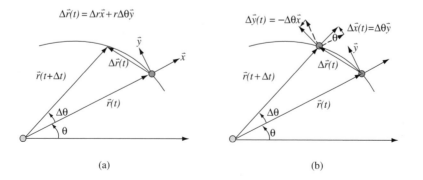

Figure 2.1   Vector from earth to satellite

Hence:

$$\left[\frac{d^2r}{dt^2} - r\left(\frac{d\theta}{dt}\right)^2\right] + \frac{\mu}{r^2} = 0 \tag{2.8}$$

$$\left[r\frac{d^2\theta}{dt^2} + 2\frac{dr}{dt}\frac{d\theta}{dt}\right] = 0 \tag{2.9}$$

From Equation (2.9), we get: $\dfrac{1}{r}\dfrac{d}{dt}\left(r^2\dfrac{d\theta}{dt}\right) = 0$;

Hence

$$\left(r^2\frac{d\theta}{dt}\right) = D \text{ (Constant)} \tag{2.10}$$

Let $u = 1/r$ in Equation (2.10), we get

$$\frac{d\theta}{dt} = Du^2 \tag{2.11}$$

Hence:

$$\frac{dr}{dt} = \frac{dr}{d\theta}\frac{d\theta}{dt} = \left(-\frac{1}{u^2}\right)\frac{du}{d\theta}\frac{d\theta}{dt} = -D\frac{du}{d\theta} \tag{2.12}$$

$$\frac{d^2r}{dt^2} = \frac{dr}{dt}\left(\frac{dr}{dt}\right) = \frac{du}{d\theta}\frac{d\theta}{dt}\left(\frac{dr}{dt}\right) = -D^2u^2\frac{d^2u}{d\theta^2} \tag{2.13}$$

Substituting Equations (2.11) and (2.13) into Equation (2.8) gives:

$$\frac{d^2u}{d\theta^2} + u = \frac{\mu}{D^2} \tag{2.14}$$

Letting

$$p = \frac{D^2}{\mu} \tag{2.15}$$

and solving the second-order liner differential equation (2.14) gives the following:

$$u = \frac{1}{p} + A\cos(\theta - \theta_0)$$

Hence:

$$r = \frac{p}{1 + pA\cos(\theta - \theta_0)} \tag{2.16}$$

where A and θ_0 are constants, and adjustment can be made so that $\theta_0 = 0$.

Therefore, we can represent Kepler's first law mathematically for satellite orbits as the following equation illustrated in Figure 2.2:

$$\vec{r}(\theta) = \frac{p}{1 + e\cos(\theta)} \tag{2.17}$$

where $e = pA$.

Note that the earth is at one of the foci. Point A is the nearest point to earth, called perigee; and point B is the furthest point, called apogee. The radius of earth is $R_E = 6378\,\text{km}$. The relationships between the parameters are listed as the following:

$$r_{min} = h_A + R_E = a(1 - e) = p/(1 + e) \tag{2.18}$$

$$r_{max} = h_B + R_E = a(1 + e) = p/(1 - e) \tag{2.19}$$

$$a = (r_{min} + r_{max})/2 = (h_A + h_B)/2 + R_E = p/(1 - e^2) \tag{2.20}$$

$$b = (a^2 - c^2)^{1/2} = a(1 - e^2)^{1/2} = (r_{min}r_{max})^{1/2} = p/(1 - e^2)^{1/2} \tag{2.21}$$

$$p = b^2/a \tag{2.22}$$

$$c = ae = (r_{max} - r_{min})/2 = (h_B - h_A)/2 = ep/(1 - e^2) \tag{2.23}$$

$$e = c/a = (r_{max} - r_{min})/(r_{max} + r_{min}) = (h_B - h_A)/(h_B + h_A + 2R_E) \tag{2.24}$$

$$p = a(1 - e^2) = 2r_{max}r_{min}/(r_{max} + r_{min}) \tag{2.25}$$

2.1.4 Kepler's second law: area swept by a satellite vector

From Equations (2.10) and (2.15), we can get the following equation:

$$\frac{1}{2}r^2\frac{d\theta}{dt} = \frac{D}{2} = \frac{1}{2}\sqrt{p\mu} \quad \text{(Constant)} \tag{2.26}$$

This agrees with Kepler's second law.

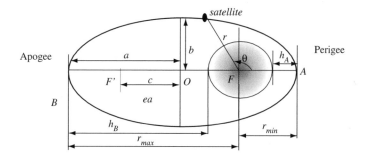

Figure 2.2 Orbit with major axis of orbit (AB) and semi-major axis of orbit (AO)

2.1.5 Kepler's third law: orbit period

Integrating from 0 to T, the period of the orbit, we get that the left-hand side of Equation 2.26 equals the area of the ellipse and the equation becomes:

$$ab\pi = \int_0^T \frac{1}{2}\sqrt{p\mu} \, dt = \frac{T}{2}\sqrt{p\mu} \tag{2.27}$$

Hence using Equation (2.22), we can rewrite Equation (2.27) into

$$T = \frac{2ab\pi}{\sqrt{p\mu}} = \frac{2\pi}{\mu} a^{\frac{3}{2}} \tag{2.28}$$

This agrees with Kepler's third law.

According to period T of a satellite completing the orbit, satellite orbits can be classified as the following types:

- Geostationary orbit, if $T = 24$ hours and $i = 0$. The orbit has the same period a sidereal day. To be more precise, a sidereal day equals 23 h 56 min 4.1 s, which totals 86 154 s. It can be calculated that the semi-major axis $a = 42\,164$ km and the satellite velocity $v = 3075$ m/s.
- Geosynchronous orbit, if $T = 24$ hours and $0 < i < 90°$.
- Non-geosynchronous orbit, if $T \neq 24$ hours. More satellites have to be used to form a constellation with a number of orbit planes and a few satellites arranged in each plane for continuous service to a coverage area.

2.1.6 Satellite velocity

Substituting Equations (2.10), (2.13) and (2.15) into Equation (2.8) gives

$$\left[\frac{d^2r}{dt^2} - \left(\frac{p\mu}{r^3}\right)^2\right] + \frac{\mu}{r^2} = 0$$

Then making use of Equation (2.25) and integrating both sides of the above equation gives:

$$\frac{1}{2}\left(\frac{dr}{dt}\right)^2 + \left(\frac{a(1-e^2)\mu}{2r^2} - \frac{\mu}{r}\right) = \int 0 \, dr = E \text{ (Constant)} \tag{2.29}$$

At perigee, there are boundary values $r = r_{min} = a(1-e)$ and $\dfrac{dr}{dt} = 0$. Substituting into Equation (2.29) gives:

$$E = \frac{a(1-e^2)\mu}{2a^2(1-e)^2} - \frac{\mu}{a(1-e)} = -\frac{\mu}{2a} \tag{2.30}$$

Using Equations (2.5), (2.25), (2.26) and (2.29), the satellite velocity v can be calculated as:

$$v^2 = \left(\frac{d\vec{r}}{dt}\right)^2 = \left(\frac{dr}{dt}\right)^2 + \left(r\frac{d\theta}{dt}\right)^2 = \left(\frac{dr}{dt}\right)^2 + \frac{a(1-e^2)\mu}{r^2}$$

$$= 2E + \frac{2\mu}{r} = \mu\left(\frac{2}{r} - \frac{1}{a}\right) \tag{2.31}$$

It can be seen from Equations (2.17), (2.18) and (2.19) that v reaches a maximum at perigee where $\vec{r} = r_{min} = a(1-e)$, and a minimum at apogee when $\vec{r} = r_{max} = a(1+e)$, and the speed can be calculated as:

$$v_{min} = \sqrt{\frac{\mu(1-e)}{a(1+e)}} \tag{2.32}$$

$$v_{max} = \sqrt{\frac{\mu(1+e)}{a(1-e)}} \tag{2.33}$$

2.2 Satellite orbit parameters

In order to define the trajectory of a satellite in space, orbital parameters are required. The shape of an orbit is described by two parameters: the semi-major axis (a) and the eccentricity (e). The position of the orbital plane in space is specified by means of other parameters: the inclination (i), the right ascension of the node (Ω) and argument of perigee (ω). The semi-major axis (a) also determines the period (T) of a satellite orbiting earth.

2.2.1 Semi-major axis (a)

This element specifies the size of the orbit (in km). It is defined as one-half of the major axis, which is the length of the chord that passes through the two foci of the orbit's ellipse. For circular orbits, the semi-major axis (a) is simply the radius of the circle. Figure 2.2 illustrates the semi-major axis and other the orbit parameters.

2.2.2 Eccentricity (e)

Eccentricity (e) determines the shape of the orbit. It is a unitless geometric constant with a value between zero and one. A pure circular orbit has an eccentricity of zero. The following values of e define the types of the satellite orbits:

- For $e = 0$, the trajectory is a circle.
- For $e < 1$, the trajectory is an ellipse.
- For $e = 1$, the trajectory is a parabola.
- For $e > 1$, the trajectory is a hyperbola.

2.2.3 Inclination of orbit (i)

The inclination (i) determines the tilt of the orbital plane with respect to the equatorial plane of the earth and is an angle measured in degrees. It is defined as the angle between the two planes shown in Figure 2.3. An orbit with an inclination of zero degrees is called an equatorial orbit; an orbit with an inclination of 90 degrees is called a polar orbit. Inclinations of less than 90 degrees correspond to direct orbits (i.e., the satellite is rotating around the North Pole heading east) and inclinations between 90 and 180 degrees correspond to retrograde orbits (i.e., the satellite is rotating around the North Pole heading west). Inclinations are limited to a maximum of 180 degrees.

According to the inclination angle i of the orbital plane, the angle between the earth's equatorial plane and the satellite's orbital plane, satellite orbits, as shown in Figure 2.4 can be classified as the following types:

- Equatorial orbit, if $i = 0$. The orbital is on the same plane as the earth equator.
- Incline orbit, if $0 < i < 90°$. The orbital plane and earth equator plane have an angle of i degrees.
- Polar orbit, if $i = 90°$. The orbital plane contains the earth pole.

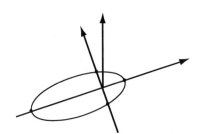
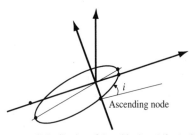

(a) Orbit plane on the equator plane ($i = 0$)   (b) Inclination of the orbit plane i ($0 < i < 90°$)

Figure 2.3   Inclination of orbit i

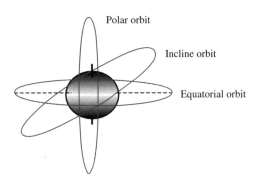

Figure 2.4   Equatorial, incline and polar orbits

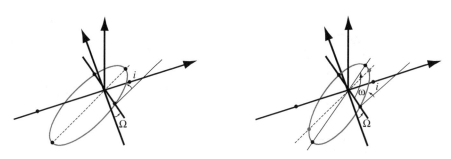

(a) Right ascension of the node Ω ($0 \leq \Omega \leq 360°$)   (b) Argument of perigee ω ($0 \leq \omega \leq 360°$)

Figure 2.5   Right Ascension of the node Ω and argument of perigee ω

2.2.4 Right ascension of the node (Ω) and argument of perigee (ω)

Right ascension of the node (Ω) determines the rotation of the orbital plane, and is an angle measured in degrees. It is defined as the angle in the equatorial plane between the lines formed by the intersection of the orbital planes and the equatorial plane as shown in Figure 2.5(a). Thus, this 'longitude' is not a normal longitude tied to the earth's surface, but is an angle measured in the equatorial plane. It is therefore also called right ascension of the ascending node (RAAN). Argument of perigee (ω) determines the rotation of perigee on the orbital plane as shown in Figure 2.5(b), and is an angle measured in degrees.

2.3 Useful orbits

According to Kepler's third law, the orbital period of a satellite is proportional to its distance from earth. Satellites in low orbits, altitudes of a few hundred to a thousand km, have orbital periods less than two hours; in contrast, the moon, at an altitude of about 380 000 km, has an orbital period of about 27 days, which is the base of the lunar month of the Chinese calendar; and earth has a orbit period of about 365 days as the base of a year.

2.3.1 Geosynchronous earth orbits

Between the extremes is an altitude that corresponds to an orbital period of one day. A satellite in a circular orbit at such an altitude revolves around earth at the same speed as earth's rotation. This altitude is 35,786.6 km, and the orbit is called a synchronous or geosynchronous orbit.

   If the orbital plane of a satellite is not coincident with earth's equatorial plane, then the orbit is said to be inclined, and the angle between the orbital plane and the equatorial plane is known as the orbit's inclination. In a geosynchronous orbit, the point on the earth directly below the satellite moves north and south in a narrow figure-eight pattern as shown in Figure 2.6 with northern and southern latitude limits corresponding to the inclination. A constellation of geosynchronous satellites is needed to provide continuous coverage of an area.

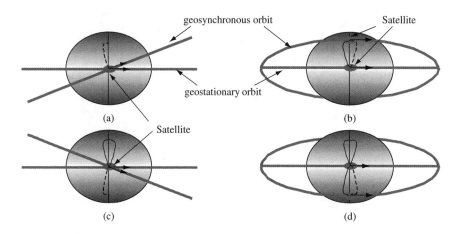

Figure 2.6   Footprints of geosynchronous satellites

2.3.2 Geostationary earth orbits (GEOs)

If inclination of a geosynchronous orbit is zero (or near zero), the satellite remains fixed (or approximately fixed) over one point on the equator. Such an orbit is known as a geostationary orbit.

An advantage of the geostationary orbit is that antennas on the ground, once aimed at the satellite, need not continue to rotate. Another advantage is that a satellite in this type of orbit continuously sees about one-third of earth.

At an altitude of 35 786.6 km above the equator, the angular velocity of a satellite in this orbit matches the daily rotation of the earth's surface, and this orbit has been widely used as a result. Propagation delay between earth station and satellite is around 0.125 ms; this leads to the widely quoted half-second round-trip latency quoted for communications via geostationary satellite.

One disadvantage of the geostationary orbit is that the gravity of the sun and moon disturb the orbit, causing the inclination to increase. The satellite's propulsion can counter this disturbance, but since the amount of fuel a satellite can carry is limited, increased inclination may remain a problem in some scenarios. The geostationary orbit's finite capacity is another disadvantage; satellites using the same frequencies must be separated to prevent mutual interference. Providing coverage of high latitudes (above 75°) is generally not possible, so full earth coverage cannot be achieved with a geostationary constellation.

2.3.3 High elliptical orbits (HEOs)

High elliptical orbits (HEOs) differ from the circular orbits. They provide coverage only when the satellite is moving very slowly relative to the ground while at apogee, furthest from the earth's surface, and power requirements in link budgets are dimensioned for this distance. Figure 2.7 illustrates a typical elliptical orbit.

These orbits are generally at an inclination of 63.4 degrees so that the orbit is quasi-stationary with respect to the earth's surface. This high inclination enables coverage of high

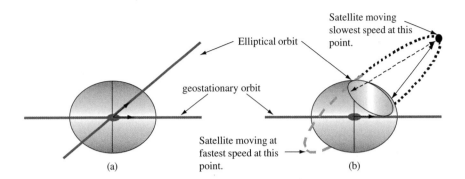

Figure 2.7 A typical high elliptical orbit

latitudes, and Russian use of Molnya and Tundra elliptical orbits for satellite television to the high-latitude Russian states is well known.

As elliptical orbits are the exception rather than the general rule, and generally provide targeted selection, rather than general worldwide coverage, we will not consider them further here.

2.3.4 Notations of low earth orbit (LEO) satellite constellations

LEO satellites move faster than the rotation of earth. Therefore, they appear to continuously circulate around the earth. Figure 2.8 illustrates the orbit and footprint of an LEO satellite. A constellation of satellites is needed to provide global coverage.

Satellite orbit plane and the point on the orbit can allocate a satellite position. For a constellation of satellites, there are simple notations or rules to describe it for a global coverage. There are two forms of notations used to describe satellite constellations in the literature – Walker notation and Ballard notation:

- Walker notation ($N/P/p$): this notation refers to: (number of satellites per plane/number of planes/number of distinct phases of planes to control spacing offsets in planes).
- Ballard notation (NP, P, m): this notation refers to: (total number of satellites NP, number of planes P, harmonic factor m describing phasing between planes).

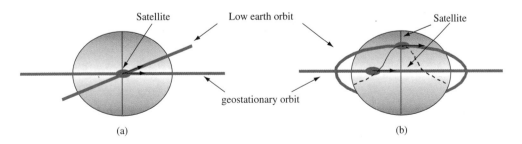

Figure 2.8 Footprint of a LEO satellite

The Walker notation is more commonly seen, although the Ballard notation can more accurately describe possible offsets between planes.

2.3.5 Orbital perturbations

There are many subtle effects that perturb earth satellite orbits, invalidating the simple orbits predicted by two-body gravity equations. Some of the factors that perturb the orbit are:

- Earth's oblateness: the earth bulges at the equator, which leads to a much more complex gravity field than the spherically symmetric field of a 'point' gravity source.
- Solar and lunar effects: these effects of the sun and moon are the most influential gravitational forces on earth satellites besides the earth's own field.
- Atmospheric drag: the friction that a satellite encounters as it passes through the diffuse upper layers of the earth's atmosphere.
- Solar radiation pressure: solar radiation pressure is caused by collisions between the satellite and photons radiating from the sun, which are absorbed or reflected.

2.3.6 Satellite altitude and coverage

The higher the altitude, the longer the distance one is able to see. The longer the distance, the higher the transmission power is required for communications. Figure 2.9 illustrates these simple relationships.

It can be seen that the GEO satellite has the highest altitude covering the largest area, the LEO satellite has the lowest altitude covering the smallest area and the MEO is in between. The GEO satellite also provides a fixed and continuous coverage of the area, but the LEO and MEO satellites will gradually move away from the coverage areas. These imply that the LEO and MEO satellites provide advantages for small and lower terminals at the cost of complicated constellations of satellite systems. Such complications also incur the high cost of deployment and operation of the constellations.

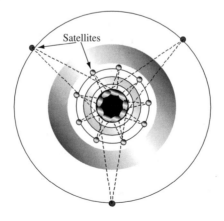

Figure 2.9   Relationships between altitude and coverage

Though research and development of constellation satellites in recent years have achieved excellence in technical aspects, economically, it will take time to exploit the full benefit of the constellations by reducing costs of the systems and creating revenues from new services and applications.

Satellite networking is to provide coverage of the earth, particularly the areas beyond coverage of terrestrial networks. Therefore, in this section we should take the earth-centric point of view to discuss the relationships between satellite networking and the earth.

2.3.7 Antenna gain and beam-width angle

In radio communication, an antenna is a very important part of the transmission link. It helps to concentrate radiation power toward the receiving antenna, but the receiver receives only a small amount of power. Most of the power is spread onto a wide area. Figure 2.10 illustrates a typical antenna pattern determined by the size of the antenna and transmission frequency used.

The maximum gain of an antenna is expressed as:

$$G_{max} = (4\pi/\lambda^2)\eta A$$

where $\lambda = c/f$ and the velocity of light $c = 3 \times 10^8$ m/s and f is the frequency of the electromagnetic wave. The geometric surface of the antenna $A = \pi D^2$ with a diameter D.

In direction of θq with respect to bore sight, the value of gain (relative to isotropic antenna) is:

$$G(\theta)dBi = G_{max,dBi} - 12(\theta - \theta_{3dB})$$

The angular beam width of the radiation pattern is:

$$\theta_{3dB} = 70(\lambda/D) = 70[c/(fD)]$$

2.3.8 Coverage calculations

The height of the satellite determines coverage of global beam antenna and the distance of earth stations from the edge of the coverage to the satellite. Figure 2.11 shows the relation between elevation angle β and altitude h_E.

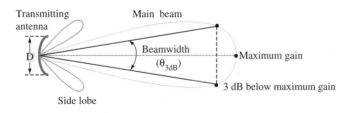

Figure 2.10 Antenna radiation pattern

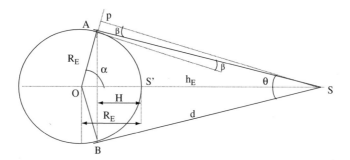

Figure 2.11   Relation between elevation angle and altitude

In Figure 2.11, OPS is a right-angled triangle. We can calculate as the following:

$$Sp = (h_E + R_E) \sin a \tag{2.34}$$

$$Op = (h_E + R_E) \cos a$$

$$Ap = Sp \tan \beta \tag{2.35}$$

As we also have $Ap = AS \sin \beta$ together with Equations (2.34) and (2.35), we can get: $AS = Sp \tan \beta / \sin \beta = (h_E + R_E) \sin a / \cos \beta$.

For a special case when $\beta = 0$, $AS = (h_E + R_E) \sin a$. We can also calculate $\cos a = R_E/(h_E + R_E)$, then, $\sin^2 a = (1 - \cos^2 a) = (1 - R_E/(h_E + R_E))^2$. Therefore:

$$(AS)^2 = (h_E + R_E)^2 - R_E^{\,2} \tag{2.36}$$

We can also calculate directly, as OAS becomes a right-angled triangle when $\beta = 0$

$$(AS)^2 + R_E^{\,2} = (h_E + R_E)^2$$

This gives the same result as Equation (2.36).

The maximum coverage area can be calculated as:

$$\text{Coverage} = 2\pi R_E H = 2\pi R_E^{\,2}(1 - R_E/(h_E + R_E))$$

2.3.9 Distance and propagation delay from earth station to satellite

Two angles are used to locate the satellite from any point on the surface of the earth:

1. Elevation angle (β): the elevation angle is the angle between the horizon at the point considered and the satellite, measured in the plane containing the point considered, the satellite and the centre of the earth.
2. Azimuth angle (α): the azimuth angle is the angle measured in the horizontal plane of the location between the direction of geographic north and the intersection of the plane containing the point considered, the satellite and the centre of the earth.

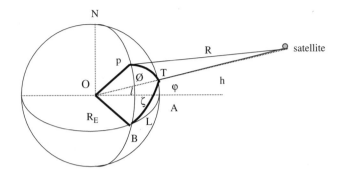

Figure 2.12 Distance between earth station and satellite

The distance from the centre of the earth to the satellite is: $r = h + R_E$.
The distance between earth station and satellite can be calculated as:

$$R^2 = R_E^2 + r^2 - 2R_E r \cos \emptyset$$

$$\tan \beta = [\cos \emptyset - (R_E/r)]/\sin \emptyset$$

$$\sin \alpha = (\sin L \cos \varphi)/\sin \emptyset$$

where $\cos \emptyset = \cos L \cos \varphi \cos l + \sin \varphi \sin l$, for GEO, we have $\varphi = 0$, then $\cos \varphi = 1$ and $\sin \varphi = 0$.

The propagation delay from an earth station to a satellite can be calculated as:

$$T_p = R/c$$

where the velocity of light $c = 3 \times 10^8$ m/s.

Therefore, one-way propagation delay from one station to another station is:

$$T_p = (R_1 + R_2)/c$$

where R_1 and R_2 are the distances from earth stations to the satellite.

2.4 Satellite link characteristics and modulations for transmissions

The basic transmission signal components include the carrier wave and modulating signal. The carrier wave is a continuous sinusoidal wave, which contains no information. The modulating signal is the message signal to be transmitted over the carrier. It can modulate (change) the amplitude, frequency or phase of the carrier wave, leading to different modulation schemes: amplitude modulation (AM), frequency modulation (FM) and phase modulation (PM). In the receiving end, the demodulator can separate the message signal

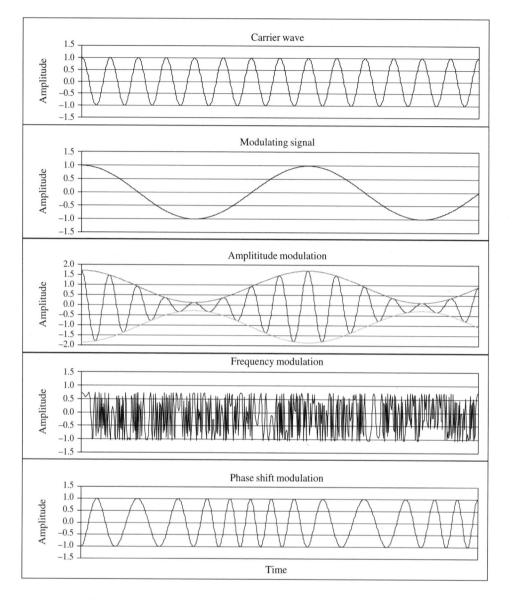

Figure 2.13   Carrier waves, modulating signals and modulated signals

from the carrier wave by a demodulation process depending on the modulation scheme used in transmission. Figure 2.13 illustrates the different modulating processes. Modulations enable transmission of message signals at the carrier frequency. They can be used to provide multiple accesses to the radio frequency bands at the frequency domain.

Beside the modulation signal, the satellite propagation channel conditions may also cause changes to the amplitude, frequency or phase of the carrier wave. Therefore, it can cause transmission errors. It is also important to have an error correction coding scheme to recover as much as possible errors occurred during the transmission.

2.4.1 Satellite link characteristics

Unlike cable, the quality of the satellite link cannot be controlled. The satellite link may cause propagation impairments depending on the following factors:

- Operating frequency: as signal attenuation by gas absorption, the severity of tropospheric impairments increases with frequency.
- Antenna elevation angle and polarisation: the length of the propagation path passing through the troposphere varies inversely with elevation angle. Accordingly, propagation losses, noise and depolarisation also increase with the decreasing elevation angle. Rain attenuation is slightly polarisation sensitive. Depolarisation is also polarisation sensitive, with circular polarisation being the most susceptible.
- Earth station altitude: because less of the troposphere is included in paths from higher altitude sites, impairments are less at high altitude.
- Earth station noise temperature: this is the level of sky noise temperature to system noise temperature, thus the effect of sky noise on the downlink signal-to-noise-ratio.
- Local meteorology: the amount and nature of the rainfall in the vicinity of the earth station are the primary factors in determining the frequency and most propagation impairments.
- Figure of merit (G/T): a figure of merit expresses the efficiency of a receiver; G is the overall gain in decibels and T is the noise temperature, expressed in decibels relative to $10°$ Kelvin. Hence G/T is expressed in decibels, with the appended symbol dB/K to indicate the Kelvin scale of temperature.

The free-space loss is the major power loss due to long propagation of the satellite link. Though it is larger than all the other losses, the other losses can also play important part in adding several dB. At frequencies of 10 GHz and above, losses due to atmospheric absorption and rain can be significant. At these frequencies, electromagnetic waves interact and resonate with molecules of atmospheric gases to cause signal attenuation. The most important resonant attenuation occurs at 22.235 GHz due to water vapour and between 53 to 65 GHz due to oxygen. Loss at other frequencies is usually small (less than 1 dB). These atmospheric losses can be calculated and included in the link equation to determine its impact on the overall quality.

At lower frequencies, less than 1 GHz, losses due to multipath fading and scintillation are predominant. Faraday rotation due to the total electron count in the atmosphere becomes significant, but, using proper polarisation, these losses can be controlled in high-gain communications.

2.4.2 Modulation techniques

As an example, we can describe the carrier wave mathematically as:

$$c_r(t) = A_c \cos(2\pi f_c t)$$

where A_c is carrier amplitude and f_c is carrier frequency.

We can describe the amplitude-modulated wave as:

$$s(t) = [A_c + k_a m(t)] \cos(2\pi f_c t)$$

where $m(t)$ is the signal and k_a is the amplitude sensitivity of the modulator.
   We can describe the frequency-modulated wave as:

$$s(t) = A_c \cos[2\pi(f_c + k_f m(t))t]$$

where k_f is the frequency sensitivity of the modulator.
   We can describe the phase-modulated wave as:

$$s(t) = A_c \cos[2\pi f_c t + k_p m(t)]$$

where k_p is the phase sensitivity of the modulator.
   In the frequency-modulated wave, let $\theta_f(t) = 2\pi(f_c + k_f m(t))t$, it can be seen that $m(t)$
causes the change of frequency $\Delta f = k_f m(t)\Delta t$ which is equivalent to the change of phase
$\Delta\theta_f = 2\pi(f_c + \Delta f)\Delta t = 2\pi[f_c + k_f m(t)]\Delta t$, hence:

$$\frac{d\theta_f(t)}{dt} = 2\pi f_c + 2\pi k_f m(t) \text{ and } \theta_f(t) = 2\pi f_c t + 2\pi k_f \int_0^t m(t)dt,$$

i.e., the frequency-modulated wave is generated by using the carrier frequency and integration
of the message signal.
   In the phase-modulated wave, $\theta_p(t) = 2\pi f_c t + k_p m(t)$, i.e., the phase-modulated wave
is generated by using the carrier frequency and the message signal. Therefore, the phase-
modulated wave can be deduced from the frequency-modulated wave, and vice versa. Phase
modulation and frequency modulation are also called angle modulation.

2.4.3 Phase shift keying (PSK) schemes for satellite transmissions

As the satellite link conditions may change during transmission for digital transmission, the
altitude of the transmission may also change with the link condition. It is difficult to use AM.
FM is difficult to implement and not efficient on bandwidth utilisation. Compared with the
AM and FM schemes, PM has the advantages of FM and is easy to implement. Therefore,
for satellite transmission, PM is used, and many different PM schemes have been developed
to trade off among power, frequency and implementation efficiency.
   The simplest modulation scheme is BPSK, which is capable of transmitting one bit in a
carry frequency period. The higher the modulation order, the larger the bandwidth efficiency
is. QPSK, for example, is capable of transmitting two bits in the carry frequency period for
the same E_b/N_0 value, but at the cost of error performance.
   It is called coherent demodulation or coherent detect if the local oscillator is exactly
coherent or synchronous, in both frequency and phase, with the carrier wave used for
modulation. Otherwise, it is called non-coherent demodulation or non-coherent detect where
a different technique is used for demodulation using match filters.

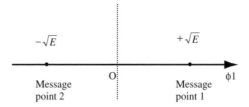

Figure 2.14 Signal-space diagram for coherent BPSK

2.4.4 Binary phase shift keying (BPSK)

In a coherent binary PSK (BPSK) system, the pair of signals used to represent binary symbols 1 and 0, respectively, is defined as:

$$
\begin{cases}
s_1 = \sqrt{\dfrac{2E_b}{T_b}} \cos(2\pi f_c t) \\[4mm]
s_2 = \sqrt{\dfrac{2E_b}{T_b}} \cos(2\pi f_c t + \pi) = -\sqrt{\dfrac{2E_b}{T_b}} \cos(2\pi f_c t)
\end{cases}
$$

These can be represented in a signal-space diagram with a basis function of unit energy:

$$
\phi_1 = \sqrt{\frac{2}{T_b}} \cos(2\pi f_c)
$$

as shown in Figure 2.14.

2.4.5 Quadrature PSK (QPSK)

Similarly, in a coherent quadrature PSK (QPSK) system, there are four signal vectors defined as:

$$
s_i = \begin{bmatrix} \sqrt{E} \cos\!\left((2i-1)\dfrac{\pi}{4}\right) \\[3mm] -\sqrt{E} \cos\!\left((2i-1)\dfrac{\pi}{4}\right) \end{bmatrix} \quad i = 1, 2, 3, 4.
$$

These can be represented in a signal-space diagram with two orthogonal basis functions of unit energy

$$
\phi_1 = \sqrt{\frac{2}{T_b}} \cos(2\pi f_c) \text{ and } \phi_2 = \sqrt{\frac{2}{T_b}} \sin(2\pi f_c)
$$

as shown in Figure 2.15.

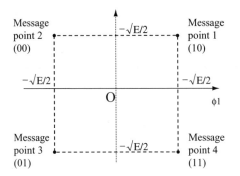

Figure 2.15   Signal-space diagram for coherent QPSK

2.4.6 Gaussian-filtered minimum shift keying (GMSK)

To improve the bandwidth efficiency and error performance, two techniques can be applied: minimum phase shift and shaping the rectangular pulse through a Gaussian filter.

Let W denote the 3 dB baseband bandwidth of the pulse-shaping filter. The transfer function $H(f)$ and impulse response $h(t)$ of the pulse filter are defined as the following respectively:

$$H(f) = \exp\left(-\frac{\log 2}{2}\left(\frac{f}{W}\right)^2\right)$$

and

$$h(f) = \sqrt{\frac{2\pi}{\log 2}} \exp\left(-\frac{2\pi^2}{\log 2}W^2 t^2\right)$$

The responses of a Gaussian filter to a rectangular pulse of unit amplitude and duration T_b (centred in the origin) are given by:

$$g(t) = \int_{T_b/2}^{T_b/2} h(t-\tau)d\tau = \sqrt{\frac{2\pi}{\log 2}}W\int_{T_b/2}^{T_b/2} \exp\left(-\frac{2\pi^2}{\log 2}W^2(t-\tau)^2\right)d\tau$$

The pulse response $g(t)$ constitutes the frequency shaping pulse of the GMSK modulator, with the dimensionless time-bandwidth product WT_b playing the role of a design parameter.

As WT_b is reduced, the time spread of the frequency shaping pulse is correspondingly increased. The limiting condition $WT_b = \infty$ corresponds to the case of ordinary MSK, and when WT_b is less than unity, increasingly more of the transmit power is concentrated inside the pass band of the GMSK signal.

2.4.7 Bit error rate (BER): the quality parameter of modulation schemes

Inherently, bit errors occur in satellite channels during transmission. The bit error rate (BER) depends on the signal-to-noise ratio (S/N) at the receiver. Thus for an acceptable level of error rate, a certain minimum signal-to-noise ratio must be ensured at the receiver and hence maintained at the transmitter.

The relationship between C/N and the bit error rate of the channel is a measure of performance for a digital link. This is computed from the carrier-to-noise density ratio, C/N_0 ratio, for a particular modulation scheme by:

$$\frac{E_b}{N_0} = \begin{cases} C/N_0 - 10\log_{10}(DataRate) \\ C/N - 10\log_{10}(DataRate/Bandwidth) \end{cases} \tag{2.37}$$

The data rate over bandwidth ratio, R/B, is called the spectrum efficiency or bandwidth efficiency of the modulation. For a given bandwidth, the E_b/N_0 value has to be large enough to achieve the transmission bit rate with a good error performance in terms of bit error ratio or probability of bit error.

There are error functions to compute symbol error rates. The number of bits per symbol is $\log_2(M)$, where M is coded levels of modulation scheme. The bit error rate p_b is related to the symbol error rate P_s by:

$$p_b = \frac{P_s}{\log_2(M)} \tag{2.38}$$

Theoretically, error performance can be calculated using the Gaussian probabilities as the following:

$$P(X > \mu x + \sigma_x) = Q(y) = \int_y^\infty \frac{1}{\sqrt{2\pi}} e^{-z^2/2} dz$$

$$Q(0) = 1/2, Q(-y) = 1 - (y), \text{when } y \geq 0.$$

$$erfc(y) \equiv \frac{2}{\sqrt{\pi}} \int_y^\infty e^{-z^2} dz = 2Q(\sqrt{2}y)$$

Table 2.1 shows the error performance common for some popular modulation schemes. Figure 2.16 shows some results calculated using the formulas in the table.

2.4.8 Satellite networking in the physical layer

In the context of the protocol reference model, satellite networking starts from the physical layer. The physical layer accepts frames from the link layer, then transmits the frame in the form of a bit stream to its peer entity via the satellite. Depending on the implementation of the satellite communication payload, there can be a transparent satellite simply forwarding the radio signal from an uplink to a downlink, or there can be an onboard processor (OBP) processing the digital signal then forward to the downlink. It is also possible to have even more complicated payloads including switching or routing functions.

Table 2.1 Modulation methods

Modulation scheme	P_E (symbol)
Coherent QPSK	$\frac{1}{2}erfc(\sqrt{E_b/N_0})$
Coherent BPSK	
Coherent MPSK	
MSK	$\frac{1}{2}erfc(\sqrt{E_b/N_0})$
GMSK	$\frac{1}{2}erfc(\sqrt{\alpha E_b/2N_0})$ where α is a constant depending on the time-bandwidth product WT_b
Coherent BFSK	$\frac{1}{2}erfc(\sqrt{E_b/2N_0})$
No coherent DPSK	$\frac{1}{2}\exp(-E_b/N_0)$
No coherent BFSK	$\frac{1}{2}\exp(-E_b/2N_0)$

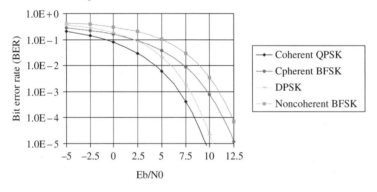

Figure 2.16 Noise performance of modulation schemes

Here the discussion focuses on the physical layer dealing with transmission and receiving the bit streams and radio signals over satellite. Figure 2.17 shows the physical layer functions of satellite networking in the context of the protocol reference model. It can be seen that a user terminal generates a bit stream. The encoders process the stream with error correction encoding function and channel coding function. The modulator uses the encoded signal to modulate the carry to transmit the signal over the satellite link. In the other side of the satellite network, a reverse processing takes place before getting the bit stream to the other user terminal. Within the satellite network, the processing is transparent to users, which can include different OBP functions or even inter-satellite links.

In a wired network, a digital bit stream can be encoded into baseband signals and transmitted directly along the wire. However, satellite uses radio links for transmission, hence

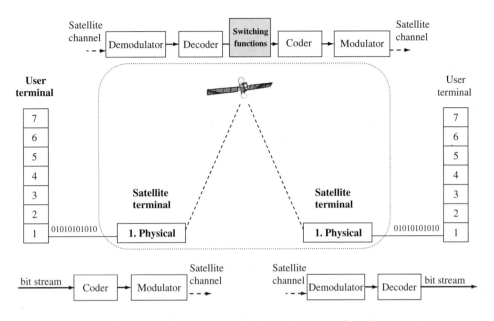

Figure 2.17 Block diagram of physical layer functions of satellite networks

modulation is required so that the signal can be transmitted over a radio channel or carrier. In addition, error correction coding is used before the channel coding to correct possible transmission errors, hence improving transmission quality by reducing probability of error.

2.5 Forward error correction (FEC)

FEC techniques try to introduce some redundancy in the transmitted data, such that when a receiver receives the data, it uses this redundancy to detect and correct errors if there are any caused by transmission, as illustrated in Figure 2.18. FEC codes consist of a wide range of classes. We give brief introductions to only some of them including linear block codes, cyclic codes, trellis and convolutional codes and turbo codes.

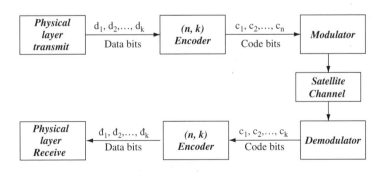

Figure 2.18 Forward error correction (FEC) coding

2.5.1 Linear block codes

Block codes are 'memory-less' codes that map k input binary signals to n output binary signals, where $n > k$ for redundancy.

Let $m = [m_0, m_1, \ldots, m_{k-1}]$ be the message bits, $b = [b_0, b_1, \ldots, b_{n-k-1}]$ and $\mathbf{P}$ be the $k \times (n-k)$ coefficient matrix known to transmitter and receiver, we can generate the parity bits as the following:

$$b = m\mathbf{P} \tag{2.39}$$

If the code word $c = [b\!:\!m]$ is transmitted, we will be able to detect some errors or even correct some errors by making use of Equation (2.39).

An example of a linear coder is the Hamming coder (n, k), where block length is $n = 2^r - 1$, number of message bits is $k = 2^r - r - 1$ (or $k + r - 1 = 2^r$), and number of parity bits $n - k = r$.

The Bose–Chaudhuri–Hocquenghem (BCH) code is a class of linear block code with the parameters: block length is $n = 2^m - 1$, number of message bits is $k \geq (n - mt)$, and minimum distance is $d_{min} \geq 2t + 1$, where m is any integer number and $t = (2^m - 1)/2$ is the maximum number of detectable errors.

The Reed–Solomon code (RS code) is a subclass of non-binary BCH codes. An RS coder of (n, k) is used to code m bit symbols into blocks consisting of $n = (2^m - 1)$ symbols $= m(2^m - 1)$ bits. A t-error correcting RS code has the parameters of: block length is $n = (2^m - 1)$ symbols, message size is k symbols, parity check size is $(n - k) = 2t$ symbols, and minimum distance is $d_{min} \geq 2t + 1$ symbols.

2.5.2 Cyclic codes

Let $g(X)$ be the polynomial of least degree of $(n - k)$, also called the generator polynomial of cyclic code, defined as the following:

$$g(X) = 1 + \sum_{i=1}^{n-k-1} g_i X^i + X^{n-k}$$

Let $m(X) = m_0 X^1 + m_1 X^2 + \cdots + m_{k-1} X^{k-1}$, and $b(X) = b_0 X^1 + b_1 X^2 + \cdots + b_{n-k-1} X^{n-k-1}$, we can divide $X^{n-k} m(X)$ by the generator polynomial $g(X)$ to obtain the remainder $b(X)$ and add $b(X)$ to $X^{n-k} m(X)$ to get $c(X)$.

Cyclic coding is often used for error check purposes because of its ability to detect error burst, also called cyclic redundancy check (CRC) code. Table 2.2 shows some useful CRC codes.

Binary (n, k) CRC codes are capable of detecting the following error patterns:

- All error bursts of length $n - k$ or less.
- A fraction of error bursts of length equal to $n - k + 1$ that the fraction equals to or greater than $1 - 2^{-(n-k-1)}$.

Table 2.2 Cyclic redundancy check (CRC) code

Code	Generator polynomial $g(X)$	$n-k$
CRC-12 code	$1 + X + X^2 + X^3 + X^{12} + X^{12}$	12
CRC-12 code (USA)	$1 + X^2 + X^{15} + X^{16}$	16
CRC-ITU code	$1 + X^5 + X^{12} + X^{16}$	16

- All combinations of $(d_{min} - 1)$ or fewer errors, where d_{min} is the minimum distance of a linear block code. The distance is defined as the number of locations of a pair of code vectors with their respective differing, also called the Hamming distance.
- All error patterns with an odd number of errors if the generator polynomial g(X) for the code has an even number of non-zero coefficients.

2.5.3 Trellis coding and convolutional codes

Trellis codes use 'memory' by remembering $(K - 1)$ input signals immediately preceding the target block of L input signals. These $(K - 1) + L = (K + L - 1)$ input binary signals are used in the generation of $n[(K - 1) + L]$ output binary signals corresponding to L input signals. Therefore the code rate is $L/[n(K + L - 1)]$.

Convolutional coding is a subset of trellis codes. The convolutional encoder can be associated to a finite state machine storing $(k - 1)$ message bits. At time j, the portion of the message sequence contains the most recent k bits $(m_{j-k+1}, m_{j-k+2}, \cdots, m_{j-1}, m_j)$, where m_j is the current bit. A convolutional decoder takes into account such memory, when trying to estimate the most likely sequence of data that produce the received sequence of code bits. This is called maximum likelihood method for decoding convolutional codes. In 1967, Andrew Viterbi developed a technique of decoding convolutional codes using this method that has since become the standard for decoding convolutional codes.

2.5.4 Concatenated codes

Linear block coders are more effective to correct burst of errors and convolutional coders are more effective on random errors, however, they can produce burst errors if there are too many random errors. In 1974, Joseph Odenwalder combined these two coding techniques to form a concatenated code.

The arrangement is that a block code is used as internal code first, then a convolutional code follows second as external code for encoding; on decoding, external convolutional code follows first, then the internal block code follows second.

Performance can be further enhanced if interleaving techniques between the two coding stages mitigate any burst that might be too long for the block code to deal with effectively.

Interleaving techniques are input-output mapping functions that permute the order of a stream of bits or symbols so that the position of the interleaved bit stream is independent from the original bit stream, and a burst of errors can be randomised into single random errors spreading into the bit stream when de-interleaved. A device or a function block of an interleaving technique is often called an interleaver.

2.5.5 Turbo codes

Turbo codes are the most powerful FEC, developed in 1993 by Claude Berrou. They enable communication transmissions closer to the Shannon limit. A turbo code consists of two coders and one interleaver so that the extrinsic information is used recursively to maximise the probability that the data is decoded correctly. Each of the two codes can be any of the existing coders. Without going into the detail of turbo codes, we will only illustrate the concepts of the turbo coder and decoder using Figures 2.19 and 2.20, respectively.

The encoder is simple and straightforward. The decoder is more complicated, where the extrinsic information is used recursively. The most convenient representation for this concept is to introduce the soft estimation of $x = [d_1, d_2, d_3, d_4]$ in decoder 1, expressed as the log-likelihood ratio:

$$l_1(d_i) = \log\left(\frac{P(d_i = 1|x, y, \tilde{l}_2(x))}{P(d_i = 0|x, y, \tilde{l}_2(x))}\right) \quad i = 1, 2, 3, 4$$

$$l_2(d_i) = \log\left(\frac{P(d_i = 1|x, z, \tilde{l}_1(x))}{P(d_i = 0|x, z, \tilde{l}_1(x))}\right) \quad i = 1, 2, 3, 4$$

$$l_1(x) = \sum_1^4 l_1(d_i), \tilde{l}_1(x) = l_1(x) - \tilde{l}_2(x)$$

$$l_2(x) = \sum_1^4 l_2(d_i), \tilde{l}_2(x) = l_2(x) - \tilde{l}_1(x)$$

where $\tilde{l}_2(x)$ is set as 0 in the first iteration. An estimation of the message $x' = [d'_1, d'_2, d'_3, d'_4]$ is calculated by hard limiting that log-likelihood ratio $l_2(x)$ at the out put of decoder 2, as the following

$$\hat{x} = sign(l_2(x))$$

where the sign function operates on each element of $l_2(x)$ individually.

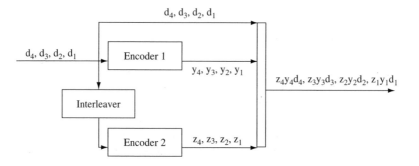

Figure 2.19   Block diagram of turbo encoder

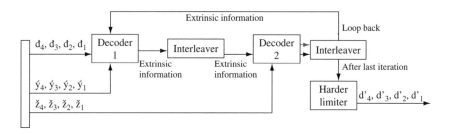

Figure 2.20 Block diagram of turbo decoder

2.5.6 Performance of FEC

The receiver can decode the data in most cases even it has been corrupted during transmission, making use of FEC techniques. The receiver may not be able to recover the data if there are too many bits corrupted, since it can only tolerate a certain level of errors. We have seen that the E_b/N_0 is the parameter affecting the error performance of satellite transmission for given codes and bandwidth resources. The FEC enables satellite links to tolerate higher transmission errors than the uncoded data in terms of error performance. This is very useful as sometimes satellite transmission alone may be difficult to achieve a certain level of performance due to limited transmission power at certain link conditions.

Let take an example: assume R is the information rate, the coded data rate R_c, as defined for a (n, k) block code, where n bits are sent for k information bits is $R_c = (R\,n/k)$. The relationship of required power between the coded and uncoded data for the same bit error rate is:

$$(C/R_c)/N_0 = (k/n)(C/R)/N_0 = (k/n)(E_b/N_0)$$

These codes, at the expense of larger required bandwidth or larger overhead (reduced throughput), provide a coding gain to maintain the desired link quality at the same available E_b/N_0. Without going through detailed mathematical analysis, we will only give a brief description using Figure 2.21.

2.6 Multiple access techniques

Considering that satellite communications use multiple access schemes on a shared medium. The access scheme refers to the sharing of a common channel among multiple users of possible multi-services. There are three principal forms of multiple access schemes as shown in Figure 2.22:

- frequency division multiple access (FDMA);
- time division multiple access (TDMA); and
- code division multiple access (CDMA).

Multiplexing is different from multiple access: it is a concentration function which shares the bandwidth resource from the same places while and multiple access shares the same resource from different places as shown in Figure 2.23.

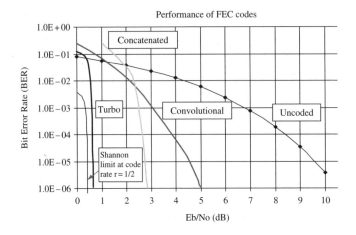

Figure 2.21   Comparison of FEC codes

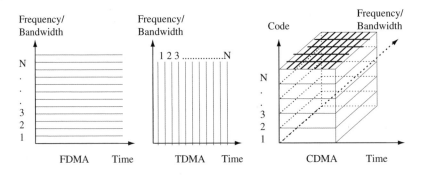

Figure 2.22   Multiple access techniques: FDMA, TDMA and CDMA

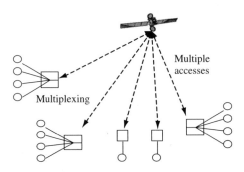

Figure 2.23   Comparison between the concepts of multiplexing and multiple access

2.6.1 Frequency division multiple access (FDMA)

FDMA is a traditional technique, where several earth stations transmit simultaneously, but on different frequencies into a transponder.

FDMA is attractive because of its simplicity for access by ground earth stations. Single channel per carrier (SCPC) FDMA is commonly used for thin-route telephony, VSAT systems and mobile terminal services for access networks. Multiplexing a number of channels to share a carrier for transit networks also uses FDMA. It is inflexible for applications with varying bandwidth requirements.

When using multiple channels per carrier for transit networks, FDMA gives significant problems with inter-modulation products (IMPs), and hence a few dB of back-off from saturation transmission power is required to overcome the problem of non-linearity at high power. The resultant reduction in EIRP may represent a penalty, especially to small terminals.

2.6.2 Time division multiple access (TDMA)

In TDMA, each earth station is allocated a time slot of bandwidth for transmission of information. Each time slot can be used to transmit synchronisation and control and user information. The synchronisation is achieved by using the reference burst time. TDMA is more convenient for digital processes and transmission. Figure 2.24 shows a typical example of TDMA.

Only one TDMA carrier accesses the satellite transponder at a given time, and the full downlink power is available for access. TDMA can achieve efficiencies in power utilisation and also in bandwidth utilisation if the guard time loss is kept at minimum when using more accurate timing techniques. This is widely used for transit networks due to high bandwidth utilisation at high transmission speed.

Clearly TDMA bursts transmitted by ground terminals must not interfere with each other. Therefore each earth station must be capable of first locating and then controlling the burst time phase during transmission. Each burst must arrive at the satellite transponder at a prescribed time relative to the reference burst time. This ensures that no two bursts overlap and that the guard time between any two bursts is small enough to achieve high transmission

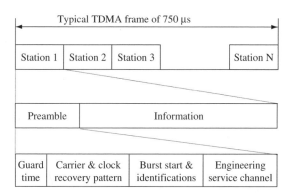

Figure 2.24   A typical example of satellite TDMA scheme

efficiency but large enough to avoid collision between time slots, since there is no clock capable of keeping time perfectly.

Synchronisation is the process of providing timing information at all stations and controlling the TDMA bursts so that they remain within the prescribed slots. All this must operate even though each earth station is fixed in relation with GEO satellites, because GEO satellites are located at a nominal longitude and typically specified to move within a 'window' with sides of 0.002 degrees as seen from the centre of the earth. Moreover, the satellite altitude varies as a result of a residual orbit eccentricity. The satellite can thus be anywhere within a box of $(75 \times 75 \times 85)\,\text{km}^3$ in space.

The tidal movement of the satellite causes an altitude variation of about 85 km, resulting in a round trip delay variation of about $500\,\mu\text{s}$ and a frequency change of signals known as the Doppler effect.

2.6.3 Code division multiple access (CDMA)

CDMA is an access technique employing the spread spectrum technique, where each earth station uses a unique spreading code to access the shared bandwidth. All theses codes are orthogonal to each other. To accommodate a large number of users, the code consists of a large number of bits resulting in wide-band signals from all users. It is also known as spread spectrum multiple access (SSMA). A feature of spread spectrum is that operation is possible in the presence of high levels of uncorrelated interference, and this is an important anti-jamming property in military communications.

The wide-band spreading function is derived from a pseudo-random code sequence, and the resulting transmitted signal then occupies a similar wide bandwidth. At the receiver, the input signal is correlated with the same spreading function, synchronised to the signal, to reproduce the originating data. At the receiver output, the small residual correlation products from unwanted user signals result in additive noise, known as self-interference.

As the number of users in the system increases, the total noise level will increase and degrade the bit-error rate performance. This will give a limit to the maximum number of simultaneous channels that can be accommodated within the same overall frequency allocation. CDMA allows gradual degradation of performance with increasing number of connections.

2.6.4 Comparison of FDMA, TDMA and CDMA

A brief comparison of FDMA, TDMA and CDMA is provided in Table 2.3. In satellite networking, we are more concerned the properties concerning efficient utilisation of bandwidth and power resources; ultimately the capacity that the multiple access techniques can deliver.

2.7 Bandwidth allocation

Multiple access schemes provide mechanisms to divide the bandwidth into suitable sizes for the required applications and services. Bandwidth allocation schemes provide mechanisms to allocate the bandwidth in terms of transmission bandwidth and time.

Bandwidth allocation schemes can be typically categorised into three classes: fixed assignment access; demand assignment multiple access (DAMA) adaptive access; and random

Table 2.3 Comparison of main multiple access method properties

Characteristic	FDMA	TDMA	CDMA
Bandwidth utilisation	Single channel per carrier (SCPC)	Multiple channels per carrier – partial allocation	SCPC, partial or full allocation
Interference rejection	Limited	Limited with frequency hopping	Can suppress interference, up to noise limit
Inter-modulation effects	Most sensitive (most back-off required)	Less sensitive (less back-off required)	Least sensitive (least back-off required)
Doppler frequency shift	Bandwidth limiting	Burst time limiting	Removed by receiver
Spectrum flexibility	Uses least bandwidth per carrier	Moderate bandwidth use per carrier	Largest demand for contiguous segment
Capacity	Basic capacity available	Can provide capacity improvement through hopping	Capacity indeterminate due to loading unknowns

access. These techniques can be used to meet the needs of different types of user traffic requirements in terms of time durations and transmission speeds. These schemes can be used individually or in combination, depending on applications.

2.7.1 Fixed assignment access

With fixed assignment, a terminal's connection is permanently assigned a constant amount of bandwidth resources for the lifetime of the terminal or for a very long period of time (years, months, weeks or days). This means that when the connection is idle, the slots are not utilised (i.e. they are wasted). For example, for transit networks, network bandwidth resources are allocated using fixed assignment based on long-term forecasts on traffic demands.

2.7.2 Demand assignment

Demand assignment allocates bandwidth resources only when needed. It has two variables: time duration and data rate. The time can be fixed or variable. For a given time duration, the data rate can be fixed or variable. With fixed rate allocation, the amount of bandwidth resources is fixed, which means that it is not very efficient if data rate changes over a wide range.

With variable rate allocation, the allocated bandwidth resources change with the changing data rate. If the changing patterns are unknown to the system, it is also difficult to meet the traffic demand. Even if signalling information is used, the propagation delay in the satellite networks makes it difficult to response to short-term demands.

Normally this scheme is used for demands of short period time and limited variation in terms of hours and minutes.

It also allows bandwidth allocation depending upon the instantaneous traffic conditions. To accommodate a combination of traffic types, bandwidth resources can be partitioned into several sections, each operating under its own bandwidth allocation schemes. The system observes the traffic conditions and makes adjustments dynamically according to the traffic conditions. This is also called the dynamic allocation scheme or adaptive allocation scheme.

2.7.3 Random access

When bandwidth demands are very short such as a frame data bits, it becomes impractical and there is too much overhead for any allocation scheme to make efficient use of bandwidth resources. Therefore, random access is the obvious option.

It allows different terminals to transmit simultaneously. Because the transmission is very short, the transmission has a very high success rate for low traffic load conditions. The transmissions may collide with each other. The chance of collisions increases with the increase of traffic load conditions. When the transmission is corrupted during transmission due to collision (or transmission), data has to been re-transmitted. The system also needs packet error or loss correction by observing transmitted data or acknowledgements from the receiver.

Such a scheme is based on the contention scheme. The contentions have to be resolved to increase the chance of success. Normally if there is any collision, the transmitting terminals back off their transmission for random period of times and increases the back-off to a longer period if collision occurs again until the contention is resolved. Back-off effectively reduces traffic load gradually to a reasonable operational level.

Random access can achieve a reasonable throughput, but cannot give any performance guarantees for individual terminals due to the nature of random access. Typical examples of random access schemes are aloha and slotted aloha. It can also work with the other schemes.

2.8 Satellite networking issues

After discussing the connections between ground earth stations and satellites, we now discuss how to link the satellites into networks. For transparent satellites, a satellite can be considered as a mirror 'bending' the link in the sky to connect ground earth stations together. For satellites with on-board processing (OBP) or on-board switching (OBS), a satellite can be considered as a node in the sky. Without losing generality, we will consider satellites as network nodes in the sky.

2.8.1 Single hop satellite connections

In this type of configuration, any end-to-end connection is routed through a satellite only once. Each satellite is set up as an 'island' to allow network nodes on the ground to be interconnected with any other ground station via the island. The topology of satellite networks forms a star, where the satellite is in the centre as shown in Figure 2.25.

2.8.2 Multi-hop satellite connections

In this type of configuration, an end-to-end connection is routed through the satellite network more than once, through the same satellite or different satellites. In the former case, it is widely used in very small aperture terminal (VSAT) networks where the signal

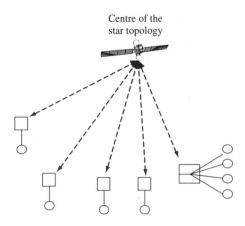

Figure 2.25   Single hop topology with satellite at the centre

between two terminals is too weak to make a direct communication and a large ground
hub is used to boost the signal between the communicating terminals. In the latter case, one
hop may not be far enough to reach remote terminals, therefore more hops are used for the
connections. The topology of the satellite network forms a star with a ground hub at the
centre of the star or multiple stars where the hubs are interconnected to link the satellites
together as shown in Figure 2.26.

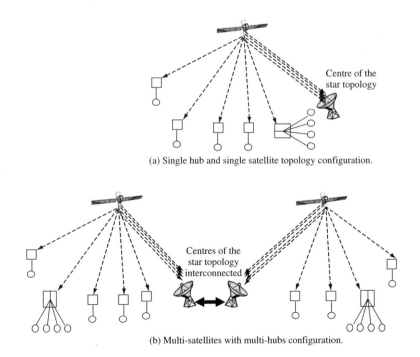

Figure 2.26   Multiple hops topology with hub at the centre

2.8.3 Inter-satellite links (ISL)

To reduce the earth segment of the network connections, we introduce the concept of inter-satellite links. Without ISL, the number of ground earth stations will increase to link more satellites together, particularly for LEO or GEO constellations where the satellites continuously moving across the sky. The topology of the network also changes with the movement of the constellation.

As the positions between satellites are relatively stable, we can link the satellite constellations together to form a network in the sky. This allows us to access the satellite sky network from the earth with fewer stations needed to link all the satellites into a network as shown in Figure 2.27.

Another advantage of using ISL is that satellites can communicate directly with each other by line of sight, hence decreasing earth–space traffic across the limited air frequencies by removing the need for multiple earth–space hops. However, this requires more sophisticated and complex processing/switching/routing on-board satellites to support the ISL. This allows completion of communications in regions where the satellite cannot see a ground gateway station, unlike the simple 'bent-pipe' satellites, which act as simple transponders.

For circular orbits, fixed fore and aft ISL in the same plane have fixed relative positions. For satellites in different orbit planes, the ISL have changing relative positions, because the line-of-sight paths between the satellites will change angle and length as the orbits separate and converge between orbit crossings, giving rise to:

- high relative velocities between the satellites;
- tracking control problems as antennas must slew around; and
- the Doppler shift effect.

In elliptical orbits, a satellite can see that the relative positions of satellites 'ahead' and 'behind' appear to rise or fall considerably throughout the orbit, and controlled pointing of the fore and aft intra-plane links are required to compensate for this, whereas inter-plane cross-links between quasi-stationary apogees (quasi GEO constellation) can be easier to maintain.

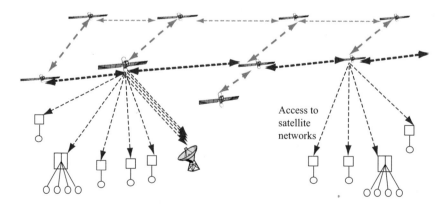

Figure 2.27   Satellite networks with inter-satellite links

We can see that it is a trade-off between complexity in the sky or on earth, i.e. it is possible to design a satellite constellation network without ISL, or with ISL of a very small number of earth stations or a moderate number of earth stations to increase the connectivity between the satellite network and ground network.

2.8.4 Handovers

Whereas the handovers (also called handoffs) of communications are well understood in the terrestrial mobile networks, the handovers in non-geostationary satellite networks add additional complexity to satellite network designs, due to relative movements between the satellites and between the satellites and ground earth stations.

Handover is needed to keep the links from source to destination connections. Satellite coverage moves along with the satellite and links must be handed over from one satellite to the next satellite (inter-satellite handover). For multi-beam satellites, handover is also needed between spot beams (beam handover or intra-satellite handover) and eventually to the next satellite (inter-satellite handover) as shown in Figure 2.28. When the next beam or satellite has no idle circuit to take over the handed-over links, the links get lost which can force termination of connection-oriented services; this event is referred to as a handover failure. Premature handover generally results in unnecessary handover and delayed handover results in increased probability of forced termination. Handover can be initiated based on the signal level strength and/or distance measurements position.

Two handover scenarios for satellite handovers are possible: intra-plane satellite handover and inter-plane satellite handover.

Intra-plane satellite handover assumes that the subscriber moves from beam to beam within the coverage area of satellite S. The gateway knows the subscriber is approaching the boundary between satellite S and satellite T because it knows the subscriber's location area code and the satellite's locations. The gateway will send a message to the trailing satellite S to prepare to handover the subscriber and another message to the leading satellite T in the same plane to prepare to accept the subscriber. The gateway will then send a message to the

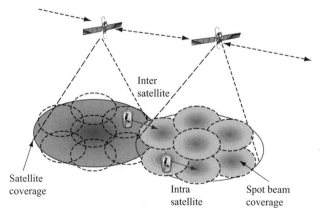

Figure 2.28 Concepts of inter-satellite beam and intra-satellite beam handovers

station via satellite S to resynchronise with the new satellite T. The handover is completed when the satellite sends a message to the station informing it of which new frequency to use. The gateway is the intelligent entity in this handover case.

Inter-plane satellite handover is the same as intra-plane satellite handover except that instead of handing over the connection to a satellite in the same orbit plane, it is handed over to a satellite in a different plane. The reason of performing a handover to a satellite in another plane is if no satellite in the same plane is able to cover the subscriber or if there are no available channels in the satellite of the same plane to perform a handover. Another reason can be that the satellite in a different plane can provide better service due to space diversity, as lower altitude satellites have more problems with shadowing than higher altitude satellites.

The time necessary for launching and executing the handover must be very short. In addition, the handovers should not degrade quality of service for the connections.

With the satellites' orbital velocity, and the dimension of coverage, the time to cross the overlap area covered by satellites is relatively short. However, due to the characteristics of the satellite constellation, a terminal can be covered by at least two satellites. This offers the possibility of optimising the handover, with respect to the quality of service of each connection, and serving a greater number of connections.

With the development of terminal technologies and integration with GPS functions, it is possible that satellite terminals will also be able to provide more assistance to the handover processes.

2.8.5 Satellite intra-beam and inter-beam handovers

Beam handover has two scenarios: intra-beam handover and inter-beam handover.

Intra-beam handover assumes that the subscriber is in beam A using frequency 1 and is associated with satellite S. As the beam approaches another geographic region, frequency 1 may no longer be available. There are two possible reasons for this. The first is government regulations, i.e. the particular set of frequencies is not available in the approaching region. Another reason is interference, which may be caused when satellite S moves too close to another satellite using the same frequency. In this case, even though the subscriber is still within beam A (satellite S), the satellite will send a message to the portable unit to change to frequency 2 in order to maintain the communication link. The satellite is the intelligent entity in this handover case.

Inter-beam handover scenario allows gateway earth stations (GES) or terminal earth station (TES) to continually monitor the radio frequency (RF) power of frequency 1 used in beam A. They also monitor the RF power of two adjacent candidate handover beams, B and C, via the general broadcast channel (information channel). The station determines when to hand over based on the RF signal strength. If the beam B signal becomes stronger than the signal used in beam A, the station will initiate a handover request to the satellite to switch the user to beam B. The satellite assigns a new frequency 3 to the station because two adjacent beams cannot use the same frequency (typically 3-, 6- and 12-beam patterns are used for efficient frequency reuse and coverage purpose). Inter-beam handover can be extremely frequent, if the beams are small and/or satellites move fast. There can also be an intelligent entity in this handover case.

2.8.6 Earth fixed coverage vs. satellite fixed coverage

The handover problem is considered according to the constellation. A satellite constellation can be designed as earth fixed coverage (EFC) or satellite fixed coverage (SFC) as shown in Figure 2.29. In EFC, each coverage area of satellite beams is fixed in relation to earth, therefore relatively it allows a longer period of time for handover. In contrast, each coverage area of SFC is moving along the satellite, hence it is fixed in relation to the satellite but moving in relation to earth. There is a relatively short period of time for handover, because the overlap between two-satellite coverage can be very small and moving away very fast.

The problems that occur in EFC constellations are due to the exaggerated difference in propagation delays in the radio signal of each satellite. The difference, due to different satellite locations, results in the loss of sequence, loss or duplication of coverage according to the position of satellites relative to earth units.

The benefit of multi-beam satellites is that each satellite can serve its entire coverage area with a number of high-gain scanning beams, each illuminating a single small area at a time. Narrow beamwidth allows efficient reuse of the spectrum and resulting high system capacity, high channel density and low transmitter power. However, if this small beam pattern swept the earth's surface at the velocity of the satellite, a terminal would have a very short period of time for communication before the next handover procedure. As in the case of terrestrial cellular systems, frequent hand-offs result in inefficient channel utilisation, high processing costs and lower system capacity.

In EFC, each satellite manages channel resources (frequencies and time slots associated with each coverage area) in the current serving area. As long as a terminal remains within the same earth fixed coverage, it maintains the same channel assignment for the duration of a call, regardless of how many satellites and beams are involved. Channel reassignments become the exception rather than the rule, thus eliminating much of the frequency management and hand-off overhead.

A database contained in each satellite defines the type of service allowed within each coverage area. Small fixed beams allow satellite constellations to avoid interference to or from specific geographic areas and to contour service areas to national boundaries. This would be

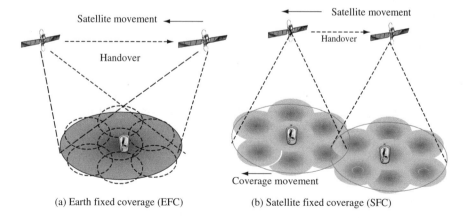

(a) Earth fixed coverage (EFC) (b) Satellite fixed coverage (SFC)

Figure 2.29 Satellite constellations of earth fixed coverage and satellite fixed coverage

difficult to accomplish with large beams or beams that move with the satellite. Active antennas are normally used to fix the beams onto earth while the satelites are flying at high speed.

2.8.7 Routing within constellation of satellites network

In addition to ISL and links between satellites and earth stations, routing finds paths to provide end-to-end connections by making use of the links. Clearly routing affects directly the utilisation of the network resources and quality of service provided by the connections.

The routing methods within constellations depend on the constellation design. The topology of a LEO constellation of satellites network is dynamic. The network connectivity between any two points is also dynamic. The satellites move with time above a rotating earth. Each satellite keeps the same position relative to other satellites in its orbital plane. Its position and propagation delay relative to earth terminals and to satellites in other planes change continuously but predictably. In addition to changes in network topology, as traffic flows through the network, routes are also changing with time. All of these factors affect the routing from source to destination of connections or packets.

The maximum delay between two end points, including the hops across satellite is constrained by real-time propagation delays. These constraints limit the hop count in systems utilising ISL. Satellite failure can create islands of communication within the LEO network. The network routing algorithm must accommodate these failures.

Due to the satellite orbital dynamics and the changing delays, most LEO systems are expected to use some form of adaptive routing to provide end-to-end connectivity. Adaptive routing inherently introduces complexity and delay variation. In addition, adaptive routing may result in packets being out of order. These out-of-order packets will have to be reordered by the receiver.

As all satellite nodes and ISLs have the same characteristics, it is convenient to separate the satellite part and terrestrial part of the routing. This allows different routing algorithms to be used effectively and they can be transparently adapted for the network characteristics.

Routing algorithms can be distributed or centralised. In centralised routing algorithms, all satellites report information about constellation command and control, which then calculates routing graphs and passes information back to the satellites for connection or packet routing.

In distributed routing algorithms, all satellites exchange network metrics (such as propagation delay, traffic load conditions, bandwidth availability and node failures, etc.) and each satellite tries to calculate its own routing graphs. QoS parameters may also be taken into account, such as delay and bandwidth requirements. The routing algorithms should also be able to trade off between QoS for user applications and efficiency for network resource utilisations.

Due to the motion of the satellites and user terminals, both the start and end points of the route may change with time and also the ISL path. Therefore satellite network routing is relatively more complicated than terrestrial network routing.

2.8.8 Internetworking

Internetworking is the final stage for satellite networking and provides connectivity directly to the user terminals or terrestrial networks. In addition to physical layer connections in terms of bandwidth and transmission speed, higher layer protocols also need to be taken into account.

According to possible differences between protocols used in satellite networks, terrestrial networks and satellite terminals, the following techniques can be used for internetworking:

- Protocol mapping is a technique used to translate the functions and packet headers between different protocols.
- Tunnelling is a technique used to treat one protocol as data to be transported in the tunnelling protocol. The tunnelled protocol is processed only at the end of the tunnel.
- Multiplexing and de-multiplexing are techniques used to multiplex several data streams into one stream and to de-multiplex one data stream into multiple streams.
- Traffic shaping is a technique used to shape the characteristics of traffic flows such as speeds and timings to be accommodated by the transport network.

2.8.9 Satellite availability and diversity

The total availability of the satellite network (A_{total}) is dependent on the availability of the satellite ($A_{satellite}$), the availability of the satellite link ($A_{propagation}$) and the availability of the satellite resources ($A_{congestion}$).

$$A_{total} = A_{satellite} \times A_{propagation} \times A_{congestion}$$

From a dependability point of view, a portion of a network connection should have the following properties:

- The fraction of time during which it is in a down state (i.e. unable to support a connection) should be as low as possible.
- Once a connection has been established, it should have a low probability of being either terminated because of insufficient data transfer performance or prematurely released due to the failure of a network component.

Availability of a network connection portion is defined as the fraction of time during which the connection portion is able to support a connection. Conversely, unavailability of a portion is the fraction of time during which the connection portion is unable to support a connection (i.e. it is in the down state). A common availability model is depicted in Figure 2.30.

	Satellite Link Unavailable	Satellite Link Available
Satellite Link in use	Unavailable State (2)	Available State (1)
Satellite Link not in use	Unavailable State (4)	Available State (3)

Figure 2.30 Satellite network availability model

The model uses four states corresponding to the combination of the ability of the network to sustain a connection in the available state and the actual use of the connection. Two independent perspectives are evident from the model:

- The service perspective, where availability performance is directly associated with the performance perceived by the user. This is represented in Figure 2.30 by states 1 and 2, even in the case of an on/off source since the user is only concerned with the connection availability performance while attempting to transmit packets.
- The network perspective, where availability performance is characterised independently of user behaviour. All four states in Figure 2.30 are applicable.

There are two availability parameters defined as the following:

- Availability ratio (AR): defined as the portion of time that the connection portion is in the available state over an observation period, whether the connection is in use or not.
- Mean time between outages (MTBO): defined as the average duration of a time interval during which the connection is available from the service perspective. Consecutive intervals of available time during which the user attempts to use are concatenated.

Diversity is technique used to improve satellite link availability. There are different types of diversity. Here we discuss only two types of diversity:

- Earth-to-space diversity uses more than one satellite at once for communication. This allows an improvement in physical availability, by decreasing the impact of shadowing due to buildings obstructing the path between the ground terminal and satellite and by providing redundancy at the physical or data-link level. Diversity is also exploited for soft handovers, i.e., the old connection is closed only after successful establishment of a new connection.
- In-orbit network diversity provides redundancy for failures in links and satellites. It is only possible due to the large number of satellites in the constellation with close spacing.

As this can affect routing across the ISL mesh, it can have a considerable effect on end-to-end delivery.

Further reading

[1] Haykin, S., *Communication Systems*, 4th edition, John Wiley & Sons, Inc., 2001.
[2] ITU, *Handbook on Satellite Communications*, 3rd edition, John Wiley & Sons, Inc., 2002.

Exercises

1. Use the laws of physics including Kepler's laws and Newton's laws to explain the features of satellite orbits.
2. Use Newton's laws to calculate GEO satellite orbit parameters.

Exercises (*continued*)

3. Design a constellation of quasi GEO satellites to provide coverage over the North Pole region.
4. Calculate the free-space loss of GEO satellite links.
5. Explain different types of modulation techniques and why the phase shift modulation technique is more suitable for satellite transmission.
6. Explain the important error correction coding schemes.
7. Explain how turbo code achieves performance close to the Shannon limit.
8. Explain the differences between the concepts of multiple access and multiplexing.
9. Explain the different bandwidth resources allocation schemes.
10. Discuss the satellite networking design issues.
11. Explain the concept of quality of service (QoS) at the physical layer in terms of bit error rate (BER) and the techniques to improve QoS.
12. Explain the quality of satellite networking in terms of availability and the techniques used to improve satellite availability.

3

ATM and Internet Protocols

This chapter aims to provide an introduction to the ATM and Internet protocols in the context of the basic protocol layering principles. It discusses internetworking between the ATM and Internet networks. It also provides the basic knowledge to help readers better understand the following chapters on satellite internetworking with terrestrial networks, ATM over satellite and Internet over satellite. When you have completed this chapter, you should be able to:

- Understand the concepts of ATM protocol and technology.
- Identify the functions of ATM adaptation layers (AAL) and the type of services they provide.
- Describe how ATM cells can be transported by different physical layer transmissions.
- Know the ATM interfaces and networks.
- Explain the relationships between traffic management, quality of service (QoS) and traffic policing functions.
- Describe the generic cell rate algorithm (GCRA).
- Knows the functions of the Internet protocol (IP).
- Understand the transmission control protocol (TCP) and user datagram protocol (UDP) and their use.
- Appreciate the concepts of internetworking between Internet and ATM.

3.1 ATM protocol and fundamental concepts

ATM is a fast packet-oriented transfer mode based on asynchronous time division multiplexing and it uses fixed-length (53 bytes) cells, each of which consists of an information field (48 bytes) and a header (5 bytes) as shown in Figure 3.1. The header is used to identify cells belonging to the same virtual channel and thus used in appropriate routings. Cell sequence integrity is preserved per virtual channel.

Satellite Networking: Principles and Protocols   Zhili Sun
© 2005 John Wiley & Sons, Ltd

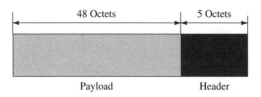

Payload   Header

Figure 3.1   ATM cell

The B-ISDN protocol reference model consists of three planes: user plane for transporting user information; control plane responsible for call control, connection control functions and signalling information; and management plane for layer management functions and plane management functions. There is no defined (or standardised) relationship between OSI layers and B-ISDN ATM protocol model layers, however, the following relations can be found. The physical layer of ATM is almost equivalent to layer 1 of the OSI model and it performs bit-level functions.

The ATM layer is equivalent to the upper layer 2 and lower layer 3 of the OSI model. The ATM adaptation layer performs the adaptation of OSI higher layer protocols. Figure 3.2 illustrates the hierarchy of the ATM protocol stack.

Higher Layer Functions		
Convergence Sublayer	CS	**AAL**
Segmentation and Reassembly	SAR	
Generic Flow Control Cell header generation/extraction Cell VPI/VCI Translation Cell Multiplexing and Demultiplexing		**ATM**
Cell rate decoupling HEC header generation/verification Cell delineation Transmission frame adaption Transmission frame generation/recovery	TC	**Physical Layer**
Bit timing Physical Media	PM	

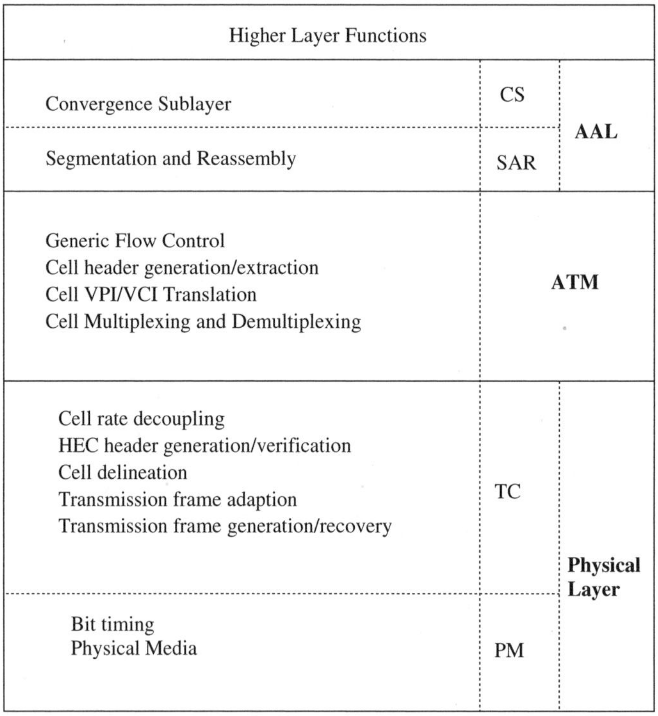

Figure 3.2   Functions of the ATM protocol stack

The number of 53 bytes is not only an unusual number (not even an even number), but also a relatively small number for a network layer packet. There are a few trade-offs involved here, including packetisation delay, packet overhead and functionality, queuing delay in switch buffers, and political compromise.

3.1.1 Packetisation delay

Standard digital (PCM) voice uses a constant stream of bits at 64 kbit/s. The voice is sampled 8000 per second. Each sample is coded using eight bits – 8000 eight-bit samples per second results in a data rate of 64 kbit/s.

To fill a cell of 40 bytes of payload, the first voice sample sits around in the partially filled cell for 40 sample times and then the cell is sent into the network. That first voice sample is therefore 5 milliseconds old before the cell even gets sent. This is called 'packetisation delay' and it is very important for real-time traffic such as voice.

In satellite communications, the delay is in the order of 250 milliseconds in each direction. Such a long delay may cause someone to experience problems in telephony communications, because the delay can interfere with normal conversational interactions. Even lower delays in the order of 10 to 100 milliseconds may cause problems due to echo in a voice network and analogue-to-digital conversion. To keep delay to a minimum, small cells are desirable.

However, there must be some overhead on the cell so that the cell can be forwarded to the right place and processed correctly. Using a five-byte header, the percentage of the bandwidth that is used by overhead can be very high. If the cell is too small it will loss efficiency. The key is to try to balance the delay characteristics with the efficiency. A five-byte header with a 48-byte payload results in less than 10% overhead as shown in Figure 3.3.

3.1.2 Queuing delay

Delay is important. Delay variation is also important. Delay variation is the amount of delay difference that cells experience as they traverse the network.

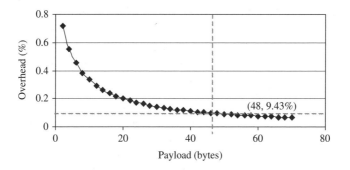

Figure 3.3 Trade-off between delay and cell payload efficiency

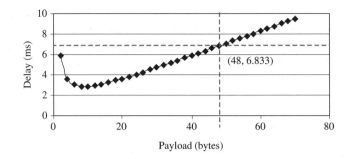

Figure 3.4   Delay due to packetisation and queuing

For example, consider a high-speed link with a 100-byte message to be transmitted. Further assume that this link is shared with 100 other streams of data. The best case for queuing delay is that there is no other data to send when the message arrives. Just send it, and effectively the queuing is almost zero. The worst case would be to wait for each of the other 100 streams to send their 100-byte message first.

Consider this worst case. If the payload is very small, one has to send so many cells that efficiency is quite poor as shown in Figure 3.4. If the cells is too large, the amount of time you have to wait for all these other cells to go, increases.

When the cells become bigger and bigger, the time to wait for a cell before one can get access to the link goes up, and it essentially goes up linearly. It can be seen that small cell sizes can have a large delay due to a large number of cells need to process, and delay will also become large if the payload becomes large as it takes time to process large cell.

3.1.3 Compromise solution between North America and Europe

Naturally, the important question of picking a cell size involves extensive analysis of technical concerns. In Europe, in fact, one of the major concerns was the packetisation delay because Europe consists of many small countries hence distance is small within each of the telephone networks. Thus, they do not need to deploy very much echo cancellation technology.

In North America, the distance within the countries is large, causing telephone companies to deploy echo cancellation technology. Thus, North Americans generally favoured a large cell with 64 octets of payload and a five-octet header, while Europeans generally favoured a smaller cell of 32 octets of payload and a four-octet header. One of the big differences was the concern about how to handle voice, since telephony services generate the revenue and major traffic in the telecommunication networks.

A political compromise was made, resulting in 48 octets for the payload with intensive averaging between 64 and 32 octets. Of course, this turned out to be too large to avoid the use of echo cancellers while failing to preserve the efficiency of 64 octets. Since Internet traffic exceeds telephony traffic, the size of the ATM cell is not important any more, however, the principles used to achieve optimisation and compromise to be acceptable at a global scale still are.

3.2 ATM layer

The ATM layer is the core of the ATM protocol stack. There are two different forms of header format for an ATM cell: one for the user network interface (UNI) between user terminal equipment and network node inter connections and the other for the network node interface (NNI) as illustrated in Figure 3.5.

3.2.1 The GFC field

The generic flow control (GFC) field occupies the first four bits in the header. It is only defined on in the UNI. It is not used in the NNI, which is the interface between switches. The GFC field can be used for flow control or for building a multiple access so that the network can control ATM cell flows from user terminals into the network.

3.2.2 The VPI and VCI fields

The important fields for routing in the header are the virtual path identifier (VPI) and virtual channel identifier (VCI) fields. A number of virtual connections through an ATM switch are shown in Figure 3.6. Within the switch, there has to be a connection table (or routing table) and that connection table associates a VPI/VCI and port number with another port number and another VPI/VCI.

When a cell comes into the switch, the switch looks up the value of the VPI/VCI from the header. Assume that the incoming VPI/VCI is 0/37. Because the cell came in on port one, the switch looks in the port one entries and discovers that this cell has to go to port three. When being sent out on port three, the VPI/VCI value is changed to 0/76, but the information content remains the same.

The VPI/VCI values change for two reasons. First, if the values were unique, there would only be about 17 million different values for use. As networks get very large, 17 million connections will not be enough for an entire network. Second, it is impossible to guarantee that in each newly established connection has a unique value in the world.

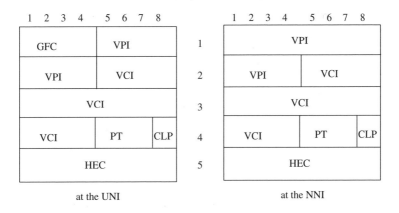

Figure 3.5 The ATM cell header format at the UNI and NNI

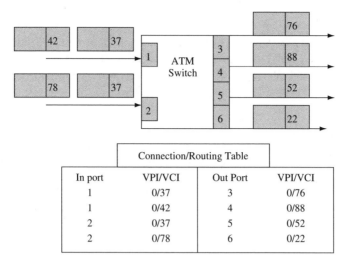

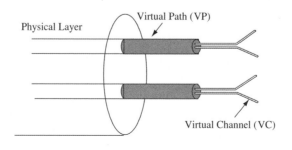

Figure 3.6   Connection/routing table in ATM switch

It is interesting to note that both of these considerations are becoming quite important in the Internet, where a limited number of TCP/IP addresses are available. If the address space were made large enough to serve as universal addresses, the overhead in comparison to the payload in the cell would become unacceptable.

Consequently, the VPI/VCI value is only meaningful in the context of the given interface. In fact, in this example '37' is used on in both interfaces, however, but there is no ambiguity because they are considered in the context of different physical interfaces. There is a separate entry for 37 for port two, which of course goes to a different destination.

So the combination of the VPI/VCI values allows the network to associate a given cell with a given connection, and therefore it can be routed to the right destination. The idea to have two values to identify a channel within the physical layer is illustrated in Figure 3.7. A virtual path is a bundle of virtual channels. The VPI is eight bits, providing up to 256 different bundles. Of course, the individual virtual channels have unique VCI values, but the VCI values may be reused in different virtual paths.

Figure 3.7   Concept of VP and VC in physical layer

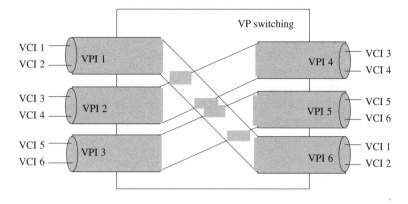

Figure 3.8   Example of VP switching

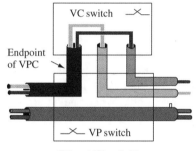

VC and VP switching

Figure 3.9   Example of VC and VP switching

ATM allows two different ways of getting connections to an ATM network shown in Figures 3.8 and 3.9. These two figures show how the network can support a 'bundle' of connections and how to switch the 'bundle' of connections and individual connection within it.

3.2.3 The CLP field

By default the one-bit cell loss priority (CLP) field is set as 0 as high priority. Cells with this bit set to 1 should be discarded before cells that have the bit set to 0. Consider reasons that why cells may be marked as expendable. First, this may be set this by the terminal. This may be desirable if, for example, in a wide area network (WAN) with a price drop for these low-priority cells. This could also be used to set a kind of priority for different types of traffic when one were aware to over use a committed service level. The ATM network can also set this bit for traffic management purposes in the traffic contract.

3.2.4 The PT field

The payload type (PT) identifier has three bits in it. The first bit is used to distinguish data cells from cells of operation, administration and maintenance (OMA). The second bit is called the congestion experience bit. This bit is set if a cell passes through a point in the network that is experiencing congestion, this bit is set. The third bit is carried transparently by the network. Currently, its only defined use is in one of the ATM adaptation layer type 5 (AAL5) for carrying IP packets.

3.2.5 The HEC field

The last eight-bit header error check (HEC) field is needed because if a cell is going through a network and the VPI/VCI values have errors, it will be delivered to the wrong place. As a security issue, it was deemed useful to put some error checking on the header. Of course, the HEC is also used, depending on the physical medium, e.g. in SONET, to delineate the cell boundaries.

HEC actually has two modes. One is a detection mode where if there is an error with the CRC calculation, the cell is discarded. The other mode allows the correction of one-bit errors. Whether one or the other mode is used depends on the actual medium in use. If fibre optics is used, one-bit error correction may make a lot of sense because typically the errors are isolated. It may not be the right thing to do if errors tend to come in bursts in the medium, such as copper and wireless link. When one-bit error correction is used, it increases the risk of a multiple-bit error being interpreted as a single-bit error, mistakenly 'corrected' and sent someplace. So the error detection capabilities drop when the correction mode is used.

Notice that the HEC is recalculated link by link because it covers the VPI and VCI values which change as ATM cells are transported through the network.

3.3 ATM adaptation layer (AAL)

AAL is divided into two sublayers as shown in Figure 3.2: segmentation and reassembly (SAR) and convergence sublayers (CS).

- *SAR sublayer*: this layer performs segmentation of the higher layer information into a size suitable for the payload of the ATM cells of a virtual connection, and at the receive side it reassembles the contents of the cells of a virtual connection into data units to be delivered to the higher layers.
- *CS sublayer*: this layer performs functions like message identification and time/clock recovery. It is further divided into a common part convergence sublayer (CPCS) and a service-specific convergence sublayer (SSCS) to support data transport over ATM. AAL service data units are transported from one AAL service access point (SAP) to one or more others through the ATM network. The AAL users can select a given AAL-SAP associated with the QoS required to transport the AAL-SDU. Five AALs have been defined, one for each class of service.

	Class A	Class B	Class C	Class D
Timing relation	required		not required	
Bit rate	constant	variable		
Connection mode	connection-oriented			connection-less

Examples: **A - Circuit emulation, CBR Video**
B - VBR video and audio
C - CO data transfer
D - CL data transfer

Figure 3.10   Service classes and their attributes

The role of the AAL is to define how to put the information of different types of services into the ATM cell payload. The services and applications are different and therefore require different types of AAL. It is important to know what kinds of services are required.

Figure 3.10 illustrates the results of the ITU-T's efforts for defining service classes.

- Class A has the following attributes: end-to-end timing, constant bit rate and connection oriented. Thus, Class A emulates a circuit connection on top of ATM. This is very important for initial multimedia applications because virtually all methods and technologies today that carry video and voice assume a circuit network connection. Taking this technology and moving it into ATM requires a supporting circuit emulation service (CES).
- Class B is similar to class A except that it has a variable bit rate. This might be performing video encoding but not playing at a constant bit rate. The variable bit rate really takes advantage of the burst nature of the original traffic.
- Classes C and D have no end-to-end timing and have variable bit rates. They are oriented toward data communications, and the only difference between the two is connection-oriented versus connection-less.

3.3.1 AAL1 for class A

Figure 3.11 shows AAL type 1 (AAL1) for Class A, illustrating the use of the 48-byte payload. One byte of the payload must be used for this protocol.

Convergence sublayer indication (CSI) consists of one bit. It indicates the existence of an eight-bit pointer if CSI = 1 and no existence if CSI = 0. Sequence number (SN) can be used

Figure 3.11   AAL 1 packet format for Class A

for cell loss detection and providing time stamps using adaptive clock methods. Sequence number protection (SNP) protects the CN by using CRC.

There are a number of functions here, including detecting lost cells and providing time stamps to support a common clock between the two end systems. It is also possible that this header could be used to identify byte boundaries by emulating a connection and identifying subchannels within the connection.

The primary objective for the adaptive clock method is to obtain clock agreement, making sure to be able to play out the original information stream. For example, in a 64 kbit/s voice service, the transmitter collects voice samples, fills up cells and sends those cells into the network at about once every 5.875 milliseconds (transmits 47 octets at a speed of one octet every 125 microseconds). The receiver is shown in Figure 3.12. The receiver plays out the original bit stream at 64 kbit/s. This is where we see the impact of variation and delay.

Using the adaptive clock method, the receiver establishes a buffer based on the characteristics of the connection at 64 kbits. It establishes a watermark and then collects some cells up to about the watermark. Then the receiver unwrap the bits from the payload and plays them out as a stream of bits at 64 kbit/s.

If the play out is too fast, the buffer becomes empty because the cells will be arriving a little bit too slow compared to rate of emptying them. Thus, we will have a buffer starvation problem. If it is a little bit too slow, the buffer will start to fill, and eventually it will overrun the buffer. Then cells get lost. The solution is that the receiver observes the fill of the buffer relative to the watermark. If it starts to get empty, it slows the (output) clock down because the clock is going a little fast. If it starts to get too full, it speeds the (output) clock up. This way, the receiver's output clock rate stays centred around the transmitter's clock.

The size of the buffer must be a function of how variable the arrival rate is for the cells. If the cells arrive in bursts, a large buffer is required. The larger the burst is the larger the buffer size is required. There is a lot of delay variation when the cells traverse the network. Bigger buffers also cause a larger delay. Cell delay variation (CDV) is a very important factor in QoS, thus it is an important parameter in traffic management.

Another important factor is the effect of losing a cell. Part of the protocol is a sequence number, which is not meant to maintain the sequence of the cells, but to detect loss. If a cell is lost, the receiver should detect the loss and essentially put in a substitute cell. Otherwise, the clock rate becomes unstable.

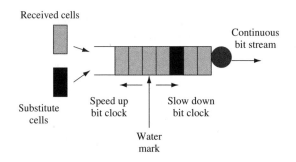

Figure 3.12   Illustration of adaptive clock method

It is interesting to note that with this kind of scheme, we can maintain a circuit-like connection of virtually any speed over ATM. As it is so important in supporting telephony service, AAL 1 is called a telephony circuit emulator.

3.3.2 AAL2 for class B

AAL type 2 (AAL2) is being defined for Class B, but it is not fully developed. This AAL is important, because it will allow ATM to support the burst nature of traffic to be exploited for packet voice, packet video, etc. Figure 3.13 illustrates the functions and frame format of the AAL2.

3.3.3 AAL3/4 for classes C and D

In AAL type 3/4 (AAL3/4), the protocol puts a header before and a trail after the original data, then the information is chopped into 44-byte chunks. The cell payloads include two bytes of header and two bytes of trailer, so this whole construct is exactly 48 bytes. Figure 3.14 illustrates the functions and frame format of the AAL3/4.

The header functions include the common part identifier (CPI) field of one byte, which identifies the type of traffic and certain values that are to be implemented in the other fields of the headers and trailers. The beginning tag (Btag) field of one byte is used to identify all the data associated with this session. The buffer allocation size (BAsize) of two bytes defines the size of the buffer in the receiver for the data. The alignment field (AL) is filler to 32-bit align the trailer. The end tag (Etag) is used with the Btag in the header to correlate all traffic associated with the payload. The length field specifies the length of the payload in bytes.

Note that there is a CRC check on each cell to check for bit errors. There is also an MID (message ID). The MID allows the multiplexing and interleaving of large packets on a single virtual channel. This is useful when the cost of a connection is very expensive since it helps to guarantee high utilisation of that connection.

3.3.4 AAL5 for Internet protocol

The other data-oriented adaptation layer is AAL type 5 (AAL5). It was designed particularly for carrying IP packet using the full 48 bytes of the ATM payload. Here, the CRC is appended at the end and the padding is such that this whole construct is exactly an integral number of 48-byte chunks. This fits exactly into an integral number of cells, so the construct

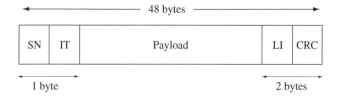

Figure 3.13   AAL 2 packet format for Class B

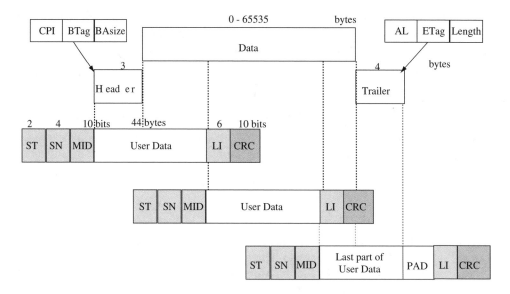

Figure 3.14 AAL 3/4 packet format for Classes C & D

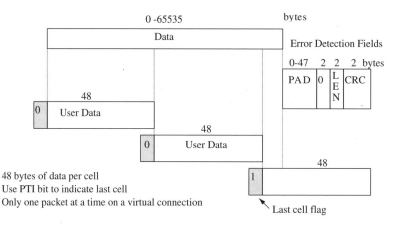

Figure 3.15 AAL 5 format for Internet protocol

is broken up into 48-byte chunks and put into cells. Figure 3.15 illustrates the functions and frame format of the AAL5.

To determine when to reassemble and when to stop reassembling, remember the third bit for PT in the ATM header. This bit is zero except for the last cell in the packet (when it is one).

A receiver reassembles the cells by looking at the VPI/VCI and, for a given VPI/VCI, reassembles them into the larger packet. This means that a single VPI/VCI may support only one large packet at a time. Multiple conversations may not be interleaved on a given connection. This is attractive when connections are cheap.

3.4 The physical layer

The first requirement for interpretability of the terminal equipment with the ATM network and network nodes with network nodes within the network is to transmit information successfully at the physical level over physical media including fibre, twisted pairs, coaxial cable, terrestrial wireless and satellite links.

As shown in Figure 3.2 the physical layer (PL) is divided into two sublayers: the physical medium (PM) and transmission convergence (TC) sublayers.

3.4.1 The physical medium (PM) sublayers

The PM sublayer contains only the PM-dependent functions (such as bit encoding, the characteristics of connectors, the property of the transmission media, etc.). It provides bit transmission capability including bit alignment, and performs line coding and also conversions of electrical, optical and radio signals if necessary. Optical fibre has been chosen as the physical medium for the ATM and coaxial and twisted pair cables and radio wireless links including satellite can also be used. It includes bit-timing functions such as the generation and reception of waveforms suitable for the medium and also insertion and extraction of bit-timing information.

3.4.2 The transmission convergence (TC) sublayer

In an ATM network, a terminal needs to have a cell to send data into the network. To keep the network receiving ATM cells correctly, the terminal still has to send an 'empty' cell into the network if there is nothing to send, because the ATM also makes use of the features of the HEC field and fixed size of the ATM cells for framing. One of the functions of the TC sublayer is to insert empty cells for transmission and remove empty cells when they get to the destination in order to keep the cell streams constant.

Because of the different kinds of details in the coupling between the fibre and other physical media, the TC sublayer differs, depending on the physical layer transmission of the ATM cells. The TC sublayer mainly has five functions as shown in Figure 3.2.

- The lowest function is generation and recovery of the transmission frame.
- The next function, i.e. transmission frame adaptation, takes care of all actions adapting cell flow according to the used payload structure of the transmission system in the sending direction. It extracts the cell flow from the transmission frame in the receiving direction. The frame can be a synchronous digital hierarchy (SDH) envelope or an envelope according to ITU-T Recommendation G.703.
- The cell delineation function enables the receiver to recover the cell boundaries from a stream of bits. Scrambling and descrambling are performed in the information field of a cell before the transmission and after reception respectively to protect the cell delineation mechanism.
- The HEC sequence generation is performed in the transmit direction and its value is recalculated and compared with the received value and thus used in correcting the header errors. If the header errors cannot be corrected, the cell is discarded.
- Cell-rate decoupling inserts the idle cells in the transmitting direction in order to adapt the rate of the ATM cells to the payload capacity of the transmission system. It suppresses

all idle cells in the receiving direction. Only assigned and unassigned cells are passed to the ATM layer.

3.4.3 ATM cell transmissions

As the ATM is a protocol defining an asynchronous mode, the ATM cells have to be transmitted over network technologies. In the ITU-T I-series standards, a target solution and evolutional solution are defined for public ATM networks at transmission speeds of 155.520 Mbit/s or higher. For lower bit rates, the ATM Forum defined transmission methods over existing standard transmission technologies. The ITU-T is responsible for public ATM network specifications. The ATM Forum is not an international standardisation organisation. It is an international non-profit organisation, formed in 1991, with the objective of accelerating the use of ATM products and services through a rapid convergence of interoperability specifications, and promotes industry cooperation and awareness. It is responsible for private ATM network specifications by adopting the ITU-T ATM standards if available or proposing one if not available.

3.4.4 Target solution for ATM transmission

Figure 3.16 shows the target solution recommended by the ITU-T I-series standards. It suggested a new transmission scheme at the physical layer so that the physical layer transmits ATM cells directly, but only provides 26/27 ATM cells to the ATM layer so that the 1/27 cell can be used for supporting operation, management and administration (OMA) functions. The choice of the 1/27 cell used for OMA is to make the new scheme compatible with evolutional approaches using the SDH standards for ATM cell transmissions. The physical layer transmission is 155.520 Mbit/s, which is the same as the SDH standards physical layer transmission speed. The ATM layer is 149.760 Mbit/s, which is the same as the SDH payload.

3.4.5 ATM over synchronous digital hierarchy (SDH)

The ITU-T defined the evolutional approach to transmit ATM cells over SDH before the future target solution. The essential feature of SDH is to keep track of boundaries of streams

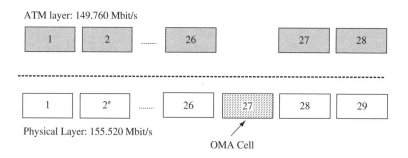

Figure 3.16   The ITU-T target solution for ATM cell transmission

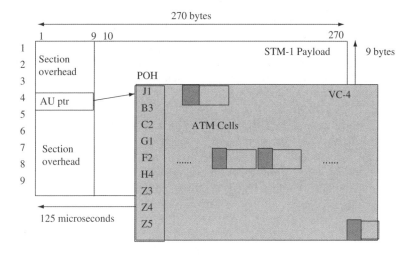

Figure 3.17 SDH STM-1 frame

that do not really depend on the particular medium. Although it was originally designed for transmission over fibre, it can in fact operate over other media.

The SDH mode type 1 (STM-1) frame is compatible to the synchronous optical network (SONET) synchronous transport signal optical carrier 3 (STS-3C) frame at 155 Mbit/s as shown in Figure 3.17. The bytes are transmitted across the medium a row at a time, wrapping to the next row. It takes nominally 125 microseconds to transmit all nine rows forming the SDH STM-1 frame.

The first nine bytes of each row have various overhead functions. For example, the first two bytes are used to identify the beginning of the frame so that the receiver can lock onto this frame.

In addition, although not shown here, there is another column of bytes included in the 'synchronous payload envelope', which is additional overhead, with the result that each row has 260 bytes of information. Consequently, 260 bytes per row × 9 rows × 8 bits divided by 125 microseconds, equals 149.76 Mbit/s of payload, which is the same as the target solution.

The STM-1 in the international carrier networks will be the smallest package available in terms of the SDH. The bit rates for SDH STM-4 are four times the bit rates of the STM-1.

SDH also has some nice features for getting to higher rates – like 622 Mbit/s – it becomes basically a recipe of taking four of these STM-1 structures and simply interleaving the bytes to get to 622 Mbit/s (STM-4). There are additional steps up to 1.2 gigabits, 2.4 gigabits, etc. And at least in theory, the recipe makes it simple to get to a speed interface from low speed ones.

Using the header error check (HEC) of the ATM cell delineates the cells within the SDH payload (VC-4 container). The receiver, when it is trying to find the cell boundaries, takes five bytes to check if they form a header or not. It does the HEC calculation on the first four bytes and matches that calculation against the fifth byte. If it matches, the receiver then counts 48 bytes and tries the calculation again. And if it finds that calculation correct several times in a row, it can probably safely assume that it has found the cell boundaries. If it fails, it just slides the window by one bit and tries the calculation again.

This kind of process must be used because we don't really know what is in the 48 bytes of payload, but the chances that the user data would contain these patterns separated by 48 bytes is essentially zero for any length of time.

For empty cells, the HEC is calculated by first calculating the CRC value, then performing an 'exclusive or' operation of the CRC value with a bit pattern called the coset, resulting in a non-zero HEC. Thus, the HEC is unique from the zeros in the empty cells, and the HEC may still be used for cell delineation. At the receiving end, another 'exclusive or' operation is performed, resulting in the original CRC for comparison.

The payload in an STM-1 frame is 135,563 Mbit/s, assuming that the entire cell payload may carry user information.

3.4.6 ATM over DS1

Digital signal level 1 (DS1) is the primary rate offered by the public carriers in North America. It is also specified by the ATM Forum to carry ATM traffic. The standard DS1 format consists of 24 consecutive bytes with a single overhead bit inserted for framing. There is a fixed pattern for these overhead bits to identify the framing bits and the frame structure, as shown in Figure 3.18.

Once the pattern has been identified, we know where the bytes within the DS1 physical layer payload are. Now, the question is how to find the cell boundaries.

The cells are going to be put into these physical layer payload bytes. Notice that there are only 24 bytes in each of these blocks, so the cell is actually going to extend across multiple blocks. There could be 24 bytes of a cell in the first block, 24 bytes of the same cell in the second block and then the remaining five bytes of the cell in the third block. However, the cell actually can fall anywhere on the byte boundaries.

Use the same mechanism as with SDH. Keep looking at five-byte windows and doing the CRC calculation, use the HEC approach. The actual payload that can be transported within a DS1 is 1.391 Mbit/s.

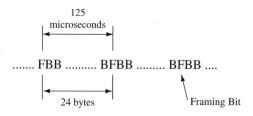

- (24 byte × 8 bit/byte)/125 microsecond = 1.536 Mbit/s of payload
- Cell delineation by HEC detection
- Cell payload = 1.536 Mbit/s × (48/53) = 1.391 Mbit/s

Figure 3.18   DS1 frame structure of 1.544 Mbit/s

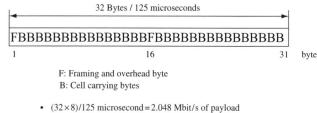

FBBBBBBBBBBBBBBBBFBBBBBBBBBBBBBBB

1   16   31   byte

F: Framing and overhead byte
B: Cell carrying bytes

- (32×8)/125 microsecond = 2.048 Mbit/s of payload
- Cell delineation by HEC detection

Figure 3.19   E1 frame structure of 2.048 Mbit/s

3.4.7 ATM over E1

The 2.048 Mbit/s interface will be particularly important in Europe, where this speed (E1) is the functional equivalent of North American DS1 interfaces. Note that in contrast to the DS1 format, there are no extra framing bits added. In fact, the 2.048 Mbit/s rate is an exact multiple of 64 kbit/s.

The basic E1 frame consists of a collection of 32 bytes, recurring every 125 microseconds. Instead of using framing bits, this format uses the first (Byte 0) and seventeenth (Byte 16) for framing and other control information. The receiver uses the information within the framing bytes to detect the boundaries of the physical layer blocks, or frames. The remaining 30 bytes are used to carry ATM cells. Consequently, the physical layer payload capacity for the E1 interface is 1.920 Mbit/s (see Figure 3.19).

Just as in SDH and DS1, as previously discussed, the HEC is used to find the cell boundaries.

3.5 ATM interfaces and ATM networking

ATM provides a well-defined interface for networking purposes between users and network, between network nodes (switches), and between networks.

3.5.1 User–network access

Two elements can be used to describe a reference configuration of the user–network access of B-ISDN: functional groups and reference points. Figure 3.20 gives the reference configuration. The B-NT1 and B-NT2 are broadband network terminators. The B-NT2 provides an interface allowing other type of TE rather than the broadband TE to be connected to the broadband network.

B-NT1 functions are similar to layer 1 of the OSI reference model and some of the functions are:

- line transmission termination;
- transmission interface handling; and
- OAM functions.

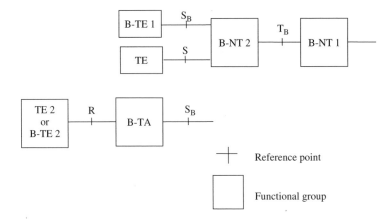

Figure 3.20   B-ISDN reference configuration

B-NT2 functions are similar to layer 1 and higher layers of the OSI model. Some functions of B-NT2 are:

- adaptation functions for different interface media and topology;
- multiplexing and de-multiplexing and concentration of traffic;
- buffering of ATM cells;
- resource allocation and usage parameter control;
- signalling protocol handling;
- interface handling;
- switching of internal connections.

B-TE1 and B-TE2 are broadband terminal equipment. B-TE1 can be connected directly to the network from the reference S_B and T_B. B-TE2 can only be connected to the network via a broadband adapter.

B-TA is broadband terminal adapter. It allows the B-TE2, which cannot be connected directly, to be connected to the broadband network.

S_B and T_B indicate reference points between the terminal and the B-NT2 and between B-NT2 and B-NT1 respectively. Reference point characteristics are:

- T_B and S_B: 155.520 and 622.080 Mbit/s;
- R: allow connection of a TE2 or a B-TE2 terminal.

3.5.2 Network node interconnections

In Figure 3.21, first consider the private ATM network in the upper left corner. The interface between the terminal and the switch is referred to as the private user-to-network interface (UNI). The interface to the public network is a public UNI. Now, these two interfaces are quite similar. For example, the cell size is the same; the cell format is the same. There are some differences, though. For example, the public UNI interface is likely to be a DS3 interface early on, but it's very unlikely that a DS3 would be deployed across campus. Consequently, there are some differences at the physical layer.

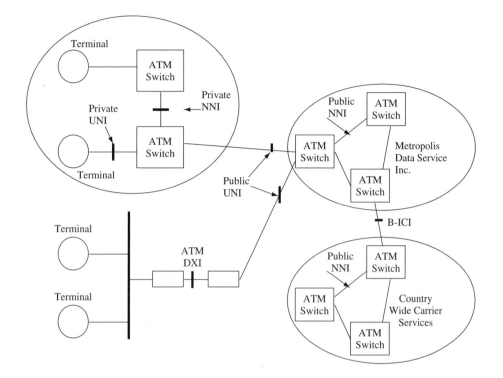

Figure 3.21 ATM interfaces network nodes interconnections

Within a private ATM network, there is the issue of connecting multiple switches together into an ATM network. This is referred to as the network node interface (NNI). In some ways, the NNI is misnamed because it is really more than an interface. It is a protocol that allows multiple devices to be interconnected in somewhat arbitrary topologies and still work as one single network.

There is a corresponding protocol in the public arena called the public NNI. It has basically the same function, but, because of the context of the problem that is being addressed, it ends up in detail to be quite different.

The ATM Forum specifies the private NNI (PNNI) protocol. The ITU specifies the public NNI. One of the major differences is that in the case of the public NNI, there is a strong dependence on the signalling network.

The B-ICI specifies how two carriers can use ATM technology to multiplex multiple services onto one link, thereby exchanging information and cooperating to offer services.

3.5.3 ATM DXI

The ATM data exchange interface (DXI) allows a piece of existing equipment – in this case, a router – to access the ATM network without having to make a hardware change. The hardware impact is in a separate channel service unit/data service unit (CSU/DSU).

Typical physical layers for the DXI are e.g. V35 or the high-speed serial interface (HSSI). Since this is a data-oriented interface, the frames are carried in HDLC frames. All that is required is a software change in the router and the CSU-DSU to perform the 'slicing' segmentation and reassembly (SAR) function.

The CSU-DSU takes the frames, chops them up into cells, does traffic shaping if required by the traffic contract, and ends up with a UNI.

3.5.4 B-ICI

The broadband inter-carrier interface (B-ICI), in its initial version, is a multiplexing technique. It specifies how two carriers can use ATM technology to multiplex multiple services onto one link, thereby exchanging information and cooperating to offer services.

The services specified in the B-ICI are: cell relay service, circuit emulation service, frame relay and SMDS. Users of the carrier network don't 'see' this interface, but it is important because it will help provide services across carriers.

3.5.5 Permanent virtual connections vs. switched virtual connections

The connections involve routing through a switch only. How to get a connection established through a network?

One technique is called a permanent virtual connection (PVC). This will be done through some form of service order process. Conceptually, there is some sort of network management system that communicates to the various devices what the VCI-VPI values are and what the translations are. For example, the network management system tells the switch what entries to make in its connection table.

There are some environments for which this is most reasonable. If there are a small number of devices attached to the ATM network, and these devices tend not to move around very much, this behaves much as telephone network private lines. This tends to make a lot of sense when there is a large community of interest between two locations. Because it takes a while to set up these connections, and to leave them up, but not to try to tear them down and set them up in a very dynamic fashion. That is why these are called permanent virtual connections.

A second technique for establishing a connection through a network is called a switched virtual connection (SVC). This allows a terminal to set up and tear down connections dynamically.

The way SVC operates is that one of the VPI/VCI values is predefined for the signalling protocol to control the connections. The value is VPI-0/VCI-5, and this connection is terminated by the call processing function. Of course, the 'receiving' terminal also has VPI-0/VCI-5 terminating at the call processing function for this (or another) switch.

A protocol called the 'signalling protocol' is used on the VPI-0/VCI-5 connection to communicate with the switch, passing information to allow the connection to be set up or to be torn down (or to even be modified while it i's in existence). The result is dynamic connection configuration. Further, these connections will probably be established in less than a second.

Note that the connection that is set up for actual information transfer should not use VPI-0/VCI-5. The other connection passing around the call processing function does not interact with the call processing functions within the switch.

3.5.6 ATM signalling

The signalling capability for ATM networks has to satisfy the following functions.

- Set up, maintain and release ATM virtual channel connections for information transfer.
- Negotiate the traffic characteristics of a connection (CAC algorithms are considered for these functions).

Signalling functions may also support multi-connection calls and multi-party calls. A multi-connection call requires the establishment of several connections to set up a composite call comprising various types of traffic like voice, video, image and data. It will also have the capability of not only removing one or more connections from the call but also adding new connections to the existing ones. Thus the network has to correlate the connections of a call. A multi-party call contains several connections between more than two end-users, such as conferencing calls.

Signalling messages are conveyed out of band in dedicated signalling virtual channels in broadband networks. There are different types of signalling virtual channels that can be defined at the B-ISDN user-to-network interface. They can be described as follows:

- A meta-signalling virtual channel is used to establish, check and release point-to-point and selective broadcast signalling virtual channels. It is bi-directional and permanent.
- A point-to-point signalling channel is allocated to a signalling endpoint only while it is active. These channels are also bi-directional and are used to establish, control and release VCC to transport user information. In a point-to-multipoint signalling access configuration, meta-signalling is needed for managing the signalling virtual channels.

3.5.7 ATM addressing

A signalling protocol needs some sort of addressing scheme. Private networks will probably use OSI NSAP type addressing, primarily because an administrative process exists. The public carriers will probably use E.164 numbers.

In order for an addressing scheme to be useful, there must be a standardised address format that is understood by all of the switches within a system. For instance, when making phone calls within a given country, there is a well-defined phone number format. When calling between countries, this format is usually modified to include information like a 'country code'.

Each call set-up message contains the information in these fields twice – once identifying the party that is being called (destination) and once identifying the calling party (source).

Figure 3.22 shows the three address formats that have been defined by the ATM Forum. The first byte in the address field identifies which of the address formats is being used. (Values for this field other than the three listed here are reserved and/or used for other functions.)

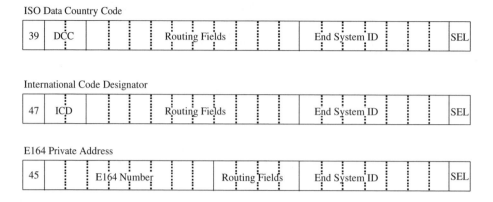

Figure 3.22 ATM address format

The three address formats are:

1. *Data country code (DCC)*. DCC numbers are administered by various authorities in each country. For instance, ANSI has this responsibility in the USA. The DCC identifies the authority that is responsible for the remainder of the 'routing fields.'
2. *International code designator (ICD)*. ICDs are administered on an international basis by the British Standards Institute (BSI).
3. *E.164 private addresses*. E.164 addresses are essentially telephone numbers that are administered by telephone carriers, with the administering authority identity code as a part of the E.164 number.

Regardless of the numbering plan used, it is very important that an ATM network implementer obtains official globally unique numbers to prevent confusion later on when ATM network islands are connected together.

Following the DCC or ICD fields – or immediately following the E.164 in the case of the E.164 format – is the 'routing field.' For DCC and IDC, this is the information that contains the address that is being called (or is placing the call).

This 'routing field' can be thought of as an address space. The term 'routing field' implies that there is more to the field than a simple address. In particular, the addressing mechanism will very probably be hierarchical to assist in the routing. In the E.164 option, the use of the 'routing field' is not defined at this time.

Each address in the routing field may refer to a particular switch, or it may even refer to a particular UNI on a switch. If it refers only to a switch, then more information will be needed to find the exact UNI that is specified. On the other hand, if it specifies a UNI, then this is sufficient to serve as a unique, globally significant address.

3.5.8 Address registration

In Figure 3.22, let's consider the case in which the first 13 bytes only specify a particular switch, as opposed to a particular UNI. In this case, the switching system must still find the appropriate UNI for the call.

This could be done using the next six bytes, called the 'end-system ID'. End systems, or terminals, could contain additional addressing information. For instance, the terminal could supply the last six bytes to the switch to identify the particular UNI. This way an entire switch could be assigned a 13-byte address, and the individual switch would then be responsible for maintaining and using the 'end-system ID'.

This mechanism might be particularly attractive to a user desiring a large 'virtual private network', so that the user would obtain 'switch addresses' from an oversight organisation and then locally administer the end-system IDs. This would have the advantage of allowing the user organisation to administer the individual addresses without involving the outside organisation. However, anyone outside the organisation desiring to call a given UNI would have to know values for both the routing field and the end-system ID.

The six bytes of the end-system ID are not specified, so its use can be left up to the manufacturers. A common anticipated use of the end-system ID is to use the six bytes (48 bits) for the unique 48-bit MAC address that is assigned to each network interface card (NIC).

Of course, both the ATM switch and the ATM terminal must know these addresses in order to route calls, send signalling messages etc. This information can be obtained automatically using the ILMI (integrated link management interface). The switch typically will provide the 13 most significant bytes (routing field) while the terminal provides the next six bytes (end-system ID).

The ATM network does not use the selector (SEL) byte, but it passes transparently through the network as a 'user information field'. Thus, the SEL can be used to identify entities in the terminal, such as a protocol stack.

3.6 Network traffic, QoS and performance issues

Network resource management concerns three aspects: the traffic to be offered (described by using traffic parameters and descriptors); the service with agreed QoS agreed upon (that the user terminals to get and the networks to provide); and the compliance requirements to check if the user terminals have got the QoS required and networks have provided the QoS expected.

To provide QoS, the ATM network should allocate network resources including bandwidth, processor and buffer space capacities to ensure good performance using congestion and flow controls, e.g., to provides particular transmission capacities to virtual channels.

Traffic management includes the following mechanisms:

- Traffic contract to specify on each virtual channel/path.
- Connection admission control (CAC) to route each virtual channel/path along a path with adequate resources and to reject set-up requests if there is not enough resource available.
- Traffic policing to mark (via cell loss priority bit) or discard ATM cells that violate the contract.
- Algorithm to check conformance to the contract or shape the traffic to confirm conform to the contract.

3.6.1 Traffic descriptors

Traffic characteristics can be described by using the following parameters known as the traffic descriptors:

- Peak cell rate (*PCR*) is the maximum rate to send ATM cells.
- Sustained cell rate (*SCR*) is the expected or required cell rate averaged over a long time interval.
- Minimum cell rate (*MCR*) is the minimum number of cells/second that the customer considers as acceptable.
- Cell delay variation tolerance (*CDVT*) tells how much variation will be presented in cell transmission times.

3.6.2 Quality of service (QoS) parameters

The QoS parameters include:

- Cell transfer delay (*CTD*): the extra delay added to an ATM network at an ATM switch, in addition to the normal delay through network elements and lines. The cause of the delay at this point is the statistical asynchronous multiplexing. Cells have to queue in a buffer if more than one cell competes for the same output. It depends on the amount of traffic within the switch and thus the probability of contention.
- Cell delay variation (*CDV*): the delay depends on the switch/network design (such as buffer size), and the traffic characteristic at that moments of time. This results in cell delay variation. There are two performance parameters associated with CDV: one-point CDV and two-point CDV. The one-point CDV describes variability in the pattern of cell arrival events observed at a single boundary with reference to the negotiated $1/T$. The two-point CDV describes variability in the pattern of cell arrival events observed at an output of a connection with the reference to the pattern of the corresponding events observer observed at the input to the connection.
- Cell loss ratio (*CLR*): the total lost cells divided by the total transmitted cells. There are two basic causes of cell loss: error in cell header or network congestion.
- Cell error ratio (*CER*): the total error cells divided by the total successfully transferred cells plus the total error cells.

3.6.3 Performance issues

There are five parameters that characterise the performance of ATM switching systems: throughput; connection blocking probability; cell loss probability; switching delay; and delay variation.

- Throughput: this can be defined as the rate at which the cells depart the switch measured in the number of cell departures per unit time. It mainly depends on the technology and dimensioning of the ATM switch. By choosing a proper topology of the switch, the throughput can be increased.

- Connection blocking probability: since ATM is connection oriented, there will be a logical connection between the logical inlet and outlet during the connection set-up phase. The connection blocking probability is defined as the probability that there are not enough resources between inlet and outlet of the switch to assure the quality of all existing connections as well as new connections.

- Cell loss probability: in ATM switches, when more cells than a queue in the switch can handle compete for this queue, cells will be lost. This cell loss probability has to be kept within limits to ensure high reliability of the switch. In internally non-blocking switches, cells can only be lost at their inlets/outlets. There is also possibility that ATM cells may be internally misrouted and erroneously reach another logical channel. This is called cell insertion probability.

- Switching delay: this is the time taken to switch an ATM cell through the switch. The typical values of switching delay range between 10 and 1000 microseconds. This delay has two parts:

    - fixed switching delay: because of internal cell transfer through the hardware.
    - queuing delay: because of the cells queued up in the buffer of the switch.

- Jitter on the delay or delay variation: this is denoted as the probability that the delay of the switch will exceed a certain value. This is called a quantile and for example a jitter of 100 microseconds at a 10^{-9} quantile means the probability that the delay in the switch is larger than 100 microsecond is smaller than 10^{-9}.

3.7 Network resource management

ATM networks must fairly and predictably allocate the resources of the network. In particular, the network must support various traffic types and provide different service levels.

For example, voice requires very low delay and low delay variation. The network must allocate the resources to guarantee this. The concept used to solve this problem is called traffic management.

When a connection is to be set up, the terminal initiating the service specifies a traffic contract. This allows the ATM network to examine the existing network utilisation and determine whether in fact a connection can be established that will be able to accommodate this usage. If the network resources are not available, the connection can be rejected.

While this all sounds fine, the problem is that the traffic characteristics for a given application are seldom known exactly. Considering a file or a web page transfer we may think we understand that application, but in reality we are not certain ahead of time how big the files going to be, or even how often a transfer is going to happen. Consequently, we cannot necessarily identify precisely what the traffic characteristics are.

Thus, the idea of traffic policing is useful. The network 'watches' the cells coming in on a connection to see if they abide by the contract. Those that violate the contract have their CLP bit set. The network has the options to discard these cells now or when the network starts to get into a congested state.

In theory, if the network resources are allocated properly, discarding all the cells with a cell loss priority bit marked will result in maintaining a level of utilisation at a good operational point in the network. Consequently, this is critical in being able to achieve the goal of ATM: to guarantee the different kinds of QoS for the different traffic types. There are many functions involved in the traffic control of ATM networks.

3.7.1 Connection admission control (CAC)

Connection admission control (CAC) can be defined as the set of actions taken by the network during the call set-up phase to establish whether a VC/VP connection can be made. A connection request for a given call can only be accepted if sufficient network resources are available to establish the end-to-end connection maintaining its required QoS and not affecting the QoS of existing connections in the network by this new connection.

There are two classes of parameters considered for the CAC. They can be described as follows:

- The set of parameters that characterise the source traffic i.e. peak cell rate, average cell rate, burstiness and peak duration etc.
- Another set of parameters to denote the required QoS class expressed in terms of cell transfer delay, delay jitter, cell loss ratio and burst cell loss etc.

Each ATM switch along the connection path in the network will be able to check if there are enough resources for the connection to meet the required QoS.

3.7.2 UPC and NPC

Usage parameter control (UPC) and network parameter control (NPC) perform similar functions at the user-to-network interface and network-to-node interface, respectively. They indicate the set of actions performed by the network to monitor and control the traffic on an ATM connection in terms of cell traffic volume and cell routing validity. This function is also known as the 'police function'. The main purpose of this function is to protect the network resources from malicious connection and equipment malfunction, and to enforce the compliance of every ATM connection to its negotiated traffic contract. An ideal UPC/NPC algorithm meets the following features:

- Capability to identify any illegal traffic situation.
- Quick response time to parameter violations.
- Less complexity and more simplicity of implementation.

3.7.3 Priority control and congestion control

The CLP (cell loss priority) bit in the header of an ATM cell allows users to generate different priority traffic flows and the low priority cells are discarded to protect the network performance for high priority cells. The two priority classes are treated separately by the network UPC/NPC functions.

Congestion control plays an important role in the effective traffic management of ATM networks. Congestion is a state of network elements in which the network cannot assure the negotiated QoS to already existing connections and to new connection requests. Congestion may happen because of unpredictable statistical fluctuations of traffic flows or a network failure.

Congestion control is a network means of reducing congestion effects and preventing congestion from spreading. It can assign CAC or UPC/NPC procedures to avoid overload situations. To mention an example, congestion control can minimise the peak bit rate available to a user and monitor this. Congestion control can also be done using explicit forward congestion notification (EFCN) as is done in the frame relay protocol. A node in the network in a congested state may set an EFCN bit in the cell header. At the receiving end, the network element may use this indication bit to implement protocols to reduce the cell rate of an ATM connection during congestion.

3.7.4 Traffic shaping

Traffic shaping changes the traffic characteristics of a stream of cells on a VP or VC connection. It spaces properly the cells of individual ATM connections to decrease the peak cell rate and also reduces the cell delay variation. Traffic shaping must preserve the cell sequence integrity of an ATM connection. Traffic shaping is an optional function for both network operators and end users. It helps the network operator in dimensioning the network more cost effectively and it is used to ensure conformance to the negotiated traffic contract across the user-to-network interface in the customer premises network. It can also be used for user terminals to generate traffic of cells conforming to a traffic contract.

3.7.5 Generic cell rate algorithm (GCRA)

The traffic contract is based on something called the generic cell rate algorithm (GCRA). The algorithm specifies precisely when a stream of cells either violates or does not violate the traffic contract. Consider a sequence of arrivals of cells. This sequence is run with the algorithm to determine which cells (if any) violate the contract.

The algorithm is defined by two parameters: the increment parameter 'I' and the limit parameter 'L'. The GCRA can be implemented by either of the two algorithms: leaky bucket algorithm or virtual scheduling algorithm. Figure 3.23 shows a flow chart of the algorithms.

The two algorithms served the same purpose: to make certain that cells are conforming (arrival within the bound of an expected arrival time) or nonconforming (arrival sooner than an expected arrival time).

3.7.6 Leaky bucket algorithm (LBA)

Sometimes referred to as a 'continuous-state leaky bucket'. Think about this as a bucket with a hole in it. To make this a little more concrete, assume that 'water' is being poured into the bucket and that it leaks out at one unit of water per cell time. Every time a cell comes into the network that contains data for this connection, I units of water are poured into the bucket. Of course, then the water starts to drain out. Figure 3.24 shows the leaky bucket illustrating the GCRA.

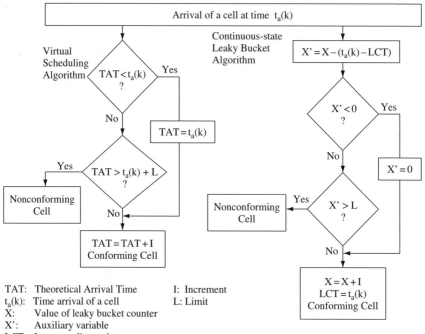

TAT: Theoretical Arrival Time   I: Increment
$t_a(k)$: Time arrival of a cell   L: Limit
X: Value of leaky bucket counter
X':   Auxiliary variable
LCT: Last compliance time

Figure 3.23   Generic cell rate (GCRA) algorithm

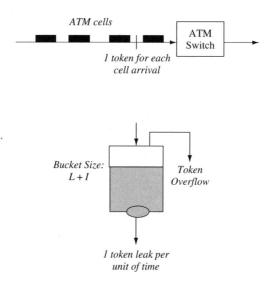

Figure 3.24   Leaky bucket algorithm (LBA)

The size of the bucket is defined by the sum of the two parameters $(I + L)$. Any cell that comes along that causes the bucket to overflow when I units have poured in violates the contract.

If the bucket was empty initially, a lot of cells can go into the bucket, and the bucket would eventually fill up. Then it would be better to slow down. In fact, the overall rate that can be handled is the difference between the size of I and the leak rate. I affects the long-term cell rate L short-term cell rate because it affects the size of the bucket. This controls how cells can burst through the network.

Let's consider the leaky bucket algorithm with a smooth traffic example. In Figure 3.25, the cell times are separated left to right equally in time. The state of the bucket just before the cell time is represented by t−, and the state of the bucket just afterwards is represented by t+.

Assume the bucket is empty and a cell comes in on this connection. We pour one-and-a-half units of water into the bucket. (Each cell contains one-and-a-half units of information. This is the increment parameter I. However, we can only leak one unit per cell time.) By the time we get to the next cell time, one unit has drained out, and, of course, by carefully planning this example, another cell comes in so you put the I units in. Now the bucket is one-half plus one and a half – it's exactly full.

At the next time, if a cell came in, that cell would violate the contract because there is not enough room to put 1.5 units into this bucket. So let's assume that we are obeying the rules. We don't send a cell and this level stays the same and then it finally drains out, and of course, you can see we're back where we started.

The reason this is a 'smooth' traffic case is because it tends to be very periodic. In this case, every two out of three cell times a cell is transmitted, and we assume that this pattern goes on indefinitely. Of course, two out of three is exactly the inverse of the increment parameter, 1.5. This can be adjusted with the I and the leak rate so that the parameter can be any increment desired – 17 out of 23, 15 out of 16, etc. There is essentially full flexibility to pick the parameters to get any fine granularity of rate.

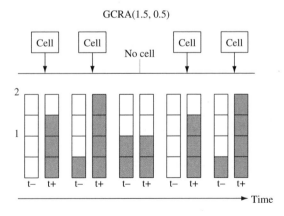

Figure 3.25   An illustration of smooth traffic coming to the leaky bucket - GCRA(1.5, 0.5)

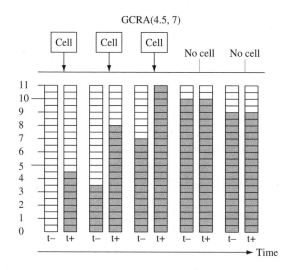

Figure 3.26 Illustration of burst traffic coming to the leaky bucket - GCRA(4.5, 7)

Now let's consider an example of more burst traffic. To make this burst, increase the limit parameter to 7, and just slow things down, the increment parameter is 4.5, so the bucket is 11.5 deep as shown in Figure 3.26.

As this example sends three cells, the information builds up and the bucket is exactly full after three cells. Now the rate is still only draining one unit of water per time but the increment is 4.5. Obviously, you're going to have to wait quite a while before you can send another cell.

If you wait a long enough for the bucket to empty completely, another burst of three cells may be accepted. This illustrates the effect of increasing the limit parameter to allow more burst type of traffic. Of course, this is especially critical for a typical data application.

3.7.7 Virtual scheduling algorithm (VSA)

In the virtual scheduling algorithm (VSA), I is the parameter used to space the time between two consecutive arrival cells. It allows the space of two cells to be smaller than I, but that must be larger than $(I - L)$. The total shift of time for a consecutive set of cells is

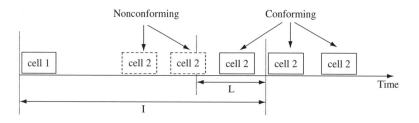

Figure 3.27 Virtual scheduling algorithm (VSA)

controlled to be less that L. Figure 3.27 illustrates the concepts of the VSA. It shows that the inter-arrival time between cell 1 and the cell 2 should be greater than or equal to I. If cell 2 arrives earlier than the inter-arrival time I but later than $(I - L)$, cell 2 is still considered as a conforming cell. Otherwise, cell 2 is considered as nonconforming cell.

3.8 Internet protocols

The developments of the Internet protocols have followed quite different paths from the ATM protocols, leading to the standards for networking. In the early years, the Internet was developed and used mainly by universities, research institutes, industry, military and the US government. The main network technologies were campus networks and dial-up terminals and servers interconnected by backbone networks. The main applications were email, file transfer and telnet.

The explosion of interest in Internet started in the mid-1990s, when the WWW provided a simple interface to ordinary users who didn't need to know anything about the Internet technology. The impact was far beyond people's imagination and entered our daily lives for information access, communications, entertainment, e-commerce, e-government, etc. New applications and services are developed every day using WWW based on the Internet.

In the meantime, the technologies and industries have started to converge so that computers, communications, broadcast, and mobile and fixed networks cannot be separated from each other any longer. The original design of the Internet could not meet the increasing demands and requirements therefore the IETF started to work on the next generation of networks. The IPv6 is the result of the development of the next generation of Internet networks. The third generation mobile networks, Universal Mobile Telecommunications Systems (UMTS), have also planned to have all-IP networks for mobile communications. Here we provide a brief introduction to the Internet protocols, and will leave further discussion to the later chapters on the next generation of Internet including IPv6 from the viewpoints of protocol, performance, traffic engineering and QoS support for future Internet applications and services.

3.8.1 Internet networking basics

Internet networking is an outcome of the evolution of computer and data networks. There are many technologies available to support different data services and applications using different methods for different types of networks. The network technologies include local area network (LAN), metropolitan area network (MAN) and wide area network (WAN) using star, bus ring, tree and mesh topologies and different media access control mechanisms.

Like ATM, the Internet is not a transmission technology but a transmission protocol. Unlike ATM, the Internet was developed to allow different technologies to be able to internetwork together using the same type of network layer packets to be transported across different network technologies.

LAN is widely used to connect computers together in a room, building or campus. MAN is a high-speed network to connect LANs together in metropolitan areas. WAN is used across a country, continent or a globe. Before the Internet, bridges were used to interconnect many different types of networks at link level by translating functions and frames formats and adapting transmission speeds between many different network technologies. Interconnecting different types of networks using different protocols together to form a larger network becomes a great challenge. The Internet protocol has taken a complete different approach from the translation between different network protocols and technologies, by introducing a common connectionless protocol in which data is carried by packets across different network technologies.

3.8.2 Protocol hierarchies

Protocol hierarchy and layering principles are also important concepts to deal with in the complexity of network design. The Internet protocols define the functions of network layers and above. Details on how to transport the network across different types of network technologies are considered as low layer functions, defined within the individual technologies, as long as the network technology is able to provide frames with payload and link layer functions capable of carrying the Internet packet across the network of the technology. On top of the network layer is the transport layer, then the application layer.

3.8.3 Connectionless network layer

The Internet network layer function is connectionless providing best-effort services. The whole network consists of many sub-networks, each of which can be of any type of network technology including LAN, MAN and WAN. User terminals can communicate directly with each other in the same sub-network using broadcast frames in shared media such as LAN, point-to-point link frames such as dialup links and multi-service frames such as WAN.

Routers are at the edge of the sub-networks and connect the sub-networks together, they can communicate with each other directly and also with user terminals in the same sub-networks. In other works, the Internet routers are interconnected together by many different network technologies. Each packet generated by source terminals carries the destination and source addresses of the terminals, and can be delivered to the destination terminal on the same sub-network or to a router on the same sub-network. The router is able to receive the packet and forward it to the next router, making use of the routing protocols, until the packet reaches its destination.

3.8.4 The IP packet format

In the Internet reference model, there is only one network layer protocol, that is the Internet protocol (IP). It is a unique protocol making use of the transmission services provided by the different types of networks below, and providing end-to-end network layer service to the transport layer protocols above.

The IP packets may be carried across different type of networks, but their IP format stays the same. Any protocol above the IP layer can only access the functions provided by the IP packet. Therefore the differences of the networks are screened out by the IP layer as shown in Figure 3.28.

Figure 3.29 shows the format of the IP packet. The following is a brief discussion of each field of the IP packet header.

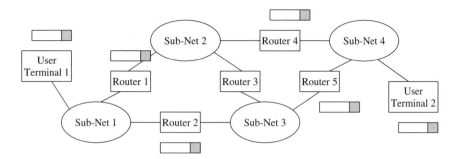

Figure 3.28 Internet packets over routers and sub-networks

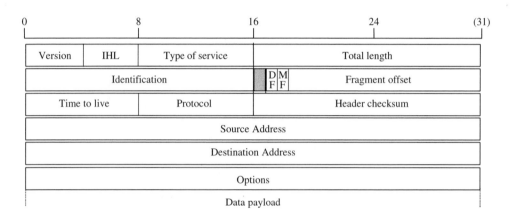

Figure 3.29 IP packet header format

- The *version* field keeps track of which version of the protocol the datagram belongs to. The current version is 4, also called IPv4. IPv5 is an experimental version. The next version to be introduced into the Internet is IPv6, the header has been changed dramatically. We will discuss this later.
- The *IHL* field is the length of the header in 32-bit words. The minimum value is 5 and maximum 15, which limits the header to 60 bytes.
- The *type of service* field allows the host to tell the network what kind of service it wants. Various combinations of delay, throughput and reliability are possible.
- The *total length* includes both header and data. The maximum value is 65 535.
- The *identification* field is needed to allow the destination host to determine which datagram a newly arrived fragment belongs to. Every IP packet in the network is identified uniquely.
- *DF*: don't fragment. This tells the network not to fragment the packet, as a receiving party may not be able to reassemble the packet.
- *MF*: more fragment. This indicates that more fragment is to come as part of the IP packet.
- The *fragment offset* indicates where in the current datagram this fragment belongs.
- The *time to live* is a counter used to limit packet lifetime to prevent the packet staying in the network forever.

Table 3.1 Option fields of the IPv4 packet header

Options	Descriptions
Security	Specifies how secret the datagram is
Strict source routing	Gives complete path to follow
Loose source routing	Gives a list of routers not be missed
Record route	Makes each router append its IP address
Time stamp	Makes each router append its address and time stamp

- The *protocol* field indicates the protocol data in the payload. It can be TCP or UDP. It is also possible to carry data of other transport layer protocols.
- The *checksum* field verifiers the IP header only.
- The *source* and *destination addresses* indicate the network number and host number.
- *Options* are variable length. Five functions are defined: security, strict routing, loose source routing, record route and time stamp (see Table 3.1).

3.8.5 IP address

The IP address used in the source and destination address fields of the IP packet is 32 bits long. It can have up to three parts. The first part identifies the class of the network address from A to E, the second part is the network identifier (net-id) and the third part is the host identifier (host-id). Figure 3.30 shows the formats of the IPv4 addresses.

In class A and B addresses, there are a large number of host-id. The hosts can be grouped into subnets each of which is identified by using the high-order host-id bits. A subnet mask is introduced to indicate the split between net-id + sub-net-id and host-id.

Similarly, there is a large number of net-id in the class C addresses. Some of the lower order bits of the net-id can be grouped together to form a supernet. This is also called classless inter domain routing (CIDR) addressing. Routers do not need to know anything within the supernet or the domain.

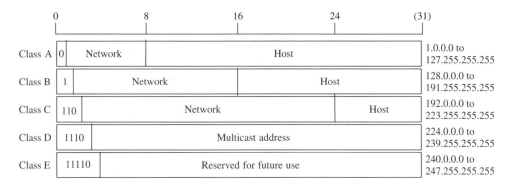

Figure 3.30 IP address formats

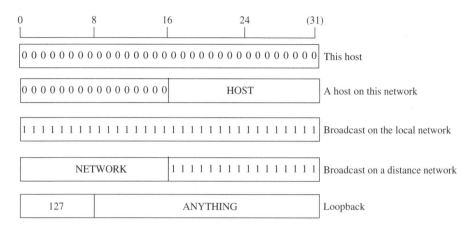

Figure 3.31   Special IP addresses

Class A, B and C addresses identify the attachment point of the hosts. Class D addresses identify the multicast address (like radio channel) but not an attachment point in the network. Class E is reserved for future use. There are also some special addresses shown in Figure 3.31.

3.8.6 Mapping between Internet and physical network addresses

An Internet address is used to identify a sub-network in the context of Internet. Each address consists of two parts: one identifies uniquely a sub-network and the other a host computer. The physical address is used to identify a network terminal related to the transmission technologies. For example, we can use a telephone number to identify individual telephones in the telephony networks, and an Ethernet address to identify each network interface card (NIC) uniquely for Ethernet networks.

Each host (computer, PC, or workstation), by installing an Ethernet NIC, will have the unique Ethernet address worldwide. A host can send data to another host or to all hosts in the Ethernet by broadcasting using the other hosts' addresses or Ethernet broadcasting address.

Each host also has a unique IP address in the Internet. All the hosts in the Ethernet have the same network identifier (net-id) forming a sub-network. The sub-networks can be connected to the Internet by using routers. All routers exchange information using routing protocols to find out the topology of the Internet and calculate the best router to be used for forwarding packets to their destinations.

Clearly, the host can send a packet to another host within the same sub-network. If the other host is outside of the sub-network, the host can send the packet to a router. The router can forward the packet to the next one until the packets reach their destinations or send to the host if the router is on the destination network. Therefore, the Internet can be seen as a network of interconnected routers by using many different network transmission technologies. However, the transmissions of the Internet packets between the routers need to use the native addresses and data frames of the network technologies. As the native address identifies access points to the network technology and the Internet address identifies the host, a mapping is required to specify the identified host attached to the network access point together forming a part of the sub-net.

A network manager can set up such a mapping manually for small networks, but it is preferable to have network protocols to map them automatically in a global scale.

3.8.7 ARP and RARP

Address resolution protocol (ARP) is a protocol used to find the mapping between the IP address and network address such as an Ethernet address. Within the network, a host can ask for the network address giving an IP address to get the mapping. If the IP address is outside the network, the host will forward the IP address to a router (it can be a default or proxy).

Reverse address resolution protocol (RARP) is the protocol used to solve the reverse problem, i.e., to find the IP address giving a network address such as Ethernet. This is normally resolved by introducing a RARP server. The server keeps a table of the address mapping. An example of using RARP is when a booting machine does not have an IP address and needs to contact a server to get an IP address to be attached to the Internet.

3.8.8 Internet routing protocols

Each router in the Internet has a routing table showing the next router or default router to forward packets to for all the destinations. As the Internet becomes larger and larger it is impractical or impossible to configure the routing table manually, although in the early days and for small networks manual configuration of network was carried out for convenience but was error prone. Protocols have to be developed to configure the Internet automatically and dynamically.

A part of the Internet owned and managed by a single organisation or by a common policy can form a domain or autonomous system (AS). The interior gateway routing protocol is used for IP routing within the domain. Between domains, the exterior gateway routing protocol has to be used as political, economic or security issues often need to be taken into account.

3.8.9 The interior gateway routing protocol (IGRP)

The original routing protocol was called the routing information protocol (RIP), which used the distance vector algorithm. Within the domain, each router has a routing table of the next router leading to the destination network. The router periodically exchanges its routing table information with its neighbour routers, and updates its routing table based on the new information received.

Due to its slow convergence problem, a new routing protocol was introduced in 1979, using the link state algorithm. The protocol was also called the link state routing protocol. Instead of getting routing information from its neighbour, each router using the link state protocol collects information on the links and sends link state information of its own and received link state information of the other neighbours by flooding the network with the link state information. Every router in the network will have the same set of link state information and can calculate independently the routing table. This solved the problems of the RIP for large-scale networks.

In 1988, the IETF began work on a new interior gateway routing protocol, called open shortest path first (OSPF) based on the link state protocol, which became a standard in 1990. It is also based on algorithms and protocols published in open literatures (this is the reason the word 'open' appears in the name of the protocol), and is designed to support: a variety of distance metrics, adaptive to changes in topology automatically and quickly; routing based on type of service and real-time traffic; load balancing; hierarchical systems and some levels of security; and also deals with routes connected to the Internet via a tunnel.

The OSPF supports three kinds of connections and networks including point-to-point lines between two routers, multicast networks (such as LAN), and multi-access networks without broadcasting (such as WAN).

When booting, a router sends a HELLO message. Adjacent routers (designated routers in each LAN) exchange information. Each router periodically floods link state information to each of its adjacent routers. Database description messages include the sequence numbers of all the link state entries, sent in the Internet packets. Using flooding, each router informs all the other neighbour routers. This allows each router to construct the graph for its domain and compute the shortest path to form a routing table.

3.8.10 The exterior gateway routing protocol (EGRP)

All an interior gateway protocol has to do is move packets as efficiently as possible. Exterior gateway routers have to worry about politics a great deal. EGRP is fundamentally a distance vector protocol, but with additional mechanisms to avoid the problems associated with the distance vector algorithm. Each EGRP router keeps track of the exact path used to solve the problems of distance vector. EGRP is also called Board Gateway Protocol (BGP).

3.9 Transport layer protocols: TCP and UDP

The transport layer protocols appear on the hosts. When a packet arrives in a host, it decides which application process to handle the data, e.g. email, telnet, ftp or WWW. There are also additional functions including reliability, timing, flow control and congestion control. There are two protocols at the transport layer within the Internet reference model.

3.9.1 Transmission control protocol (TCP)

TCP is a connection-oriented, end-to-end reliable protocol. It provides reliable inter-process communication between pairs of processes in host computers. Very few assumptions are made as to the reliability of the network technologies carrying the Internet packets. TCP assumes that it can obtain a simple, potentially unreliable datagram service from the lower level protocols (such as IP). In principle, TCP should be able to operate above a wide spectrum of communication systems ranging from hard-wired LAN and packet-switched networks and circuit-switched networks to wireless LAN, wireless mobile networks and satellite networks.

3.9.2 The TCP segment header format

Figure 3.32 illustrates the TCP segment header. The functions of the fields are the following:

- *Source port* and *destination port* fields, each of which has 16 bits, specify source and destination port numbers to be used by the process as addresses so that the processes in the source and destination computers can communicate with each other by sending and receiving data from the addresses.
- *Sequence number* field consists of 32 bits. It identifies the first data octet in this segment (except when SYN control bit is present). If SYN is present the sequence number is the initial sequence number (ISN) and the first data octet is ISN +1.
- *Acknowledgement number* field consists of 32 bits. If the ACK control bit is set this field contains the value of the next sequence number the sender of the segment is expecting to receive. Once a connection is established this is always sent.
- *Data offset* field consists of four bits. The number of 32-bit words in the TCP header. This indicates where the data begins. The TCP header (even one including options) is an integral number of 32 bits long.
- *Reserved* field of six bits for future use (must be zero by default).
- *Control bits* consist of six bits (from left to right) for the following functions:

    - URG: urgent pointer field indicator;
    - ACK: acknowledgement field significant;
    - PSH: push function;
    - RST: reset the connection;
    - SYN: synchronise sequence numbers;
    - FIN: no more data from sender.

- *Window* field consists of 16 bits. The number of data octets beginning with the one indicated in the acknowledgement field, which the sender of this segment is willing to accept.

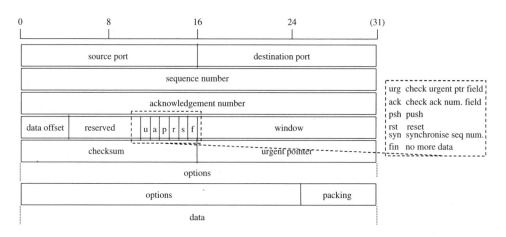

Figure 3.32 The TCP segment header

- *Checksum* field consists of 16 bits. It is the 16-bit one's complement of the one's complement sum of all 16-bit words in the header and text. If a segment contains an odd number of header and text octets to be checksummed, the last octet is padded on the right with zeros to form a 16-bit word for checksum purposes. The pad is not transmitted as part of the segment. While computing the checksum, the checksum field itself is replaced with zeros.
- *Urgent pointer* field consists of 16 bits. This field communicates the current value of the urgent pointer as a positive offset from the sequence number in this segment.
- *Options* and *padding* fields have variable length. The option allows additional functions to be introduced to the protocol.

To identify the separate data streams that a TCP may handle, the TCP provides the port identifier. Since port identifiers are selected independently by each TCP they might not be unique. To provide for unique addresses within each TCP, IP address and port identifier are used together to create a unique socket throughout all sub-networks in the Internet.

A connection is fully specified by the pair of sockets at the ends. A local socket may participate in many connections to different foreign sockets. A connection can be used to carry data in both directions, i.e., it is 'full duplex'.

The TCP are free to associate ports with processes however they choose. However, several basic concepts are necessary in any implementation. Well-known sockets are a convenient mechanism for a priori associating socket addresses with standard services. For instance, the 'telnet-server' process is permanently assigned to a socket number of 23, FTP-data 20 and FTP-control 21, TFTP 69, SMTP 25, POP3 110, and WWW HTTP 80.

3.9.3 Connection set up and data transmission

A connection is specified in the system call OPEN by the local and foreign socket arguments. In return, the TCP supplies a (short) local connection name by which the user refers to the connection in subsequent calls. There are several things that must be remembered about a connection. To store this information we imagine that there is a data structure called a transmission control block (TCB). One implementation strategy would have the local connection name be a pointer to the TCB for this connection. The OPEN call also specifies whether the connection establishment is to be actively pursued or passively waited for.

The procedures used to establish connections utilise the synchronisation (SYN) control flag and involve an exchange of three messages. This exchange has been termed a three-way handshake. The connection becomes 'established' when sequence numbers have been synchronised in both directions. The clearing of a connection also involves the exchange of segments, in this case carrying the finish (FIN) control flag.

The data that flows on the connection may be thought of as a stream of octets. The sending process indicates in each system call SEND that the data in that call (and any preceding calls) should be immediately pushed through to the receiving process by setting of the PUSH flag.

The sending TCP is allowed to collect data from the sending process and to send that data in segments at its own convenience, until the push function is signalled, then it must send all unsent data. When a receiving TCP sees the PUSH flag, it must not wait for more data from the sending TCP before passing the data to the receiving process. There is no necessary relationship between push functions and segment boundaries. The data in any particular segment may be the result of a single SEND call, in whole or part, or of multiple SEND calls.

3.9.4 Congestion and flow control

One of the functions in the TCP is end-host based congestion control for the Internet. This is a critical part of the overall stability of the Internet. In the congestion control algorithms, TCP assumes that, at the most abstract level, the network consists of links for packet transmission and queues for buffering the packets. Queues provide output buffering on links that can be momentarily oversubscribed. They smooth instantaneous traffic bursts to fit the link bandwidth.

When demand exceeds link capacity long enough to cause the queue buffer to overflow, packets must get lost. The traditional action of dropping the most recent packet ('tail dropping') is no longer recommended, but it is still widely practised.

TCP uses sequence numbering and acknowledgements (ACKs) on an end-to-end basis to provide reliable, sequenced, once-only delivery. TCP ACKs are cumulative, i.e., each one implicitly ACKs every segment received so far. If a packet is lost, the cumulative ACK will cease to advance.

Since the most common cause of packet loss is congestion in the traditional wired network technologies, TCP treats packet loss as an indicator of network congestion (but such an assumption is not applicable in wireless or satellite networks where packet loss is more likely to be caused by transmission errors). This happens automatically, and the sub-network need not know anything about IP or TCP. It simply drops packets whenever it must, though some packet-dropping strategies are fairer than others.

TCP recovers from packet losses in two different ways. The most important is by a retransmission timeout. If an ACK fails to arrive after a certain period of time, TCP retransmits the oldest unacknowledged packet. Taking this as a hint that the network is congested, TCP waits for the retransmission to be acknowledged (ACKed) before it continues, and it gradually increases the number of packets in flight as long as a timeout does not occur again.

A retransmission timeout can impose a significant performance penalty, as the sender will be idle during the timeout interval and restarts with a congestion window of one following the timeout (slow start). To allow faster recovery from the occasional lost packet in a bulk transfer, an alternate scheme known as 'fast recovery' can be introduced.

Fast recovery relies on the fact that when a single packet is lost in a bulk transfer, the receiver continues to return ACKs to subsequent data packets, but they will not actually acknowledge (ACK) any data. These are known as 'duplicate acknowledgements' or 'dupacks'. The sending TCP can use dupacks as a hint that a packet has been lost, and it can retransmit it without waiting for a timeout. Dupacks effectively constitute a negative acknowledgement (NAK) for the packet whose sequence number is equal to the acknowledgement field in the incoming TCP packet. TCP currently waits until a certain number of dupacks (currently three) are seen prior to assuming a loss has occurred; this helps avoid an unnecessary retransmission in the face of out-of-sequence delivery.

In addition to congestion control, the TCP also deals with flow control to prevent the sender overrunning the receiver. The TCP 'congestion avoidance' (RFC2581) algorithm is the end-to-end system congestion control and flow control algorithm used by TCP. This algorithm

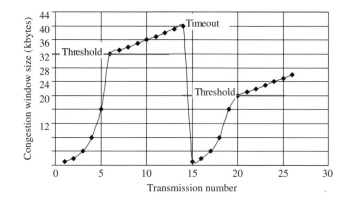

Figure 3.33   Congestion control and avoidance

maintains a congestion window (cwnd) between the sender and receiver, controlling the amount of data in flight at any given point in time. Reducing cwnd reduces the overall bandwidth obtained by the connection; similarly, raising cwnd increases the performance, up to the limit of the available bandwidth.

TCP probes for available network bandwidth by setting cwnd at one packet and then increasing it by one packet for each ACK returned from the receiver. This is TCP's 'slow-start' mechanism. When a packet loss is detected (or congestion is signalled by other mechanisms), cwnd is set back to one and the slow-start process is repeated until cwnd reaches one half of its previous setting before the loss. Cwnd continues to increase past this point, but at a much slower rate than before to avoid congestion. If no further losses occur, cwnd will ultimately reach the window size advertised by the receiver. Figure 3.33 illustrates an example of the congestion control and congestion avoidance algorithm.

3.9.5 User datagram protocol (UDP)

The UDP is defined to make available a datagram mode of the transport layer protocol. This protocol assumes that the Internet protocol (IP) is used as the underlying protocol.

This protocol provides a procedure for application programs to send messages to other programs with a minimum of protocol mechanism. The protocol provides connectionless service and does not provide any guarantee on delivery, duplicate protection and order of delivery, or even make any effort to recover any lost data. Therefore, it makes the protocol very simple and particularly useful for real-time data transportation.

Figure 3.34 illustrates the UDP datagram header format. The functions of the fields of the UDP datagram header are discussed here.

- *Source port* field is an optional field, when meaningful, it indicates the port of the sending process, and may be assumed to be the port to which a reply should be addressed in the absence of any other information. If not used, a value of zero is inserted.
- *Destination port* field has a meaning within the context of a particular Internet destination address.

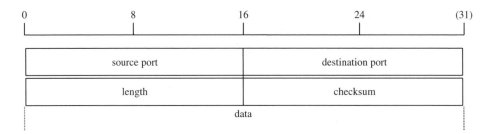

Figure 3.34 The UDP datagram header format

- *Length* field indicates the length in octets of this user datagram including its header and the data. (This means the minimum value of the length is eight.)
- *Checksum* is the 16-bit one's complement of the one's complement sum of a pseudo header of information from the IP header, the UDP header.
- The data, padded with zero octets at the end (if necessary) to make a multiple of two octets.

The major uses of this protocol are the Internet name server, and the trivial file transfer, and recently for real-time applications such as VoIP, video streaming and multicast where retransmission of lost data is undesirable. The well-known ports are defined in the same way as the TCP.

3.10 IP and ATM internetworking

Since there are vast numbers of computers and network terminals interconnected by using LANs, MANs and WANs and the Internet protocols operating on these networks, a key to success will be the ability to allow for interoperability between these network technologies and ATM. A key to success of future Internet is its ability to support QoS and to provide a uniform network view to higher layer protocols and applications.

There are, however, two fundamentally different ways of running Internet protocols across an (overlay mode) ATM network as shown in Figure 3.35. In one method, known as native

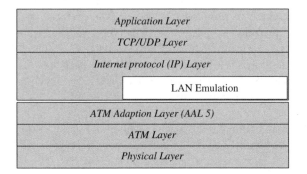

Figure 3.35 Protocol stacks for LAN emulation and classical IP over ATM

IP over ATM (or classic IP over ATM) mode operation, address resolution mechanisms are used to map Internet addresses directly into ATM addresses, and the Internet packets are then carried across the ATM network.

The alternative method of carrying network layer packets across an ATM network is known as LAN emulation (LANE). As the name suggests, the function of the LANE protocol is to emulate a local area network on top of an ATM network. Specifically, the LANE protocol defines mechanisms for emulating either an IEEE 802.3 Ethernet or an 802.5 token ring LAN.

3.10.1 LAN emulation (LANE)

LAN emulation means that the LANE protocol defines a service interface for higher layer (that is, network layer) protocols, which is identical to that of existing LANs, and that data sent across the ATM network are encapsulated in the appropriate LAN MAC packet format. It does not mean that any attempt is made to emulate the actual media access control protocol of the specific LAN concerned (that is, CSMA/CD for Ethernet or token passing for 802.5). In other words, the LANE protocols make an ATM network look and behave like an Ethernet or token ring LAN – albeit one operating much faster than a real such network.

The rationale for doing this is that it requires no modifications to higher layer protocols to enable their operation over an ATM network. Since the LANE service presents the same service interface of existing MAC protocols to network layer drivers, no changes are required in those drivers. The intention is to accelerate the deployment of ATM, since considerable work remains to be done in fully defining native mode operation for the plethora of existing network layer protocols.

It is envisaged that the LANE protocol will be deployed in two types of ATM-attached equipment:

- *ATM network interface cards (NIC)*: ATM NIC will implement the LANE protocol and interface to the ATM network, but will present the current LAN service interface to the higher level protocol drivers within the attached end system. The network layer protocols on the end system will continue to communicate as if they were on a known LAN, using known procedures. They will, however, be able to use the vastly greater bandwidth of ATM networks.
- *Internetworking and LAN switching equipment*: the second class of network gear that will implement LANE will be ATM-attached LAN switches and routers. These devices, together with directly attached ATM hosts equipped with ATM NIC, will be used to provide a virtual LAN service, where ports on the LAN switches will be assigned to particular virtual LANs, independent of physical location. LAN emulation is a particularly good fit to the first generation of LAN switches that effectively act as fast multi-port bridges, since LANE is essentially a protocol for bridging across ATM. Internetworking equipment, such as routers, will also implement LANE to allow for virtual LAN internetworking.

The LANE protocols operate transparently over and through ATM switches, using only standard ATM signalling procedures. ATM switches may well be used as convenient platforms upon which to implement some of the LANE server components, but this is independent of the cell relay operation of the ATM switches themselves. This logical decoupling is one of the great advantages of the overlay model, since it allows ATM switch designs to proceed independently of the operation of overlying internetworking protocols, and vice versa.

The basic function of the LANE protocol is to resolve MAC addresses into ATM addresses. By doing so, it actually implements a protocol of MAC bridge functions using ATM; hence the close fit with current LAN switches. The goal of LANE is to perform such address mappings so that LANE end systems can set up direct connections between themselves and forward data. The element that adds significant complexity to LANE, however, is supporting LAN switches – that is, LAN bridges. The function of a LAN bridge is to shield LAN segments from each other.

3.10.2 LANE components

The LANE protocol defines the operation of a single emulated LAN (ELAN). Multiple ELANs may coexist simultaneously on a single ATM network. A single ELAN emulates either Ethernet or token ring, and consists of the following entities:

- *LAN emulation client (LEC)*: a LEC is the entity in an end system that performs data forwarding, address resolution and other control functions for a single end-system within a single ELAN. A LEC also provides a standard LAN service interface to any higher layer entity that interfaces to the LEC. In the case of an ATM NIC, for instance, the LEC may be associated with only a single MAC address, while in the case of a LAN switch; the LEC would be associated with all MAC addresses reachable through the ports of that LAN switch assigned to the particular ELAN.
- *LAN emulation server (LES)*: the LES implements the control function for a particular ELAN. There is only one logical LES per ELAN, and to belong to a particular ELAN means to have a control relationship with that ELAN's particular LES. Each LES is identified by a unique ATM address.
- *Broadcast and unknown server (BUS)*: the BUS is a multicast server that is used to flood unknown destination address traffic and forward multicast and broadcast traffic to clients within a particular ELAN. The BUS to which a LEC connects is identified by a unique ATM address. In the LES, this is associated with the broadcast MAC address, and this mapping is normally configured into the LES.
- *LAN emulation configuration server (LECS)*: the LECS is an entity that assigns individual LANE clients to particular ELANs by directing them to the LES that correspond to the ELAN. There is logically one LECS per administrative domain, and this serves all ELANs within that domain.

3.10.3 LANE entity communications

LANE entities communicate with each other using a series of ATM connections. LECs maintain separate connections for data transmission and control traffic. The control connections are as follows:

- Configuration direct VCC: this is a bi-directional point-to-point VCC set up by the LEC to the LECS.
- Control direct VCC: this is a bi-directional VCC set up by the LEC to the LES.

- Control distribute VCC: this is a unidirectional VCC set up from the LES back to the LEC; this is typically a point-to-multipoint connection.

The data connections are as follows:

- *Data direct VCC*: this is a bi-directional point-to-point VCC set up between two LECs that want to exchange data. Two LECs will typically use the same data direct VCC to carry all packets between them, rather than opening a new VCC for each MAC address pair between them, so as to conserve connection resources and connection set-up latency. Since LANE emulates existing LAN, including their lack of QoS support, data direct connections will typically be UBR or ABR connections, and will not offer any type of QoS guarantees.
- *Multicast send VCC*: this is a bi-directional point-to-point VCC set up by the LEC to the BUS.
- *Multicast forward VCC*: this is a unidirectional VCC set up to the LEC from the BUS, this is typically a point-to-multipoint connection, with each LEC as a leaf.

The higher layer protocol processing within the router is unaffected by the fact that the router is dealing with emulated or physical LAN. This is another example of the value of LANE in hiding the complexities of the ATM network.

One obvious limitation of this approach, however, is that the ATM router may eventually become a bottleneck, since all inter-ELAN traffic must traverse the router. LANE has another limitation. By definition, the function of LANE is to hide the properties of ATM from higher layer protocols. This is good, particularly in the short to medium term, since it precludes the need for any changes to these protocols. On the other hand, LANE also precludes these protocols from ever using the unique benefits of ATM, and specifically, its QoS guarantees. LANE is defined to use only UBR and ABR connections, since it is these that map best to the connectionless nature of MAC protocols in LANs.

3.10.4 Classical IP over ATM

The IETF IP-over-ATM working group has developed protocols for IP transport over ATM. The transport of any network layer protocol over an overlay mode ATM network involves two aspects: packet encapsulation and address resolution. Both of these aspects have been tackled by the IETF, and are described below:

3.10.5 Packet encapsulation

The IETF has defined a method for transporting multiple types of network or link layer packets across an ATM (AAL 5) connection and also for multiplexing multiple packet types on the same connection. As with LANE, there is value to reusing the same connection for all data transfers between two nodes since this conserves the connection resource space, and saves on connection set-up latency, after the first connection set up. This is only possible, however, as long as only UBR or ABR connections are used – if the network layer requires QoS guarantees then every distinct flow will typically require its own connection.

In order to allow connection re-use, there must be a means for a node that receives a network layer packet across an ATM connection to know what kind of packet has been received, and to what application or higher level entity to pass the packet to; hence, the packet must be prefixed with a multiplexing field. Two methods for doing this are defined in RFC 1483:

- *Logical link control/sub-network access point (LLC/SNAP) encapsulation.* In this method, multiple protocol types can be carried across a single connection with the type of encapsulated packet identified by a standard LLC/SNAP header. A further implication of LLC/SNAP encapsulation, however, is that all connections using such encapsulations terminate at the LLC layer within the end systems, as it is here that the packet multiplexing occurs.
- *VC multiplexing.* In the VC multiplexing method, only a single protocol is carried across an ATM connection, with the type of protocol implicitly identified at connection set up. As a result, no multiplexing or packet type field is required or carried within the packet, though the encapsulated packet may be prefixed with a pad field. The type of encapsulation used by LANE for data packets is actually a form of VC multiplexing.

The VC multiplexing encapsulation may be used where direct application-to-application ATM connectivity, bypassing lower level protocols, is desired. As discussed earlier, however, such direct connectivity precludes the possibility of internetworking with nodes outside the ATM network.

The LLC/SNAP encapsulation is the most common encapsulation used in the IP over ATM protocols. The ITU-T has also adopted this as the default encapsulation for multiprotocol transport over ATM, as has the ATM Forum's multiprotocol over ATM group. In related work, the IP over ATM group has also defined a standard for a maximum transfer unit (MTU) size over ATM. This defines the default MTU as 9180 bytes to be aligned with the MTU size for IP over SMDS. It does, however, allow for negotiation of the MTU beyond this size, to the AAL 5 maximum of 64 kbytes, since important performance improvements can be gained by using larger packet sizes. This standard also mandates the use of IP path MTU discovery by all nodes implementing IP over ATM to preclude the inefficiency of IP fragmentation.

3.10.6 IP and ATM address resolution

In order to operate IP over ATM, a mechanism must be used to resolve IP addresses to their corresponding ATM addresses. For instance, consider the case of two routers connected across an ATM network. If one router receives a packet across a LAN interface, it will first check its next-hop table to determine through which port, and to what next-hop router, it should forward the packet. If this look-up indicates that the packet is to be sent across an ATM interface, the router will then need to consult an address resolution table to determine the ATM address of the destination next-hop router (the table could also be configured, of course, with the VPI/VCI value of a PVC connecting the two routers).

This address resolution table could be configured manually, but this is not a very scalable solution. The IP-over-ATM working group has defined a protocol to support automatic address resolution of IP addresses in RFC 1577. This protocol is known as 'classical IP over

ATM' and introduces the notion of a logical IP sub-net (LIS). Like a normal IP sub-net, an LIS consists of a group of IP nodes (such as hosts or routers) that connect to a single ATM network and belong to the same IP sub-net.

To resolve the addresses of nodes within the LIS, each LIS supports a single ATM address resolution protocol (ATMARP) server, while all nodes (LIS clients) within the LIS are configured with the unique ATM address of the ATMARP server. When a node comes up within the LIS, it first establishes a connection to the ATMARP server, using the configured address. Once the ATMARP server detects a connection from a new LIS client, it transmits an inverse ARP 53 request to the attaching client and requests the node's IP and ATM addresses, which it stores in its ATMARP table.

Subsequently, any node within the LIS wishing to resolve a destination IP address would send an ATMARP request to the server, which would then respond with a ATMARP reply if an address mapping is found. If not, it returns an ATM_NAK response to indicate the lack of a registered address mapping. The ATMARP server ages out its address table for robustness, unless clients periodically refresh their entry with responses to the servers inverse ARP queries. Once an LIS client has obtained the ATM address that corresponds to a particular IP address, it can then set up a connection to the address.

The operation of the classical model is very simple. It does, however, suffer from a number of limitations. One of these limitations is indicated by the phrase 'classical'. What this means is that the protocol does not attempt to change the IP host requirement that any packet for a destination outside the source node's IP sub-net must be sent to a default router. This requirement, however, is not a good fit to the operation of IP over ATM, and a whole class of other 'non-broadcast multi-access' (NBMA) networks, such as frame relay or X.25. In all such networks, it is possible to define multiple LIS, and the network itself could support direct communications between two hosts on two different LIS.

However, since RFC 1577 preserves the host requirements, in the context of IP over ATM, communications between two nodes on two different LIS on the same ATM network must traverse each ATM router on the intermediate hops on the path between the source and destination nodes. This is clearly inefficient, since the ATM routers become bottlenecks; this also precludes the establishment of a single connection with a requested QoS between the two nodes.

Further reading

[1] Black, U., *ATM: Foundation for Broadband Networks*, Prentice Hall Series in Advanced Communication Technologies, 1995.
[2] Comer, D.E., *Computer Networks and Internet*, 3rd edition, Prentice Hall, 1999.
[3] Cuthbert, G., ATM: broadband telecommunications solution, *IEE Telecommunication Series No.29*, 1993.
[4] Tanenbaum A., *Computer Networks*, 4th edition. Prentice Hall, 2003.
[5] RFC 791, Internet Protocol, Jon Postel, IETF, September 1981.
[6] RFC 793, Transmission control protocol, Jon Postel, IETF, September 1981.
[7] RFC 768, User datagram protocol, Jon Postel, IETF, August 1980
[8] RFC 826, An Ethernet Address Resolution Protocol, David C. Plummer, IETF, November 1982.
[9] RFC 903, A Reverse Address Resolution Protocol, Finlayson, Mann, Mogul, Theimer, IETF, June 1984.
[10] RFC 2328, OSPF Version 2, J. Moy, IETF, April 1998.
[11] RFC 2453, RIP Version 2, G. Malkin, IETF, November 1998.
[12] RFC 1771, A Border Gateway Protocol 4 (BGP-4), Y. Rekhter and T. Li, IETF, March 1995.
[13] RFC 2581, TCP Congestion Control, M. Allman, V. Paxson and W. Stevens, IETF, April 1999.

[14] RFC 1483, Multiprotocol Encapsulation over ATM Adaptation Layer 5, Juha Heinanen, IETF, July 1993.
[15] RFC 1577, Classical IP and ARP over ATM, M. Laubach, IETF, January 1994.

Exercises

1. Explain the concepts of the ATM protocol and technology.
2. Discuss the functions of ATM adaptation layers (AAL) and the type of services they provide.
3. Use a sketch to explain how to transport ATM cells using an E1 connection.
4. Explain the concepts of VP and VC switches.
5. Explain how to achieve QoS and efficient utilisation of network resources.
6. Describe the leaky bucket and virtual scheduling algorithms.
7. Explain the functions of the Internet protocol (IP).
8. Explain the transmission control protocol (TCP) and user datagram protocol (UDP) and their differences.
9. Explain the deployment scenarios of LAN emulation.
10. Explain the concept of classical IP over ATM.

4

Satellite Internetworking with Terrestrial Networks

This chapter aims to provide an introduction to satellite internetworking with terrestrial networks and related access and transit transmission networks. When you have completed this chapter, you should be able to:

- Know the basic terminologies and concepts concerning internetworking.
- Know about network traffic related to user plane, control plane and management plane.
- Describe the network hypothetical reference connection.
- Describe the differences between multiplexing and multiple access schemes.
- Understand the basic concept of traffic engineering in telephony networks.
- Understand the evolution of digital networks including PDH, SDH and ISDN.
- Identify different types of signalling schemes.
- Identify the performance objectives of satellite networks in end-to-end reference connections.
- Understand the issues of SDH over satellite.
- Understand the issues of ISDN over satellite.

4.1 Networking concepts

Telecommunication networks were originally designed, developed and optimised with respect to the speech transmission quality of narrow-band 3.1 kHz real-time telephony services.

In the early generation of data networks in wide area, people tried to fully utilise the 3.1 kHz for data communications without the additional costs of a network infrastructure.

Satellite Networking: Principles and Protocols   Zhili Sun
© 2005 John Wiley & Sons, Ltd

At that time, the transmission speed of the data terminals was relatively low. In addition to telephony services, the networks can also support the transmission of non-voice signals such as fax and modem transmission, and wholly digital data transmission. To some extent, the telecommunication networks could meet the transmission demand of data communications.

Because of the development of computers as network terminals, high-speed data networks had to be developed to meet the demand of data communications. This led to the development of different types of networks for different services. Traffic on data networks is becoming larger and larger, and the same applies to network capacity. The increase in traffic generated the opportunity to transport telephony voice services over data networks. High-capacity user terminals and network technologies enable the convergence of telephony services and data services, and also broadcasting services. A new type of network, broadband networks, has been developed to support the convergence of services and networks.

All these developments are great for new services and applications, but also bring great challenges to internetwork between these different types of networks. Due to economic reasons, new networks are 'forced' to interface with legacy networks.

It is even more of a challenge for satellite networks to interwork with all these different types of networks. One of the great problems in telephony networks is that the terminals and networks are so well engineered that any change in one party would be restricted in the other party. Modern networks try to separate the functions of user terminals from the networks so that the user terminals provide services without concerning too much about how the traffic is transported over the networks, and the networks provide different types of transport schemes with little concern about how the terminals are going to process the traffic.

We will follow the same principle to discuss satellite internetworking with terrestrial networks, i.e., what the requirements are from the terrestrial networks, and how the satellite networks will be able to meet these requirement for internetworking purposes.

Medium and large private networks consist of several interconnected multi-line telephony systems (MLTS). The terms 'corporate network' or 'enterprise network' are sometimes used to describe a large private network; in some countries these terms are used in a legal sense for a group of interconnected private networks. From the point of view of networking, there is no difference between a large private network and several smaller interconnected networks. Therefore, only the term 'private network' will be used to refer to this type of networks.

A private network can be a terminating network (one to which terminal equipment is connected). It can also provide transit connections between other networks. We will emphasise the case for terminating networks, as the transit network case is quite similar to public networks.

We will focus more on the principles of all kinds of intra- or internetwork connections rather than detailed implementations, regardless of the number of public or private networks involved, or the specific configuration in which they are interconnected.

Therefore, there is neither restriction on the network with respect to size, configuration, hierarchy, technology used, nor on the components of the network.

Although all communications networks are now digital (almost universal in Europe), radio resource management in the frequency domain still uses the same principle as analogue networks. Naturally we will focus our discussion more on digital networks because of the prevalence of digital signal transmission media and digital signal handling in switching equipment.

4.2 Networking terminology

Before going into details, we will explain the definitions of a number of terms pertinent to the relevant concepts:

- *Reference point*: is a conceptual group at a conjunction of two non-overlapping function groups. The two function groups exchange information through the reference point using the same defined conceptual group.
- *Interworking*: is a general term describing two systems or subsystems exchanging information, and covers both internetworking and service interworking aspects.
- *Internetworking*: describes the concept of the interconnection of different networks to provide services interoperable across these networks.
- *Service interworking*: describes the concept that the full or limited service of one network is converted or made available in another (similar) service of the same network or another network.
- *Interworking unit (IWU)*: is a physical entity located between reference points containing one or more interworking functions (IWF). It is used to interconnect two function groups. If they do not have common reference points, mapping or translating is required for the two function groups to communicate with each other.

4.2.1 Private network

The term 'private network' is used to describe a network which provides features only to a restricted user group in contrast to the public network (PSTN) available to the general public. In general, a private network is a terminating network and consists of several interconnected nodes (i.e. PBXs, routers, gateways), with interconnections to other networks via mainly public networks.

A private network has the following characteristics:

- It consists normally of more than one network node element, connected via public networks or leased lines or via a virtual private network (VPN).
- It provides network functions and all other features only to a single user or to a group of users, and is not accessible to the general public.
- It is not limited by geographical size or to a specific national area or region, though most of the private networks use LAN technologies in a single site.
- It has no limitation with regard to the number of extensions and access points to other networks.

4.2.2 Public network

The term 'public network' refers to networks providing transmission, switching and routing functions as well as features which are available to the general public, and are not restricted to a specific user group. In this context, the word 'public' does not imply any relation to the legal status of the network operator.

In some cases, a public network may provide a limited set of features only. In a competitive environment, a public network may be restricted to serve a limited number of customers, or restricted to specific features or functions. Generally, public networks provide access points to other networks or terminals only within a specific geographical area.

From the point of view of an end-to-end connection, a public network can function either as a 'transit network' (a link between two other networks) or as a combination of 'transit and terminating network' in cases where the public network provides connections to terminal equipment such as telephone sets, PBXs, routers or gateways.

4.2.3 Quality aspects of telephony services

In telephony networks, quality aspects take into consideration both the telephone set and different components in the network. The perception of speech transmission quality during a telephone conversation is primarily a 'subjective' judgement. The concept of 'quality' may not be considered as a unique discrete quantity, but may vary, depending on the user's expectation of sufficient 'speech transmission quality' for a 3.1 kHz telephony call for the terminal mode (e.g. handset) as well as the particular service (e.g. wireless). An end-to-end consideration is taken into account from one human's mouth to another human's ear.

For the judgement of the quality in a given configuration, and the performance of 'subjective tests', the ITU-T has developed several methods. One of the most common methods is to perform laboratory tests (e.g. 'listening-only tests'), wherein the test subjects are requested to classify the perceived quality into categories. For example, a 'quality rating' can be graded on a 1–5 grade scale as bad, poor, fair, good and excellent.

The scores are used to calculate the average value of the judgement of several test subjects for the same test configuration. The result is the so-called 'mean opinion score' (MOS), which may, theoretically, range between 1 and 5. An assessment about the speech transmission quality can also be obtained by calculating the percentage of all test persons rating the configuration as 'good or better' or as 'poor or worse'. For a given connection these results are expressed as 'percentage good or better' (%GoB) and 'percentage poor or worse' (%PoW).

Therefore, it is a complicated task to evaluate quality of services in telephony networks, and involves the collection of the necessary information on the various network components in the configuration investigated and their contribution of transmission impairments that impact the end-to-end connection speech transmission quality. The ITU-T has developed several methods and tools to evaluate QoS over telephony networks.

In digital networks, the impairment in any part of the network does not propagate from one part to the other part. Therefore, the quality of services can be evaluated for each element separately. For example, the modern network terminals are capable of buffering digitised voice or putting the voice into memory before playing out. The terminals should be given the freedom of how long to keep and how much to buffer the voice. Similarly, modern networks process the digitised voice in terms of frames or packets, and should also be given the freedom of how much time to process and what the sizes of frames or packet should be.

4.2.4 IP based network

IP based networks have been developed based on the Internet protocols transmitted over different types of network technologies including LAN, WAN and wireless and satellite networks. From a board gateway protocol (BGP) router point of view, the world consists of autonomous systems (AS) and lines connecting them. Two AS are considered connected if there is a line between boarder routers in each one. The network can be grouped into three categories: the first category is the stub networks, which have only one BGP router connecting to the outside, hence cannot be used for transit traffic. The second category is the multi-connection networks, which can be used for transit traffic except that they refuse to carry the transit traffic. Finally the third category is the transit networks, which are willing to handle third parties, possibly with some restrictions, and usually charge for their services. Each AS has a similar structure. The stub network sending traffic to and receiving traffic from backbone networks, and backbone networks to transport the traffic between the AS. Typical networks include:

- private enterprise intranets (LANs);
- Internet service provider (ISP) domains via WANs;
- public Internet (concatenated WANs).

They consist of internal routers and edge routers (e.g. between LAN and WAN). The telephony network can be used to link the routers together and link IP terminals to ISPs.

IP based networks rely on the Internet protocol (IP) and provide packet-based transport of data. Thus, a digitised speech signal will be divided into small segments for the real-time transport protocol (RTP) in the application layer, the user datagram protocol (UDP) at the transport layer protocol and then the Internet protocol (IP) at the network layer. The header of these protocol layers in general contains the following data:

- specific information for dealing with real-time applications;
- port numbers to identify the process of real-time applications;
- IP addresses for packet delivery;
- network physical addresses and frames to transport the IP packets.

Finally, at the receiving side, the speech segments are used to construct the original continuous digital speech signal. For non-real-time data services, transmission control protocol (TCP) is used at the transport layer.

4.3 Network elements and connections

Network components in an end-to-end connection can be categorised into three main groups: network terminals, network connections and network nodes.

4.3.1 Network terminals

With respect to speech transmission, terminals are all types of telephone sets, digital or analogue, wired, cordless or mobile, including the acoustical interfaces to the user's mouth and ear. These components are characterised by their send loudness rating (SLR) and receive loudness rating (RLR), which contribute to the overall loudness rating (OLR) of a connection. Other parameters, such as the side tone masking rating (STMR), the listener side tone rating

(LSTR), the design of the handset (D-factor), and the frequency response in send and receive directions and the noise floor, also contribute to the end-to-end connection rating of speech transmission quality.

In the case of wireless or IP based systems, additional distortions and delay may be added, depending on the coding and modulation algorithms used in such interfaces. However, with packet networks, there are great advantages in the terminal with memory and processing power overcoming the problems of telephony networks.

4.3.2 Network nodes

Network nodes are all types of switching equipment, such as local PBXs and switches in telephony networks and routers in Internet. These nodes may use analogue or digital switching or packet-based technology. The main impairment contributions of analogue systems are loss and noise in telecommunication networks. Where four-wire to two-wire conversions take place within or between switching equipment interfaces, signal reflections contribute to impairments as a source for echo effects. Digital switching systems contribute to the end-to-end delay, due to signal processing, and also to the amount of quantisation distortion associated with digital pads and code conversion. Packet-based routers contribute, in addition, to delay variation versus time and packet loss.

4.3.3 Network connections

Network connections use all kinds of media as the facility between network nodes and between nodes and network terminals. The physical media of these connections may be metallic (copper), fibre optics or radio. The signal form is either analogue or digital. Impairments associated with analogue signal transmission include propagation time (generally proportional to distance), loss, frequency response and noise (mainly due to longitudinal interference). Impairments due to frequency response and noise can usually be neglected for short and medium line lengths.

For digital transmission, the main impairment is caused by the propagation time via metallic, optic and radio media. For wireless sections, additional delay is introduced, depending on the coding and modulation algorithm used. Where the connection includes analogue-to-digital conversion, loss and distortion are additional impairment factors.

Multiplexing is generally used to transport several channels via one single physical media. A variety of multiplexing systems are in use in the existing networks:

- time division multiplex (TDM);
- digital circuit multiplication equipment (DCME);
- packet-based networks, connection oriented (ATM) and connectionless (Ethernet, LAN, IP, etc.).

In telephony networks, connections support either 64 kbit/s pulse code modulation (PCM), or the more recently introduced compression techniques based on low bit-rate codecs. In broadband networks, the connections will be able to support traffic at a much higher speed of video and data in addition to the telephony voice services.

4.3.4 End-to-end connections

An end-to-end connection between two user terminals can be as near as next door or as far as the other side of the world. The connection may just involve a private network or a local exchange, or a private network and a local exchange, a long-distance connection in public networks and international connections.

In telephony networks, the predominance of incoming and outgoing calls are originated or terminated only within a local calling area. We can divide traffic into local calls, national long-distance calls and international calls. Therefore, the large number of user terminals can be supported by a small number of national long-distance connections. Similarly, we can use smaller international connections to support more national calls.

The end-to-end connection may also involve different type of network technologies including cable, optical, terrestrial wireless or satellite networks. All the technologies contribute in different ways to the performance of the networks and quality of service (QoS) to the connection. Trade-off has to be made between different types of technologies so that end-to-end connection quality is acceptable to users.

For example, for an acceptable level of telephony quality, one may reasonably expect that the impairment of the connection should not affect or disturb the normal communication caused by delay, noise, echo or other disruptive factors. However, the same level of quality may not be acceptable for listening to music.

The level of acceptable quality varies also depending on considerations of economic, technical and advantage factors. In terms of economic factors, it may be concerned with the cost of use and implementation, for technical with the limit of technologies, and for advantage that people may accept lower quality for mobile, long distance and satellite network use, if it would otherwise be unavailable.

4.3.5 Reference configurations

Reference configurations provide an overview of the considered end-to-end connections and to the identification of all terminals, nodes and connections, which contribute impairments to the end-to-end QoS and performance.

Due to the variety of hierarchy, structure, routing, number and types of network technologies in a network, different networking technologies (wireless, cable and satellite) may play different roles in the reference configuration. Here we try to identify some typical reference configurations, which can be used for evaluation of QoS and performance of networks with different technologies and their roles in providing network services.

Figure 4.1 shows a basic reference configuration of a telephony network. It is generalised to include international scenarios, the public network, the private network and therefore the entire connection.

It is assumed, that the impairment allowance between the access points for calls within the national public network are allocated symmetrically with reference to the international connection, which can be considered as the virtual centre of the public network for international calls. For connections not involving an international connection, the equivalent virtual centre can be assumed to be within the portion of the highest-ranking network shown as the public network in Figure 4.1.

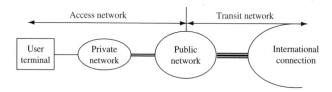

Figure 4.1 Basic configuration of access and transit networks

The private network normally connects to a local exchange (e.g. LEC), usually the lowest hierarchy and the common connection point in a public network. It is also possible to connect the private network directly to a higher hierarchy level, e.g. an international connection, bypassing the local exchange. In some cases, especially for larger private networks, bypass may permit more allocation of specific transmission parameters, e.g. delay, to the private network.

A virtual private network (VPN) although provided by the public network operator, should be considered as part of the private network. The same is valid for leased lines interconnecting private networks usually provided by public network carriers. The private network with leased lines and VPN connections has some implications on end-to-end QoS and performance.

4.4 Network traffic and signalling

Internetworking involves the following types of traffic: user traffic, signalling traffic and management traffic. User traffic is generated and consumed directly at user terminals. Signalling traffic conveys the intelligence for subscribers to interconnect with the others across the networks. Management traffic provides information in the networks for effective control of the user traffic and network resources dynamically to meet the QoS requirement of the user traffic. User traffic belongs to application layer, which consumes the major amount of network resources (such as bandwidth). The management traffic also consumes a significant amount of resources. Figure 4.2 illustrates the relationships between user, signalling and management functions.

4.4.1 User traffic and network services

User traffic is generated by a range of user services. Satellite networks can support a wide range of telecommunication services including telephony, fax, data, ISDN, B-ISDN, etc. Figure 4.3 illustrates some typical network connection and interfaces.

Telephony, fax and various low bit-rate data transmission services were originally based on analogue transmission. Nowadays, they are systematically implemented and developed based on digital technologies. In analogue transmission, network bandwidth is allocated in the frequency domain for the duration of network connection. In the digital domain, network bandwidth is allocated in the time domain. The use of time division multiplexed digital carriers, especially when combined with technologies such as adaptive differential pulse

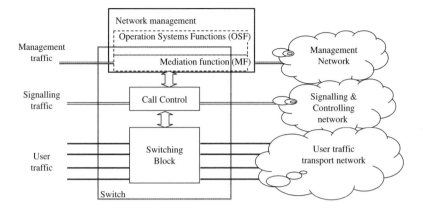

Figure 4.2   Relationships between user, signalling and management functions

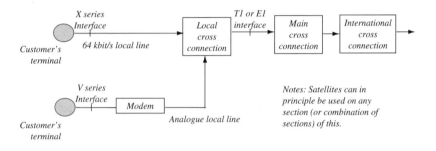

Figure 4.3   Example of network connections and interfaces

code modulation (ADPCM), low bit-rate encoding and digital speech interpolation (DSI) with digital circuit multiplication equipment (DCME), can provide increased traffic capacity in terms of a large numbers of channels on such carriers.

For ISDN service, the basic user access includes two B-channels at 64 kbit/s and a D-channel at 16 kbit/s. It can support digital voice, 64 kbit/s data in circuit and packet switched modes, telex, fax and slow-scan video. The primary access is 2.048 Mbit/s in Europe or 1.544 Mbit/s in North America and Japan. It can support fast fax, videoconference, high-speed data transmission, and high-quality audio or sound programme channels and packet-switched data services. It can also support multiplexed data streams of below 64 kbit/s. For a broadband ISDN service, the user can access at speeds as high as 155.520 Mbit/s or more. It can support integration of voice, video and data or combinations of these as multimedia services.

Satellite usage must take into account the end-to-end customer requirements as well as signalling/routing constraints of a particular network configuration. The requirements of these services may also differ depending on whether they are carried on a dedicated (leased) circuit within the main network or a switched connection.

4.4.2 Signalling systems and signalling traffic

Traditionally, telephony networks classified signalling generally into subscriber signalling and inter-switch signalling and functionally into audible-visual signalling, supervisory signalling and address signalling.

Subscriber signalling tells the local switch that a subscriber wishes to contact another subscriber by dialling the number identifying the distance subscriber. Inter-switch signalling provides information allowing switches to route the call properly. It also provides supervision of the call along its path. Signalling provides information for the network operator to charge for the use of network services.

The audible-visual signalling provides alerting (such as ring, paging and off-hook warning) and progress of the call (such as dial tone, busy tone and ring back). Supervisory signalling provides forward control from user terminal to local switch to seize, hold or release a connection and backward status including idle, busy and disconnect. Address signalling is generated from the user terminal by rotary dialling or digital dialling and used by the network to route the call.

Two trade-off factors are the signalling delay after the user dialled the number and signalling cost for setting up the call, as the network needs to reserve resources link by link until the call is set up successfully or has failed.

4.4.3 In-band signalling

In telephony networks, in-band signalling refers to signalling systems using an audio tone, or tones inside the conventional voice channel, to convey signalling information. It is broken down into three signalling categories: single frequency (SF), two frequency (TF), and multi-frequency (MF). As conventional voice channel occupies the frequency band from 300 Hz to 3400 Hz, SF and TF signalling systems utilise the 2000–3000 band where less speech energy is concentrated.

SF signalling is used almost exclusively for supervision. The most commonly used frequency is 2600 Hz, particularly in North America. On two-wire trucks 2600 Hz is used in one direction and 2400 Hz in the other. Figure 4.4(a) illustrates the concept of in-band

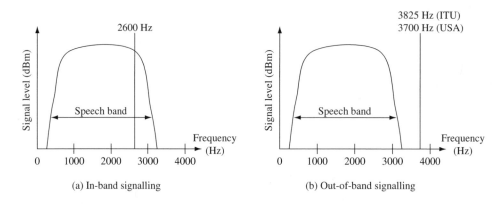

(a) In-band signalling   (b) Out-of-band signalling

Figure 4.4   Analogue network in-band signalling and out-of-band signalling

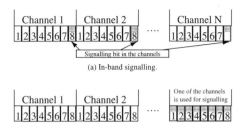

Figure 4.5 Digital network in-band signalling and out-of-band signalling

signalling of 2600 Hz within the frequency band, and Figure 4.4(b) illustrates two out-of-band signalling of 3700 Hz used in North America or 3825 Hz for ITU. Similarly in digital networks, there can also be in-band signalling and out-of-band-signalling as shown in Figure 4.5.

A two-frequency signal is used for both supervision (line signalling) and address signalling. SF and TF signalling systems are often associated with carrier (FDM) operation. In supervision line signalling 'idle' refers to the on-hook condition while 'busy' refers to the off hook condition. Thus, for such types of line signalling there are two audio tones, of which SF and TF are typical, for 'tone on when idle' and 'tone on when busy'.

You may have noticed that a major problem with in-band signal is the possibility of 'talk-down' which refers to the premature activation or deactivation of supervisory equipment by an inadvertent sequence of voice tones through the normal use of the channel. Such tones could simulate the SF tone, forcing a channel dropout (i.e., the supervisory equipment would return the channel to idle state). Chances of simulating TF tone set are less likely. To avoid the possibility of talk-down on an SF circuit, a time delay circuit or slot filters may be used to bypass the signalling tone. Such filters can cause some degradation to speech unless they are switched off during conversation. They must be switched off if the circuit is used for data transmission. Therefore, TF or MF signalling systems overcome the problem of SF. TF signalling is widely used for addressing signalling.

Multi-frequency (MF) signalling is widely used for addressing signalling between switches. It is an in-band method utilising five or six tone frequencies, two at a time, of which each have four different frequencies, forming the typical signalling of 16 buttons in the telephone set.

4.4.4 Out-of-band signalling

With out-of-band signalling, supervisory information is transmitted above 3400 Hz of the conventional voice band. In all cases, it is a single frequency system. The advantage of out-of-band signalling is that either system 'tone on' or 'tone off' may be used when idle. Talk-down cannot occur because all supervisory information is passed out of band away from the speech information portion of the channel. The preferred out-of-band frequency is 3825 Hz, whereas 3700 Hz is commonly used in the US (see Figure 4.4(b)). Out-of-band signalling is attractive, but one drawback is that when channel patching is required, signalling leads have to be patched as well.

4.4.5 Associated and disassociated channel signalling

Traditionally, signalling goes along with the traffic on the same channel it is associated with on the same media. This signalling may or may not go on the same media or path. Most often, this type of signalling is transported on a separate channel in order to control a group of channels. A typical example is the European PCM E1 where one separate digital channel supports all supervisory signalling for 30 traffic channels. It is still associated channel signalling if it travels on the same media and path as its associated traffic channels.

If the separated signalling channel follows a different path using perhaps different media, it is called disassociated signalling. See Figure 4.6. ITU-T Signalling System No. 7 (ITU-T SS7) always uses separated channels, but can be associated and disassociated. Disassociated channel signalling is also called non-associated channel signalling.

4.4.6 ITU-T signalling system No. 7 (ITU-T SS7)

ITU-T SS7 was developed to meet the advanced signalling requirements of the all-digital network based on the 64 kbit/s channels. It operates in a quite different manner than

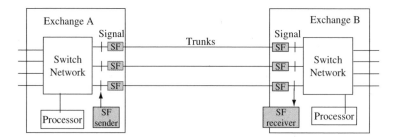

(a) Conventional associated channel signalling.

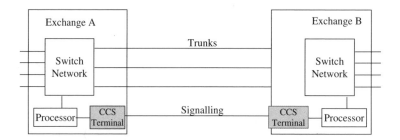

(b) Separate channel signalling with common channel signalling (CCS).

Figure 4.6   Associated and separate signalling

conventional signalling systems. Nevertheless, it must provide supervision of circuits, address signalling, call progressing and alert notification. It is a data network entirely dedicated to interswitching signalling, and can be summarised as the following:

- it is optimised for operation with digital networks where switches use stored-program control (SPC);
- it meets the requirements of information transfer for inter-processor transactions with digital communication networks for call control, remote control, network database access and management, and maintenance signalling; and
- it provides a reliable means of information transfer in the correct sequence without loss or duplication.

Since 1980, it has become known as the signalling system for ISDN. The SS No.7 network model consists of network nodes, termed signalling points (SP), which are interconnected by point-to-point signalling links, with all the links between two SPs called a link set. Messages between two SPs may be routed over a link set directly connecting the two points. This is referred to as the associated mode of signalling. Messages may also be routed via one or more intermediate points relaying the messages at the network layer. This is called the non-associated mode of signalling. It supports a special case of static routing, called quasi-associated mode, in which routing only changes in response to events such as link failures or addition of new SPs. The function of relaying messages at the network layer is called the signalling transfer points (STP).

There are certain relationships between the SS No.7 and the OSI/ISO reference model as illustrated as Figure 4.7.

It can be seen that SS No.7 has three layers corresponding to layers 1–3 of the OSI/ISO reference model within the communication networks. The application processes within a communication network invoke protocol functionality to communicate with each other in much the same way as 'end users'. The signalling system also encompasses operation, administration and maintenance (OAM) activities related to communications. Sublayer 4 of SS No.7 corresponds to OSI layer 4 upward, and consists of user parts and the signalling connection control part (SCCP).

There are three user parts: telephone user part (TUP), data user part (DUP) and ISDN user part (ISDN). Layers 1–3 together make up the message transfer part (MTP). The SCCP

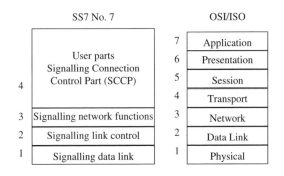

Figure 4.7   Relationship between the SS No.7 and OSI/ISO reference model

provides additional functions to the MTP for both connection-oriented and connectionless services to transfer circuit-related and non-circuit-related signalling information between switches and specialised centres in telecommunication networks via SS No.7 networks. It is situated above the MTP in level 4 with the user parts.

4.4.7 Network management

In the OSI reference model, there are five categories for network management functions defined as the following:

- configuration and name management;
- performance management;
- maintenance management;
- accounting management; and
- security management.

Configuration and name management comprise a set of functions and tools to identify and manage network objects. The functions include the ability to change the configuration of objects, assign names to objects, collect state information from objects (regularly and in emergencies) and control states of objects.

Performance management comprises a set of functions and tools to support planning and improve system performance, including mechanisms to monitor and analyse network performance and QoS parameters, and control and tune the network.

Maintenance management comprises a set of functions and tools to locate and deal with abnormal operation of the network, including functions and mechanisms to collect fault reports, run diagnostics, locate the sources of faults, and take corrective actions.

Accounting management comprises a set of functions and tools to support billing for the use of network resources, including functions and mechanisms to inform users of costs incurred, limit use of resources by setting a cost limit, combine costs when several network resources are used, and calculate the bills for customers.

Security management comprises a set of functions and tools to support management functions and to protect managed objects, including authentication, authorisation, access control and encipherment and decipherment and security logging. Please note that security management is more to provide security for the network than user information.

4.4.8 Network operation systems and mediation functions

Network management is implemented in network operation systems including user specific functions and common functions; the later are further subdivided into infrastructure functions and user generic functions.

Infrastructure functions provide underlying computer-related capabilities which support a wide range of processes. These include such services as physical communications and message passing, data storage and retrieval and human–machine interface (such as in a workstation computer with windows).

User-generic functions are general utilities in the network operation systems (NOS). They can support a number of user-specific functions. Some of the generic functions are listed in the following as examples:

- *Monitoring*: to observe the system and basic system parameters at a remote site.
- *Statistics, data distribution and data collection*: to generate and update statistics, to collect system data and to provide other functions with system data.
- *Test execution and test control*: independent of the purpose of test, whether it is done to detect a fault or to prove the correct operation of unit or an element, a test is performed in the same way. Tests are used by maintenance installation of equipment or new features, performance management and normal operations. Configuration control and protection actions might be involved if the test uses additional network resources to minimise the resources used for tests and maximise system availability during the test.
- *Configuration management*: to keep track of the actual configuration of the network and to know about valid network or network element configurations. To reconfigure the network or a network element or to support reconfiguration if it is necessary.

Network operation systems (NOS) involve four layers of management functions: business management, service management, network management and element management with business at the top of the layers and element at the bottom as shown in Figure 4.8.

- Business management includes functions necessary to implement policies and strategies with the organisation owning and operating the services and possibly also the network. These functions are influenced by still higher levels of control such as legislation or macro-economic factors and might include tariff policies and quality management strategies, which give guidance on service operation when equipment or network performance is degraded. Many of these functions may not initially be automated.
- Service management looks after particular services such telephone, data, Internet or broad-band services. The service may be implemented across several networks. The functions

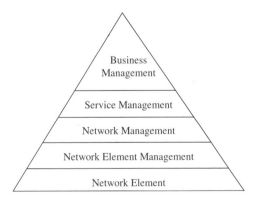

Each layer manages multiple occurrences of the layer below

Figure 4.8 Layers of management functions in network operation systems (NOS)

may include customer-related functions (e.g. subscription record, access rights, usage records and accounts) and establishment and maintenance of the facilities provided by the service itself additional to the network facilities.

- Network management provides functions to manage the network in question, including network configuration, performance analysis and statistical monitoring.
- Element management provides functions to manage a number of network elements in a region. These functions are most likely to focus on maintenance but could also include configuration capability and some statistical monitoring of the network elements. It does not cater for network wide aspects.

The mediation function (MF) acts on information passing between network element functions and the operation systems functions (OSF) to achieve smooth and efficient communication. It has functions including communication control, protocol conversion and data handling, and communication primitive functions. It also includes data storage and processing involving decision making.

4.5 Access and transit transmission networks

According to ITU-T recommendation Y.101, access network is defined as an implementation comprising those entities (such as cable plant, transmission facilities, etc.) which provide the required transport bearer capabilities for the provision of telecommunications services between the network and user equipment. Transit network can be considered as a set of nodes and links that provide connections between two or more defined points to facilitate telecommunication between them. The interface has to be well defined in terms of capacity and functionality to allow independent evolutions of user equipment and the network, and new interfaces have to be developed to accommodate new user equipment with large capacity and new functionality. The evolution of access and transit networks can be seen from analogue transmission from telephone networks, to digital transmission telephony networks, synchronous transfer mode in transit network, integration of telephony networks and data ISDN, Internet networks, broadband networks in B-ISDN, etc.

4.5.1 Analogue telephony networks

Although almost all of today's networks are digital, the connections from many residential homes to the local exchanges are still in analogue transmission. They are gradually fading away with the installation of broadband access networks such as asymmetric digital subscriber line (ADSL). ADSL is a modem technology that converts twisted-pair telephone lines into access paths for multimedia and high-speed data communications. The bit rates transmitted in both directions are different with a typical ratio of 1 to 8 between user terminal and local switch.

We discuss analogue telephony networks not because the technology itself is important for the future, but because the principles of design, implementation, control, management and operation developed with the network have been used for many years, are still very important to us today, and will continue to be important in the future. Of course these principles have to be used and developed in the new network context.

The telephony networks were well designed, well engineered and optimised for telephony services. In the context of available technologies and knowledge, the user service was telephony, the network resource was channel, and bandwidth of 4 kHz was allocated to each channel to support good acceptable quality of service.

4.5.2 Telephony network traffic engineering concept

The networks were dimensioned to provide the service to a large number of people (almost all the homes and offices today) with 4 kHz channels, taking into account factors of economics such as user demands and costs of the network to meet the demands. There were well-developed theories to model user traffic, network resource and performance of the network and grade of service.

- Traffic is described by patterns of arrivals and holding times. Traffic is measured in Erlang, named after the Danish mathematician for his contribution to telephony network traffic engineering. The Erlang is a dimensionless unit. Erlang is defined as a product of number of calls (A) and average holding time in hours (H) of these calls: $A \times H$ Erlang. One Erlang represents one call lasting for one hour or one circuit is occupied for one hour. The patterns of call arrivals and holding times are stochastic in nature, hence described by statistical methods in terms of probability distributions, means, variance, etc. Traffic varies in time in different time scales: instantaneously, hourly, daily, seasonal, trend with a gradual increase.
- The network can provide full availability of resources to meet all the traffic requirements but is expensive or has limited availability to meet most requirements economically. The network can also allow traffic to queue to wait for network resources to be available or give priority or some kinds of treatments to a portion of the traffic.
- Performance criteria allow quantitative measurement of network performance with parameters including: probability of delay, average delay, probability of delay exceeding a range of time values, number of delayed calls and number of blocked calls.
- Grade of service is one of the parameters used to measure probability of loss of calls to be achieved by the network and expected by users as acceptable quality of service.

There are well-established mathematical theories to deal with these factors in classical scenarios in terms of call arrivals and holding-time distribution, number of traffic sources, availability of circuits and handling of lost calls. Some of the mathematical formulas are simple and useful and can be summarised as the following:

- Erlang B formula to calculate the grade of service (E_B) is:

$$E_B = \frac{A^n/n}{\sum\limits_{x=0}^{n} (A^x/x!)}$$

where n is number of circuits available and A is the mean of the traffic offered in Erlang. The formula assumes an infinite number of sources, equal traffic density per source and traffic lost call cleared.

- Poisson formula to calculate the probability of lost calls or delayed calls (P) because of insufficient number of channels (n) with the traffic offered (A) is:

$$P = e^{-A} \sum_{x=n}^{\infty} \frac{A^x}{x!}$$

The formula assumes an infinite number of sources, equal traffic density per source and lost calls held.

- Erlang C formula is:

$$P = \frac{\frac{A^n}{n!} \frac{n}{n-A}}{\sum_{x=0}^{n-1} \frac{A^x}{x!} + \frac{A^n}{n!} \frac{n}{n-A}}$$

The formula assumes an infinite number of sources, lost call delayed, exponential holding times and calls served in order of arrival.

- Binomial formula is:

$$P = \left(\frac{s-A}{s}\right)^{s-1} \sum_{x=n}^{s-1} \binom{s-1}{x} \left(\frac{A}{s-A}\right)^x$$

The formula assumes a finite number of sources (s), equal traffic density per source and lost calls held.

4.5.3 Access to satellite networks in the frequency domain

In the frequency domain, we can see each signal telephony channel is allocated a bandwidth of 4 kHz to access the local exchange, or many of the single channels are multiplexed together to form the transmission hierarchy. To transmit the telephony channel over satellite, a carrier has to be generated which is suitable for satellite radio transmission on the allocated frequency band and channel signal modulating the carrier can be transmitted over satellite. At the receiving side, the demodulating process can separate the channel signal from the carrier; hence the receiver can get back the original telephony signal to be sent to a user terminal or to a network which can route the signal to the user terminal.

If a single channel modulates the carrier, we call it single carrier per channel (SCPC), i.e., each carrier carries only a single channel. This is used normally for user terminals to be connected to the network or other terminals as an access network. It is also possible to use this as a thin route to connect a local exchange to the network where the traffic density is low.

If a group of channels modulate the carrier, we call it multi channel per carrier (MCPC). This is normally used for interconnect between networks as a transit network or local exchange to the access network.

4.5.4 On-board circuit switching

If all connections between earth stations used single global beam coverage, there would be no need to have any switching functions on-board satellite. If multiple spot beams are used,

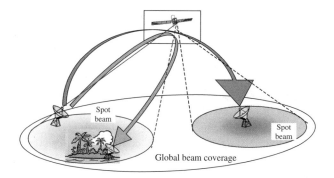

Figure 4.9   Illustration of on-board circuit switching

there are great advantages to using on-board switching, since it allows the earth stations to transmit multiple channels to several spot beams at the same time without separating these channels on the transmitting earth stations. Therefore, on-board switching will give satellite networks great flexibility and potentially save bandwidth resources.

Figure 4.9 illustrates the concept of on-board switching with two spot beams. If there is no on-board switching function, the two transmissions have to be separated at the transmission earth station by using two different bent-pipes, one of which is for connection within the spot beam and the other is for connection between the spot beams. If the same signal is to be transmitted to both spot beams, it will require two separate transmissions of the same signal; hence it will need twice the bandwidth at the uplink transmissions. It is also possible to reuse the same bandwidth in different spot beams.

By using on-board switching, all the channels can be transmitted together and will be switched on-board satellite to their destination earth stations in the different spot beams. Potentially, if the same signal is to be sent to different spot beams, the on-board switch may be able to duplicate the same signal to be sent to the spot beams without multiple transmissions at the transmitting earth station. The same frequency band can be used in the two spot beams by taking appropriate measures to avoid possible interferences.

4.6 Digital telephony networks

In the early 1970s, digital transmission systems began to appear, utilising the pulse code modulation (PCM) method first proposed in 1937. PCM allowed analogue waveforms, such as the human voice, to be represented in binary form (digital). It was possible to represent a standard 4 kHz analogue telephone signal as a 64 kbit/s digital bit stream. The potential with digital processing allowed more cost-effective transmission systems by combining several PCM channels and transmitting them down the same copper twisted pair as had previously been occupied by a single analogue signal.

4.6.1 Digital multiplexing hierarchy

In Europe, and subsequently in many other parts of the world, a standard TDM scheme was adopted whereby thirty 64 kbit/s channels were combined, together with two additional channels carrying control information including signalling and synchronisation, to produce a channel with a bit rate of 2.048 Mbit/s.

As demand for voice telephony increased, and levels of traffic in the network grew ever higher, it became clear that the standard 2.048 Mbit/s signal was not sufficient to cope with the traffic loads occurring in the trunk network. In order to avoid having to use excessively large numbers of 2.048 Mbit/s links, it was decided to create a further level of multiplexing. The standard adopted in Europe involved the combination of four 2.048 Mbit/s channels to produce a single 8.448 Mbit/s channel. This level of multiplexing differed slightly from the previous in that the incoming signals were combined one bit at a time instead of one byte at a time, i.e. bit interleaving was used as opposed to byte interleaving. As the need arose, further levels of multiplexing were added to the standard at 34.368 Mbit/s, 139.246 Mbit/s, and even higher speeds to produce a multiplexing hierarchy, as shown in Figure 4.10.

In North America and Japan, a different multiplexing hierarchy is used but with the same principles.

4.6.2 Satellite digital transmission and on-board switching

Digital signals can be processed in the time domain. Therefore, in addition to sharing bandwidth resources in the frequency domain, earth stations can also share bandwidth in the time domain. Time division multiplexing can be used for satellite transmission at any level of the transmission hierarchy as shown in Figure 4.10. Concerning on-board switching, a time-switching technique can be used often working together with circuit switching (or space switching).

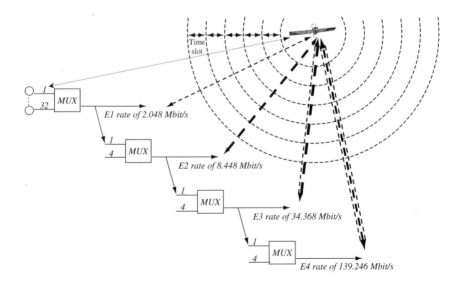

Figure 4.10   Example of traffic multiplexing and capacity requirement for satellite links

4.6.3 Plesiochronous digital hierarchy (PDH)

The multiplexing hierarchy appears simple enough in principle but there are complications. When multiplexing a number of 2 Mbit/s channels they are likely to have been created by different pieces of equipment, each generating a slightly different bit rate. Thus, before these 2 Mbit/s channels can be bit interleaved they must all be brought up to the same bit rate adding 'dummy' information bits, or 'justification bits'. The justification bits are recognised as de-multiplexing occurs, and are discarded, leaving the original signal. This process is known as plesiochronous operation, meaning in Greek 'almost synchronous' as illustrated in Figure 4.11.

The same problems with synchronisation, as described above, occur at every level of the multiplexing hierarchy, so justification bits are added at each stage. The use of plesiochronous operation throughout the hierarchy has led to adoption of the term plesiochronous digital hierarchy (PDH).

4.6.4 Limitations of the PDH

It seems simple and straightforward to multiplex and de-multiplex low bit streams to higher bit-rate streams, but in practice it is not so flexible and not so simple. The use of justification bits at each level in the PDH means that identifying the exact location of the low bit-rate stream in a high bit-rate stream is impossible. For example, to access a single E1 2.048 Mbit/s stream in an E4 139.246 Mbit/s stream, the E4 must be completely de-multiplexed via E3 34.368 and E2 8.448 Mbit/s as shown in Figure 4.12.

Once the required E1 line has been identified and extracted, the channels must then be multiplexed back up to the E4 line. Obviously this problem with the 'drop and insert' of channels does not make for very flexible connection patterns or rapid provisioning of services, while the 'multiplexer mountains' required are extremely expensive.

Another problem associated with the huge amount of multiplexing equipment in the network is one of control. On its way through the network, an E1 line may have travelled via a number of possible switches. The only way to ensure it follows the correct path is to keep careful records of the interconnection of the equipment. As the amount of reconnection activity in the network increases it becomes more difficult to keep records current and the

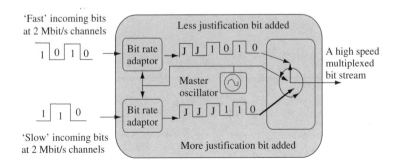

Figure 4.11 Illustration of the concept of plesiochronous digital hierarchy (PDH)

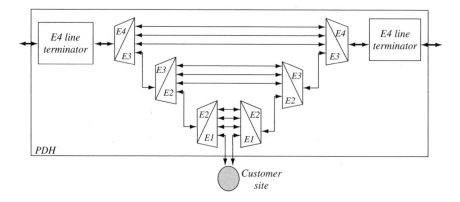

Figure 4.12   Multiplexing and de-multiplexing to insert a network node in PDH network

possibility of mistakes increases. Such mistakes are likely to affect not only the connection being established but also to disrupt existing connections carrying live traffic.

Another limitation of the PDH is its lack of performance-monitoring capability. Operators are coming under increasing pressure to provide business customers with improved availability and error performance, and there is insufficient provision for network management within the PDH frame format for them to do this.

4.7 Synchronous digital hierarchy (SDH)

PDH reached a point where it was no longer sufficiently flexible or efficient to meet the demands of users and operators. As a result, synchronous transmissions were developed to overcome the problems associated with plesiochronous transmission, in particular the inability of PDH to extract individual circuits from high-capacity systems without having to de-multiplex the whole system as shown in Figure 4.13.

Synchronous transmission can be seen as the next logical stage in the evolution of the transmission hierarchy. Concerted standardisation efforts were involved in its development. The opportunity of defining the new standard was also used to address a number of other problems. Among these were network management capability within the hierarchy, the need to define standard interfaces between equipment and international standard transmission hierarchies.

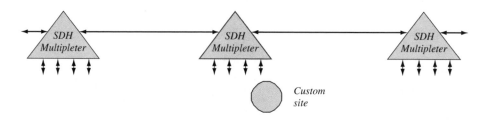

Figure 4.13   Add and drop function to insert a network node in SDH network

4.7.1 Development of SDH

The development of the SDH standards represents a significant advance in technology. Services such as videoconferencing, remote database access and multimedia file transfer require a flexible network with the availability (on demand) of virtually unlimited bandwidth. SDH overcomes the complexity of the network based on plesiochronous transmission systems.

Using essentially the same fibre, a synchronous network is able to significantly increase available bandwidth while reducing the amount of equipment in the network. In addition, the provision within the SDH for sophisticated network management introduces significantly more flexibility into the network.

Deployment of synchronous transmission systems is straightforward due to their ability to interwork with existing plesiochronous systems. The SDH defines a structure which enables plesiochronous signals to be combined together and encapsulated within a standard SDH signal. This is called backward compatible, i.e., new technology is able to interwork with legacy technology.

The sophisticated network management capabilities of a synchronous network give an improved control of transmission networks, improved network restoration and reconfiguration capabilities, and availability.

4.7.2 The SDH standards

This standards work culminated in ITU-T recommendations G.707, G.708, and G.709 covering the synchronous digital hierarchy. These were published in the ITU-T *Blue Book* in 1989. In addition to the three main ITU-T recommendations, a number of working groups were set up to draft further recommendations covering other aspects of the SDH, such as the requirements for standard optical interfaces and standard OAM functions.

The ITU-T recommendations define a number of basic transmission rates within the SDH. The first of these is 155.520 Mbit/s, normally referred to as synchronous transport module level 1 (STM-1). Figure 4.14 shows the STM-1 frame. Higher transmission rates of STM-4

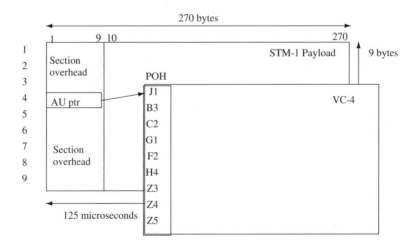

Figure 4.14   STM-1 frame of the SDH network

and STM-16 (622 Mbit/s and 2.4 Gbit/s respectively) are also defined, with further levels proposed for study.

4.7.3 Mapping from PDH to SDH

The recommendations also define a multiplexing structure whereby an STM-1 signal can carry a number of lower rate signals as payload, thus allowing existing PDH signals to be carried over a synchronous network as shown in Figure 4.15.

All plesiochronous signals between 1.5 Mbit/s and 140 Mbit/s are accommodated, with the ways in which they can be combined to form an STM-1 signal defined in Recommendation G.709.

SDH defines a number of 'containers', each corresponding to an existing plesiochronous rate. Information from a plesiochronous signal is mapped into the relevant container. Each container then has some control information known as the path overhead (POH) added to it. Together the container and the POH form a 'virtual container' (VC).

In a synchronous network, all equipment is synchronised to an overall network clock. It is important to note, however, that the delay associated with a transmission link may vary slightly with time. As a result, the location of virtual containers within an STM-1 frame may not be fixed. These variations are accommodated by associating a pointer with each VC. The pointer indicates the position of the beginning of the VC in relation to the STM-1 frame. It can be increased or decreased as necessary to accommodate the position of the VC.

G.709 defines different combinations of virtual containers which can be used to fill up the payload area of an STM-1 frame. The process of loading containers and attaching overhead is repeated at several levels in the SDH, resulting in the 'nesting' of smaller VCs within larger ones. This process is repeated until the largest size of VC is filled, and this is then loaded into the payload of the STM-1 frame (referring to Figure 4.15).

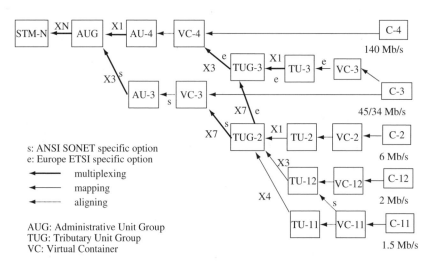

Figure 4.15 Mapping from PDH to SDH

When the payload area of the STM-1 frame is full, some more control information bytes are added to the frame to form the 'section overhead'. The section overhead bytes are so-called because they remain with the payload for the fibre section between two synchronous multiplexers. Their purpose is to provide communication channels for functions such as OAM, facilities and alignment.

When a higher transmission rate than 155 Mbit/s of STM-1 is required in the synchronous network, it is achieved by using a relatively straightforward byte-interleaved multiplexing scheme. In this way, rates of 622 Mbit/s (STM-4) and 2.4 Gbit/s (STM-16) can be achieved.

4.7.4 The benefits of SDH

One of the main benefits in the SDH network is the network simplification brought about through the use of synchronous equipment. A single synchronous multiplexer can perform the function of an entire plesiochronous 'multiplexer mountain', leading to significant reductions in the amount of equipment used. The more efficient 'drop and insert' of channels offered by an SDH network, together with its powerful network management capabilities can ease the provisioning of high bandwidth lines for new multimedia services, as well as provide ubiquitous access to those services.

The network management capability of the synchronous network enables immediate identification of link and node failure. Using self-healing ring architectures, the network will be automatically reconfigured with traffic instantly rerouted until the faulty equipment has been repaired.

The SDH standards allow transmission equipment from different manufacturers to interwork on the same link. The ability to achieve this so-called 'mid-fibre meet' has come about as a result of standards, which define fibre-to-fibre interfaces at the physical (photon) level. They determine the optical line rate, wavelength, power levels, pulse shapes and coding. Frame structure, overhead and payload mappings are also defined. SDH standards also facilitate interworking between North American and European transmission hierarchies.

4.7.5 Synchronous operation

The basic element of the STM signal consists of a group of bytes allocated to carry the transmission rates defined in G.702 (i.e. 1.5 Mbit/s and 2 Mbit/s transmission hierarchies). The following describe each level of the transmission hierarchy in SDH.

- Virtual container level n (VC-n), where $n = 1 - 4$, is built up from the container plus additional capacity to carry the path overhead (POH). For a VC-3 or VC-4 the payload may be a number of tributary units (TU) or tributary unit groups (TUG) as opposed to a simple basic VC-1 and VC-2.
- Tributary unit level n (TU-n), where $n = 1 - 3$, consists of a virtual container plus a tributary unit pointer. The position of the VC within the TU is not fixed, however, the position of the TU pointer is fixed with relation to the next step of the multiplex structure, and indicates the start of the VC.
- Tributary unit group (TUG) is formed by a group of identical TUs.

- Administration unit level n (AU-n), where $n = 3 - 4$, consists of a VC plus an AU pointer. The phase alignment of the AU pointers is fixed with relation to the STM-1 frame as a whole and indicates the positions of the VC.
- Synchronous transfer module level 1 (STM-1) is the basic element of the SDH. It is formed from a payload (made up of the AU) and additional bytes to form a section overhead (SOH). The frame format is shown in Figure 4.14 and the header is shown in Figure 4.16. The section overhead allows control information to be passed between adjacent synchronous network elements.

Within an STM-1 frame, information type repeats every 270 bytes. Thus, the STM-1 frame is often considered as a (270 byte × 9 line) structure. The first nine columns of this structure constitute the SOH area, while the remaining 261 columns are the 'payload' area.

The SOH bytes are used for communication between adjacent pieces of synchronous equipment. As well as being used for frame synchronisation, they perform a variety of management and administration facilities. The purpose of individual bytes is detailed below:

- A1, A2 are bytes for framing
- B1, B2 are parity check bytes for error detection.
- C1 identifies an STM-1 in an STM-N frame.
- D1–D12 are for data communication channels and for network management.
- E1, E2 are used for order wire channels.
- F1 is used for user channels.
- K1, K2 are used for automatic protection switching (APS) channels
- Z1, Z2 are reserved bytes for national use.

The path overhead (POH) of the VC-4 (as shown in Figure 4.14) consists of the following bytes:

- B3 BIP-8 (bit interleaved parity): provides bit-error monitoring over the path using an even bit parity code, BIP-8.

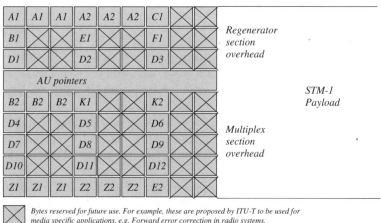

Figure 4.16   Section Overhead (SOH) of the STM-1 frame

- C2 signal label: indicates the composition of the VC-n payload.
- F2 path user channel: provides a user communication channel.
- G1 path status: allows the status of the received signal to be returned to the transmitting end of the path from the receiving end.
- H4 multiframe indicator: used for multiframe indication.
- J1 path trace: used to verify the VC-n path connection.
- Z3–Z5: provided for national use.

Synchronous transfer module level N (STM-N) is constructed by combining lower level STM signals using byte interleaving. The basic transmission rate defined in the SDH standards is 155.520 Mbit/s (STM-1). Given that an STM-1 frame consists of 2430 eight-bit bytes, this corresponds to frame duration of 125 microseconds. Two higher bit rates are also defined: 622.080 Mbit/s (STM-4) and 2488.320 Mbit/s (STM-16).

Once the STM-1 payload area is filled by the largest unit available, a pointer is generated which indicates the position of the unit in relation to the STM-1 frame. This is known as the AU pointer. It forms part of the section overhead area of the frame. The use of pointers in the STM-1 frame structure means that plesiochronous signals can be accommodated within the synchronous network without the use of buffers. This is because the signal can be packaged into a VC and inserted into the frame at any point at time. The pointer then indicates its position. Use of the pointer method was made possible by defining synchronous virtual containers as slightly larger than the payload they carry. This allows the payload to slip in time relative to the STM-1 frame in which it is contained.

Adjustment of the pointers is also possible where slight changes of frequency and phase occur as a result of variations in propagation delay and the like. The result of this is that in any data stream, it is possible to identify individual tributary channels, and drop or insert information, thus overcoming one of the main drawbacks of PDH.

4.7.6 Synchronous optical network (SONET)

In North America ANSI published its SONET standards, which were developed in the same period of time using the same principles as SDH, and can be thought of as a subset of the worldwide SDH standards, however, there are some differences.

The basic module in SONET is synchronous transport signal level 1 (STS-1), which is three times smaller than the STM-1 in terms of bit rate and frame size. It has the same bit rate of 51.840 Mbit/s as the optical carrier level 1 (OC-1). The STS-1 frame consists of (9 × 90) bytes with frame duration of 125 microseconds, of which three columns are used as transport overhead and 87 columns as STS-1 payload called envelope capacity.

4.7.7 SDH over satellite – the Intelsat scenarios

ITU-T and ITU-R standards bodies together with Intelsat and its signatories developed a series of SDH compatible network configurations with satellite forming part of the transmission link. The ITU-R Study Group 4 (SG 4) was responsible for studying the applicability of the ITU-T recommendations to satellite communication networks.

SDH was not designed for the transmission of basic rate signals. Because it is a great challenge to implement and operate a satellite network at a bit rate of 155.520 Mbit/s, various network configurations were studied to allow relevant SDH elements to operate at lower bit rate whenever there is a need to transport SDH signals over satellite. These network configurations were referred as 'scenarios'. These scenarios defined different options to support SDH over satellite, summarised as follows:

- Full STM-1 transmission (point to point) through a standard 70 MHz transponder. This required the development of an STM-1 modem capable of converting the STM-1 digital signal to an analogue format for transmitting through a standard 70 MHz transponder. While the Intelsat signatories generally supported this, there was limited confidence that this approach would yield reliable long-term results. It was considered as an engineering challenge and risk to support the required transmission quality since the carriage of an STM-1 will very closely approach the theoretical limits of a 70 MHz transponder. In addition there was no recognised need for this amount of capacity via an SDH satellite link. High bit-rate PDH IDR satellite links were generally used for submarine cable restoration (although there are some exceptions), but to develop a complete new generation of satellites for restoration of high-capacity SDH cables was not considered as a cost-effective use of satellite resources.
- Reduced rate of STM (STM-R) uplink with STM-1 downlink (point to multipoint). This scenario suggested a multi-destination system, and required considerable on-board processing of SDH signals, however, the advantage was flexible transponder usage for the network operators using the system. Most network operators did not generally favour this approach due to reliability and future proofing reasons. This approach might prevent alternative usage of the satellite transponders in the future, and additional complexity was likely to reduce the reliability and lifetime of the satellite, and increase its initial expense.
- Extended intermediate data rate (IDR). This approach has been favoured by a large number of signatories, since it retains the inherent flexibility of the satellite (regarded as a major advantage over cable systems), and would require the minimum of alterations to satellite and earth station design. Additionally, some of the management advantages of SDH are retained, including end-to-end path performance monitoring, signal labelling and other parts of the 'overhead'. The development work was centred on determining what aspects of the data communication channels could also be carried with the IDR.
  Since the bit rate of IDR is capable of supporting a range of PDH signals at a much lower bit rate than STM-1, it can be implemented with minimal rearrangement of the transponder band plans, with the possibility of mixing PDH- and SDH-compatible IDR carriers. Development work was carried out to modify existing IDR modems to be compatible with SHD at lower rates, rather than more expensive options of developing new modems (for example, for the STM-1 and STM-R options). This option is widely used in current satellite network operations.
- PDH IDR link with SDH to PDH conversion at the earth station. This is the simplest option of all to provide operators with any SDH compatibility, however, all the advantages of SHD are lost, with additional costs incurred in the SDH to PDH conversion equipment. In the early days of SDH implementation, it may be the only available method, however.

With the speed of development of new technologies, all the conversion equipment can become out of date very quickly.

4.8 Integrated services digital networks (ISDN)

Integrated services digital networks (ISDN) consist of a range of ITU-T I-series recommendations for subscriber services, user/network services and internetwork capabilities to ensure a level of international compatibility. ISDN represented the efforts by the IUT-T with the standards to integrate telephony and data networks for a wide range of services with a worldwide connectivity. The ISDN standards explain a wide range of ISDN concepts and associated principles. They also describe in detail the service and network aspects of ISDN, including service capabilities, overall network aspects and functions, user network interface (UNI) and internetwork interface with a wide range of protocols.

4.8.1 Basic rate interface (BRI)

The basic rate interface (BRI) is specified in ITU-T recommendation I.430. The recommendation defines ISDN communication between terminal equipment. The BRI comprises two B channels of 64 kbit/s each and one D channel of 16 kbit/s (2B + D).

The B channel is the basic user channel and can serve all types of traffic including digital voice, data and slow video in a circuit or packet switched mode. The D channel is primarily used for signalling required to control the B channels, but can also be used for message oriented packet data as shown in Figure 4.17. The D channel would be routed to the selected services points with the signalling (s-information), telemetry (t-information), and low speed packet switched data (p-information).

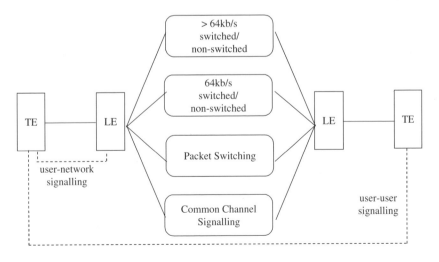

Figure 4.17 Basic architectural features of an ISDN

ISDN components include terminal equipment (TE), terminal adapters (TA), network-termination (NT) devices, line-termination (LE) equipment and exchange-termination equipment. Basic rate access may use a point-to-point or point-to-multipoint configuration between LE and TEs.

A number of reference points are specified in ISDN. These reference points define logical interfaces between functional groups such as TAs and NT1s. ISDN reference points include R (the reference point between non-ISDN equipment and a TA), S (the reference point between users' terminals and the NT2), T (the reference point between NT1 and NT2 devices) and U (the reference point between NT1 devices and line-termination equipment in the carrier network). The U reference point is relevant only in North America, where the carrier network does not provide the NT1 function. Figure 4.18 shows the ISDN reference points and functional groups.

There are three devices attached to an ISDN switch at the central office. Two of these devices are ISDN-compatible, so they can be attached through an S reference point to NT2 devices. The third device (a standard non-ISDN telephone) is attached through the R reference point to a TA. Any one of those devices could also be attached to an NT1/2 device, which would replace both the NT1 and the NT2.

In North America, the NT1 is customer premises equipment (CPE). The NT2 is a more complicated device typically found in digital private branch exchanges (PBXs), which performs layers 2 and 3 protocol functions and concentration services. An NT1/2 device also exists. It is a single device that combines the functions of an NT1 and an NT2.

4.8.2 Primary rate interface (PRI)

The primary rate interface (PRI) is defined by the physical layer protocol and also by higher protocols included LAPD. It has a full duplex point-to-point serial, synchronous configuration. The ITU-T recommendations G.703, G.704 define the electrical interfaces and the frame formats. There are two different interfaces:

- North America T1 (1.544 Mbit/s): multiplexes 24 B channels. One PRI frame has 193 bits, consisting of one framing bit plus 192 (24 × 8) bits for user channels.

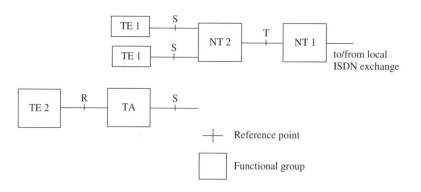

Figure 4.18 Narrowband ISDN (N-ISDN) reference points and functional groups

- Europe E1 (2.048 Mbit/s): multiplexes 32 B channels. One PRI frame has 256 (32×8) bits, consisting of 240 (30×8) bits for user traffic, eight bits for framing and synchronisation and another eight bits for signalling and controlling.

4.8.3 ISDN physical layer (layer 1)

The ISDN physical layer provides transmission capability for B channels and D channels in the form of encoded bit streams with timing and synchronisation functions. It also provides signalling capability to allow terminals and network equipment to access the D channel resources and make use of D channels to control B channels.

ISDN physical layer (layer 1) frame formats differ depending on whether the frame is outbound (from a terminal to a network) or inbound (from a network to a terminal) as shown in Figure 4.19.

The frames are 48 bits long, of which 36 bits represent data. The F bits provide synchronisation. The L bits adjust the average bit value. The E bits are used for contention resolution when several terminals on a passive bus contend for a channel. The A bit activates devices. The S bits have not yet been assigned. The B1, B2 and D bits are for users' B channels and D channels.

Multiple ISDN user devices can be physically attached to one circuit. In this configuration, collisions can occur if two terminals transmit simultaneously. ISDN therefore provides features that determine link contention. When an NT receives a D bit from the TE, it echoes it back in the next E bit position. The TE expects the next E bit to be the same as its last transmitted D bit.

Terminals cannot transmit in the D channel unless they first detect a specific number of '1' bits (indicating 'no signal') corresponding to a pre-established priority. If the TE detects a bit in the echo (E) channel, that is different from its D bit, it should stop transmission immediately. This simple technique ensures that only one terminal will transmit its D bit at a given time.

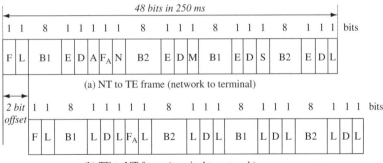

F = Framing bit, F_A = Auxiliary framing bit, L = DC balancing, E = Echo of previous D bits,
D = D channel, A = Activation bit, M = Multiframing bit, S = Spare bits
B1 = B1 channel bits, B2 = B2 channel bits, N = a binary value $N = F_A$ (NT to TE)

Figure 4.19 Frame format at T and S reference points

After a successful D message transmission, the terminal's priority is reduced by requiring it to detect more '1' bits in a row before transmitting. Terminals may not raise their priority until all the other devices on the line have had an opportunity to send a D message. Telephone connections have higher priority than all other services, and signalling information has a higher priority than non-signalling information.

4.8.4 ISDN link layer (layer 2)

Layer 2 of the ISDN signalling protocol is link access procedure, D channel (LAP-D). It is based on the LAP-B system used within X-25.

LAP-D is similar to high-level data link control (HDLC) and link access procedure balanced (LAP-B). As LAP-D's acronym indicates, it is used across the D channel to ensure that control and signalling information flows and has been received properly. LAP-D frame format is shown in Figure 4.20. Like HDLC it uses supervisory, information and unnumbered frames. The LAP-D protocol is formally specified in ITU-T Q.920 and Q.921 for signalling.

The LAP-D flag and control fields are identical to those of HDLC. The LAP-D address field can be either one or two bytes long. If the extended address bit of the first byte is set, the address is one byte long. Otherwise, the address field's length is two bytes. The first address field byte contains the service access point identifier (SAPI), which identifies the port at which LAP-D services are provided to layer 3.

The C/R indicates whether the frame contains a command or a response. The terminal end-point identifier (TEI) identifies either a single terminal or multiple terminals. A TEI of all '1' indicates a broadcast.

4.8.5 ISDN network layer (layer 3)

Two layer 3 specifications are used for ISDN signalling: ITU-T I.450 (also known as ITU-T Q.930) and ITU-T I.451 (also known as ITU-T Q.931). Together, these protocols support user-to-user, circuit-switched and packet-switched connections. A variety of call establishment, call termination, information and miscellaneous messages are specified, including

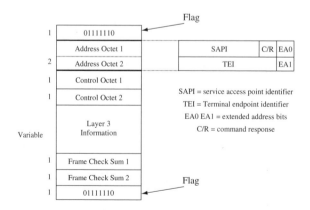

Figure 4.20 LAP-D frame structure (layer 2)

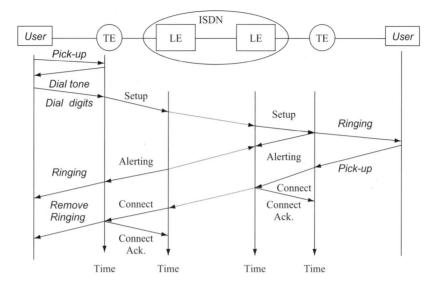

Figure 4.21 Illustration of the ISDN layer 3 signalling

SETUP, CONNECT, RELEASE, USER INFORMATION, CANCEL, STATUS and DIS-CONNECT. These messages are functionally similar to those provided by the X.25 protocol. Figure 4.21 shows the typical stages of an ISDN circuit switched call.

4.9 ISDN over satellite

Due to the availability of satellite networks, it is the natural to make use of satellite networks to extend the ISDN network for a global coverage. Though the ISDN does not restrict using any particular transmission systems, it is important from a satellite radio engineering point of view to investigate how satellite transmission systems differ from the traditional systems required to support ISDN, how satellite transmission error performance affects ISDN, and how propagation delay via satellite link affects operations of ISDN. It is the responsibility of the ITU-R SG 4 to define relevant requirements on conditions and performance for satellite links to carry ISDN channels and translate the ITU-T standards in terms that are significant for the satellite portion of the overall ISDN connections.

4.9.1 ITU-T ISDN hypothetical reference connection (IRX)

The ISDN hypothetical reference connection (IRX) is defined in the ITU-T G.821 recommendation. It is used to specify the performance requirement of the major transmission segments of the overall end-to-end connection. The distance for reference of the overall end-to-end connections is 27 500 km, which is the longest possible connection along the earth surface between subscribers (at reference point T).

Three basic segments are identified with distances that are expected to be typical distances of the portion in the overall end-to-end connections in the context of IRX, which are allocated allowable performance degradation of 30%, 30% and 40% to low-, medium- and high-grade segments.

The 30% for the low-grade segment is shared by two sides of the connections from user terminal to local exchange. Similarly, there are two medium-grade segments from local exchange to international exchange sharing the 30%. Satellite links of fixed satellite service should be equivalent to half of the high-grade segment as 20%, if used in the end-to-end ISDN connection.

In terms of distance, the high-grade segment counts for 25 000 km and the low and medium segments on one side of the connection count for 1250 km and on the other side 1250 km. Satellite link counts for 12 500 km, if used in the end-to-end ISDN connection.

4.9.2 ITU-R hypothetical reference digital path (HRDP) for satellite

ITU-R defined the hypothetical reference digital path (HRDP) in ITU-R S.521 to study the use of a fixed satellite link in a part of the ISDN HRX defined by ITU-T. As shown in Figures 4.22 and 4.23, the HRDP should consist of one earth–satellite–earth link with possibly

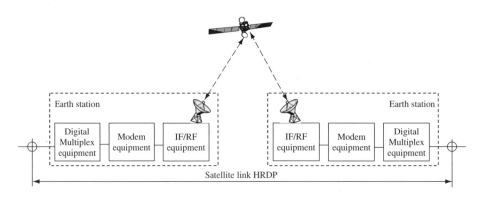

Figure 4.22   Hypothetical reference digital path (HRDP)

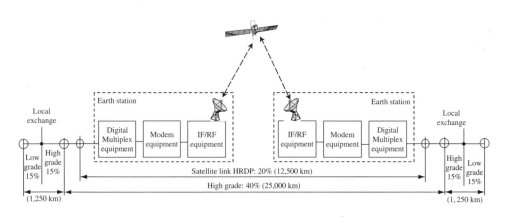

Figure 4.23   HRDP in ITU-T IRX at 64 kbit/s

one or more inter-satellite links in the space segment and interface with the terrestrial network appropriate to the HRDP. The HRDP should accommodate different types of access as single channel or TDMA, and allow for use of techniques such as digital speed interpolation (DSI) or low rate encoding (LRE) in the digital multiplex equipment.

Additionally, the earth stations should include facilities to compensate for the effects of satellite link transmission time variation introduced by satellite movements, which are of particular significance in digital transmission in the time domain such as PDH.

ITU-R HRDP uses 12 500 km from the IRX to develop performance and availability objectives. The distance has been defined by taking into account various satellite network configurations with a maximum single hop covering an equivalent terrestrial distance of approximately 16 000 km. Consequently, in the majority of cases satellite is used in international segments of the connection with two landing points usually less than 1000 km from the users. In practice, satellite network landing points should be designed as close as possible to user terminals.

4.9.3 Performance objectives

Satellite networks to support ISDH should allow end-to-end-connections to meet the performance objectives defined by the ITU-T. The ITU-R has developed recommendations for satellite to achieve the performance objectives in the end-to-end connections:

- The ITU-R S.614 on quality objectives for a 64 kbit/s ISDN circuit gives specification related to the ITU-T G.821 (see Tables 4.1 and 4.2).
- The ITU-R S.1062 on error performance for an HDRP operating at or above the primary rate gives specification related to the ITU-T G.826 (see Table 4.3).

4.9.4 Satellite network to ISDN interconnection scenarios

A satellite network interconnected to the ISDN should be capable of supporting all the ISDN services. As a minimum the satellite network needs to support ISDN circuit mode bear

Table 4.1 Quality objectives for digital telephony and 64 Kbit/s ISDN

Measurement conditions	Digital (PCM) telephony (S.522) Bit Error Rate (BER)	64 Kbit/s ISDN (S.522) Bit Error Rate (BER)
20% of any month (10 minutes mean value)	10^{-6}	—
10% of any month (10 minutes mean value)	—	10^{-7}
2% of any month (10 minutes mean value)	—	10^{-6}
0.3% of any month (1 minute mean value)	10^{-3}	—
0.05% of any month (1 second mean value)	10^{-4}	—
0.03% of any month (1 second mean value)	—	10^{-3}

Table 4.2 Overall end-to-end and satellite HRDP error performance objectives for international ISDN connections

Performance classification	Definition	End-to-end objective	Satellite HRDP objectives
Degrade seconds	Minutes intervals with BER $> 10^{-6}$ (more than 4 errors/minute)	$< 10\%$	$< 2\%$
Severely errored seconds	Minutes intervals with BER $> 10^{-3}$	$< 0.2\%$	$< 0.03\%$
Errored seconds	Minutes intervals with one or more errors	$< 8\%$	$< 1.6\%$

services, which requires adequate capacity for channels ranging from 64 kbit/s to 1920 kbit/s plus either 16 or 64 kbit/s D channel. In addition, if the satellite networks are used for data communications, it is logical to also support ISDN packet mode bear services. The satellite network should be able to support some of the ISDN supplementary services such as sub-addressing, direct dialling, multiple subscriber numbers and close groups.

The satellite network is often viewed as a part of the user network connection to ISDN through a NT2 network termination. Figure 4.24 illustrates a single node distributed ISDN customer network. The ISDN may offer at T reference point a basic or primary rate interface through network terminal NT1. NT2 forms part of the customer networks, typically using VSAT system. The NT2 can be envisaged as the node of a distributed PABX, while the S interface represents the standard for the interface between the PABX terminal equipment.

Figure 4.25 illustrates a multiple node distributed ISDN customer network. The satellite is used to interconnect several private ISDN networks (nodes), each of which consists of a earth station, a network termination NT1 and a few user terminals. In both scenarios, the private ISDN networks (nodes) are connected to a public ISDN network via a hub. In the case of VSAT, terminals can communicate with each other via the hub if it is a star configuration and with each other directly if it is a mesh configuration.

4.9.5 Routing plan

Before 1980 the ITU-T routing plan was based on a network with a hierarchical structure with descending central transit (CT) levels. Since 1980, ITU-T has made a radical change in its international routing plan. The new plan is a 'free routing structure', i.e. no hierarchical routing is required. It assumes that national telephone operators will maintain national hierarchical networks. Obviously the change was brought about by the long reach of satellite networks with which international high-usage (HU) trunks can terminate practically anywhere in the territory of a national telecommunication networks. Figure 4.26 illustrates the concepts of routing in a telecommunication network hierarchy with satellite link for the international connections. Of course, satellite can be used to replace any connection in the hierarchy.

In practice, the large majority of international telephone traffic is routed on direct circuits (i.e., no intermediate switching point) between international switching centres (ISC).

Table 4.3 Overall end-to-end and satellite HRDP error performance objectives for digital connection at primary rate or above

Performance classification	Definition	End-to-end objective		Satellite HRDP objectives	
Bit rate	—	1.5 to 15 Mbit/s	15 to 55 Mbit/s	1.5 to 15 Mbit/s	15 to 55 Mbit/s
Bit per block	—	2000–8000	4000–20000	2000–8000	4000–20000
Errored seconds (EES) ratio (SESR)	ES/t: • SES: 1 s with one or more errored blocks, and • t: available time during fixed measurement interval.	0.04	0.0075	0.014	0.0262
Severe Errored seconds (EES) ratio (SESR)	ES/t: • SES: 1 s with 30% errored blocks or one Severely Disturbed Period (SDP), • SDP: 4 continuous blocks or 1 s with BER of 10^{-2}	0.002	0.002	0.007	0.007
Background block errored (BBE) ratio (SESR)	BBE/b • BBE: an errored block not occurring as a part of SES. • b: total number of blocks during fixed measurement interval (excluding blocks during SES and unavailable time)	3×10^{-4}	2×10^{-4}	1.05×10^{-4}	0.7×10^{-4}

Note: Higher possible rates can also be found in ITU0R S.1062, including 55–160 Mbit/s and 160 to 3500 Mbit/s.

ITU-T E.171 provides the rules governing routing of connections consisting of a number of circuits in tandem. These connections have an importance in the network because:

- They are used as alternate routes to carry overflow traffic in busy periods to increase network efficiency.
- They can provide a degree of service protection in the event of failure of other routes.
- They can facilitate network management when associated with the ISC having temporary alternative routing capabilities.

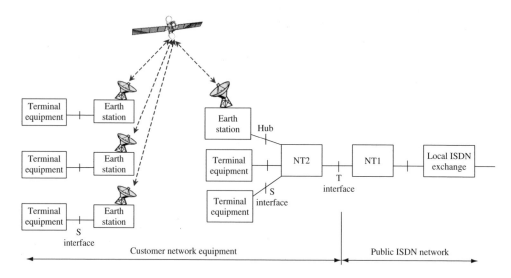

Figure 4.24   Single node distributed ISDN customer network

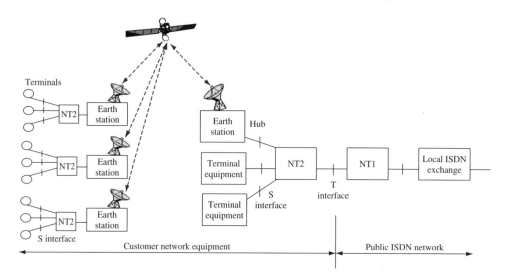

Figure 4.25   Multiple nodes distributed ISDN customer network

The rules have been designed to preserve the freedom of network operators to route their originating traffic directly or via any transit administration they choose, and to offer transit capabilities to as wide a range of destinations as possible. The governing features of this routing plan include that:

- It does not have to be hierarchical.
- Operators are free to offer whatever transit capabilities they wish, providing they conform to the recommendation.

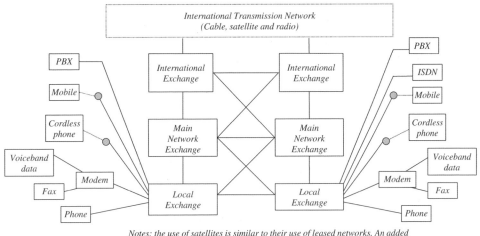

Notes: *the use of satellites is similar to their use of leased networks. An added complication which restricts their use is the need to carefully control circuit routing through a switched network to minimise the chance of picking up 2 satellite hops on a particular call.*

Figure 4.26   Switching and routing concepts in the telecommunication networks

- Direct traffic should be routed over final (fully provided) or high usage circuit groups.
- There are no more than four international circuits in tandem between the originating and terminating ISCs.
- Advantage should be taken of the non-coincidence of international traffic by the use of alternative routings and provide route diversity (referring to ITU-T E.523).
- The routing of transit switched traffic should be planned to avoid possible circular routings.

When a circuit group has both terrestrial and the satellite circuits, the choice of routing should be governed by:

- The guidance given in ITU-T G.114 (e.g., no more than 400 ms one-way propagation time).
- The number of satellite circuits likely to be utilised in the overall connection.
- The circuit which provides the better transmission quality and overall service quality.
- The inclusion of two or more satellite circuits (two or more hops over satellite) in the same connection should be avoided in all but exceptional cases when the satellite is the only possible link available for the connection.

4.10 Interworking with heterogeneous networks

Satellites offer a variety of means to accommodate different transmission rates including basic rate, primary rate and high speed IDR. Therefore, there is a significant difference between the interconnections of different networks.

Satellite networks can be used as thin routes between pairs of earth stations, as access networks to provide basic rate and primary rate and as transit networks to interconnect main networks with a capacity measured in thousands of circuits.

Transmission bit rate is the physical layer feature of the networks and is only one aspect of interworking issues. There are also high layer protocol issues. Often interworking units have to be introduced to deal with differences of functions at higher layer protocols when interconnecting different type of networks. Here we will discuss some general issues concerning interworking with heterogeneous networks.

4.10.1 Services

Different services are available in heterogeneous networks. For example, video telephony service is supported by the ISDN network and ordinary telephony is supported by plain telephony network. If a call is made from one to the other, the video information must be left out for the connection to be successful. Voice services in the video telephony terminal should be able to work as an ordinary telephone terminal because the ordinary telephony service is only a subset of video telephony. Another example is the conversion between email and fax where the interworking functions are more complicated because different terminals providing different services are involved.

Some services do not always need to be able to internetwork with each other such as file transfer, while some services may not be able to internetwork together at all. Normally service level internetwork defines the functional requirement to implement internetworking functions for heterogeneous networks.

4.10.2 Addressing

In heterogeneous networks, addressing is an important issue to be considered. We have to try to maintain the independent different networking schemes. Each address used to identify a terminal in the network must be unique.

Each internetworking unit of two networks should have two addresses, one of which is used in one network and one for the other network. A mapping between terminals and the internetworking unit must be available so that a terminal can make use of the internetworking unit when trying to connect to a terminal in the other networks. A long-haul connection may traverse many heterogeneous networks from source to destination of the connection.

Typical types of address include: Internet address, local area network address (such as Ethernet) and telephone network address (such as telephony number).

4.10.3 Routing

Routing is another important issue because the two networks can have significantly different transmission speeds, routing mechanisms, protocol functions and QoS requirements. Therefore, it is important to keep the routing independent within each network. The differences that the internetworking unit has to deal with include the protocol to access the networks, packet and frame format and size, and maintaining the QoS for the end-to-end connection requirement. In addition to the user information to be transported across different types of networks, signalling and management also have to be considered.

Typical examples of heterogeneous routing can be found in the Internet and IP telephony services where end-to-end connections traverse across LAN, MAN, telephony, mobile and satellite networks.

4.10.4 Evolution

Evolution is an important issue for all actors in the area of telecommunications because it predicts the long-range future development of both the network and services. It is different from planning, which concentrates on precise tasks and gives information and detailed figures on actions to be taken in the future. The driving force of evolution is technical and technological progress, which influence two main areas characterised by mutual interdependences: change and growth. In addition, economic considerations and conditions strongly influence evolution. One issue for future development is the transition from individual sub-networks with different capacities into a single overall network combining all components and offering even more facilities and capabilities.

Due to flexibility and adaptability, satellite systems have been used in a wide range of network topologies from simple point-to-point connections to complex multipoint-to-multipoint networks and transmission speeds from a few kbit/s to hundreds of Mbit/s. In many cases, satellite networks can present a alternative means of communication with respect to terrestrial networks, and bring advantages from both technical and economic viewpoints.

The functions of satellite networks can be the same as a traditional transmission medium, or provide advanced switching capabilities to work with all types of terrestrial networks.

In an early phase of broadband networks, satellite can be used as a flexible transport mechanism to provide an effective means to link users who cannot be reached by the broadband networks. For some users, satellites will provide an initial access to the broadband networks.

Once the broadband network is well established, satellite can be used to complement the terrestrial networks in areas where installation of other network technologies is difficult or expensive, and to provide services such as broadcast and mobile services to cover the globe.

To meet the challenge, satellite technologies are evolving, which exploit technological advantage by using higher frequency band on-board switching technologies to provide mobile and broadcasting services.

Further reading

[1] ITU-R Recommendation S.614-3, Allowable error performance for a hypothetical reference digital path in the fixed-satellite service operating below 15 GHz when forming part of an international connection in an integrated services digital network, Question ITU-R 52/4, 1986-1990-1992-1994.

[2] ITU-R Recommendation S.1062-2, Allowable error performance for a hypothetical reference digital path operating at or above the primary rate, Question ITU-R 75/4, 1994-1995-1999.

[3] ITU-T Recommendation G.107, The E-model, a computational model for use in transmission planning, 03/2003.

[4] ITU-T Recommendation G.108 Application of the E-model: a planning guide, 09/1999.

[5] ITU-T Recommendation G.821, Error performance of an international digital connection cooperating at a bit rate below the primary rate and forming part of an integrated services digital network, 08/1996.

[6] ITU-T Recommendation G.826, Error performance parameters and objectives for international constant bit rate digital paths at or above the primary rate, 02/1999.

[7] ITU-T Recommendation I.351/Y.801/Y.1501, Relationships among ISDN, Internet protocol and GII performance recommendations (1988, 1993, 1997, 2000), 01/2000.

[8] ITU-T Recommendation I.525, Interworking between networks operating at bit rates less than 64 kbit/s with 64 kbit/s-based ISDN and B-ISDN, 08/1996.

[9]   ITU-T Recommendation Y.101 Global information infrastructure terminology: terms and definitions, 03/2000.
[10] ITU-T Recommendation E.800, Terms and definitions related to quality of service and network performance including dependability, 08/94.
[11] ITU-T Recommendation G.702, Digital hierarchy data rates, 11/88.
[12] ITU-T Recommendation G.703, Physical/electrical characteristics of hierarchical digital interfaces, 11/2001.
[13] ITU-T Recommendation G.704, Synchronous frame structures used at 1544, 6312, 2048, 8448 and 44 736 kbits/s hierarchical levels, 10/98.
[14] ITU-T Recommendation G.707/Y.1322, Network node interface for the synchronous digital hierarchy (SDH), 12/2003.
[15] ITU-T Recommendation G.708, Sub STM-0 network node interface for the synchronous digital hierarchy (SDH), 06/99.
[16] ITU-T Recommendation G.709/Y.1331, Interfaces for the Optical Transport Network (OTN), 03/2003.
[17] ITU-T Recommendation I.430, Basic user-network interface – layer 1 specification, 11/95.
[18] ITU-T Recommendation I.431, Primary rate user-network interface – layer 1 specification, 03/93.
[19] ITU-T Recommendation Q.920, ISDN user-network interface data link layer - General aspects, 03/93
[20] ITU-T Recommendation Q.921, ISDN user-network interface - Data link layer specification, 09/97.
[21] Recommendation Q.930, Digital subscriber signalling system No. 1 (DSS 1) – user-network interface layer 3 – general aspects, 03/93
[22] ITU-T Recommendation Q.931, ISDN user-network interface layer 3 specification for basic call control, 05/98.
[23] ITU-R S.521-4, Hypothetical reference digital paths for systems using digital transmission in the fixed-satellite service, 01/2000
[24] ITU-R S.614-4, Allowable error performance for a satellite hypothetical reference digital path in the fixed-satellite service operating below 15 GHZ when forming part of an international connection in an integrated services digital network, 02/05.
[25] ITU-R S.1062-2 - Allowable error performance for a satellite hypothetical reference digital path operating below 15 GHZ, Draft Revision of Recommendation, 02/05.
[26] ITU-T G.114, One-way transmission time, 05/03.
[27] ITU-T E.171, International telephone routing plan, 11/88.
[28] ITU-T E.523, Standard traffic profiles for international traffic streams, 11/88.

Exercises

1. Explain the terms: interworking and internetworking.
2. Use a sketch to explain the concept of reference configuration.
3. Explain the ISDN reference points and function blocks.
4. Explain the different network traffic in user plane, control plane and management plane.
5. Explain the network hypothetical reference connection and related performance objectives for satellite.
6. Explain the basic models and parameters of traffic engineering in telephony networks.
7. Explain the principles of digital networks including PDH, SDH and ISDN.
8. Explain different types of signalling schemes and their role in the network.
9. Explain how to calculate the performance objectives of satellite networks in end-to-end reference connections.
10. Discuss the issues of SDH over satellite.
11. Discuss the issues of ISDN over satellite.

5

ATM over Satellite Networks

This chapter aims to provide an introduction to the concept of broadband satellite networking based on ATM technology. Although ATM networking is evolving towards all-IP networking, we can see that the new-generation Internet networks have started to adopt the basic principles and techniques developed for ATM networks to be able to support quality of service (QoS), class of service (CoS), fast packet switching, traffic control and traffic management. When you have completed this chapter, you should be able to:

- Know the design issues and concepts concerning ATM over satellites.
- Know the GEO satellite ATM networking and advanced satellite networking with LEO/MEO constellations.
- Describe the architecture of broadband network interconnection and terminal access.
- Describe the major roles of satellites in broadband networks with ATM over satellite networking.
- Understand the basic concept of satellite transparent and on-board switching payload for ATM networks.
- Understand ATM QoS and performance issues and enhancement techniques for satellite ATM networks.

5.1 Background

In the early 1990s, research and development in broadband communications based on ATM and fibre optic transmission cable generated a significant demand for cost-effective interconnection of private and public broadband ATM LANs (also called ATM islands), experimental ATM networks and testbeds, and for cost-effective broadband access via satellite to these broadband islands. However, there was a shortage of terrestrial networks to provide broadband connections in wide areas, particularly in more remote or rural areas where terrestrial

---

lines are expensive and uneconomical to install and operate. Satellite networking was considered as an alternative solution to 'broadband for all' to complement terrestrial broadband networks due to its flexibility and immediate global coverage. It was also expected to provide distribution and broadcasting services.

In the commercial arena, the need to provide broadband networks over satellite was also expected to increase significantly broadband services. Examples of the identified applications included linking remote office sites (e.g. oil rigs) to the enterprise backbone and providing broadband entertainment services to mobile platforms (e.g. aeroplanes, ships). Other examples included emergency and disaster relief scenarios and remote/rural medical care where the infrastructure was either disrupted or lacking.

5.1.1 Networking issues

One of the key networking issues was to provide interconnection and also access to geographically dispersed broadband islands in the context of ATM networks with the required QoS and bandwidth. Due to their global coverage and broadcasting nature, satellite networks can also be best used for broadband mobile and broadcasting services, where the major technology challenge is how to design small satellite terminals at low cost but with high-speed transmission for broadband services.

The design of satellite ATM networks was also expected to be directly compatible with the terrestrial networks. It is widely recognised that the development of B-ISDN based ATM was not revolutionary but evolutionary. This also required satellite ATM networks to be able to interconnect the ATM networks as well as existing data networks such as the LAN and MAN.

Like other packet networks, ATM is a set of protocols using asynchronous transfer mode to support broadband services; it is not a transmission technology, but can be transported over different types of transmission technologies and media including wireless, cable and satellite networks. It has been standardised by the ITU-T and the ITU-R to exploit the potential of satellite ATM networks.

By the late 1990s, the emerging WWW services and applications, based on the Internet, changed the landscape of the telecommunications and data communications networks industries. It became mandatory to support Internet protocol (IP) over ATM solutions, and also made it difficult to support QoS. This also led to the convergence of user terminals, networks and services and applications in the telecoms industry and Internet towards the next generation of Internet by taking advantage of both the IP and ATM networks.

5.1.2 Satellite services in the B-ISDN networking environment

The principal advantages of satellite systems are their wide coverage and broadcasting capabilities. There are enough satellites to provide broadband connections anywhere in the world. The cost and complexity are independent of distance. There are clear advantages to extend the broadband capabilities to rural and remote areas. Satellite links are quick and easy to install with fewer geographical constraints. They make long-distance connections more cost-effective within the coverage areas, particularly for point-to-multipoint and broadcasting services. Satellites can also be complementary to the terrestrial networks and mobile networks.

In a broadband networking environment, satellite networking can be used for user access mode and also for network transit mode. In the user access mode, the satellite system is positioned at the border of the broadband network. It provides access links to a large number of users directly or via local networks. The interfaces to the satellite system in this mode are

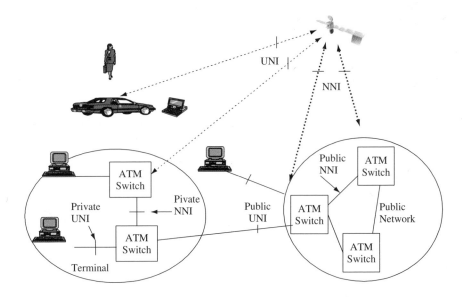

Figure 5.1   Example of user access mode via satellite ATM network

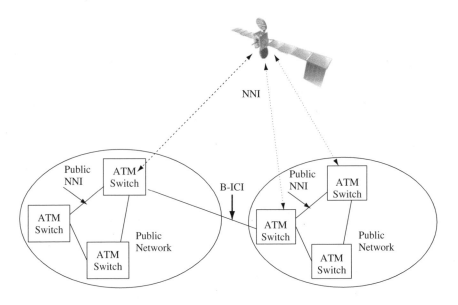

Figure 5.2   Example of network transit mode via a satellite ATM network

of the user network interface (UNI) type on one side and the network node interface (NNI) type on the other side.

In the network transit mode, the satellite systems provide high bit-rate links to interconnect the B-ISDN network nodes or network islands. The interfaces on both sides are NNI type. Figure 5.1 illustrates an example of a configuration of the satellite system for broadband network access and mobile access and Figure 5.2 shows the interconnection of broadband islands/networks.

5.2 Design issues of satellite ATM systems

The satellite networks are fundamentally different from terrestrial networks in terms of delay, error and bandwidth characteristics, and can have an adverse impact on the performance of ATM traffic, congestion control procedures and transport protocol operations.

5.2.1 Propagation delay

The propagation delay for the packets of a connection consists of the following three quantities: from the source ground terminal to satellite uplink propagation delay (t_{up}); the inter-satellite link propagation delays (t_i) (if ISL are used); and from the satellite to destination ground terminal downlink propagation delay (t_{down}).

The uplink and downlink satellite–ground terminal propagation delays (t_{up} and t_{down}, respectively) represent the time taken for the signal to travel from the source ground terminal to the first satellite in the network (t_{up}), and the time taken for the signal to reach the destination ground terminal from the last satellite in the network (t_{down}). They can be calculated as the following:

$$t_{up} = \frac{\text{distance\_from\_source\_terminal\_to\_satellite}}{\text{speed\_of\_light}}$$

$$t_{down} = \frac{\text{distance\_from\_satellite to destination\_terminal}}{\text{speed\_of\_light}}$$

The end-to-end delay also depends on LEO/MEO constellation designs. In contrast to GEO satellites, the LEO uplink and downlink propagation delay is much shorter but variable over time.

We can also note the transmission delay as t_t, the inter-satellite link delay as t_i, the on-board switching and processing delay as t_s, the buffering delay as t_q and delay due to the terrestrial networks (terrestrial tail) as t_n. The inter-satellite, on-board switching, processing and buffering delays are cumulative over the path traversed by a connection. The delay variation is caused by orbital dynamics, buffering, adaptive routing (in LEO) and on-board processing. Then, the end-to-end delay (D) can be calculated as:

$$D = t_t + t_{up} + t_i + t_{down} + t_s + t_q + t_n$$

The transmission delay (t_t) is the time taken to transmit a single data packet at the network data rate as:

$$t_t = \frac{\text{packet\_size}}{\text{data\_rate}}$$

For broadband networks with high data rates, the transmission delays become negligible in comparison to the satellite propagation delays. For example, it only takes about 212 microseconds to transmit an ATM cell at a 2 Mbit/s link. This delay is much less than the propagation delays in satellites. Compared with the propagation delays, all the t_i, t_s, t_q and t_n are very small, hence can be neglected in calculation.

The *inter-satellite link delay* (t_i) is the sum of the propagation delays of the inter-satellite links (ISL) traversed by the connection. It may be *in-plane* or *cross-plane* links. In-plane links connect satellites within the same orbit plane, while cross-plane links connect satellites in different orbit planes. In GEO systems, ISL delays can be assumed to be constant over a connection's lifetime because GEO satellites are almost stationary over a given point on the earth, and with respect to one another. In LEO constellations, the ISL delays depend on the orbital radius, the number of satellites per orbit, and the inter-orbital distance (or the number of orbits). All in-plane links in circular orbits are considered to be constant. Cross-plane ISL delays change over time, break at highest latitudes and must be reformed. As a result, LEO systems can exhibit a high variation in ISL delay.

LEO satellites have lower propagation delays due to their lower altitudes, but many satellites are needed to provide a global service. While LEO systems have lower propagation delays, they exhibit higher delay variation due to connection handovers and other factors related to orbital dynamics.

The large delays in GEO, and delay variations in LEO, affect both real-time and non-real-time applications. Many real-time applications are sensitive to the large delay experienced in GEO systems, as well as to the delay variation experienced in LEO systems. In an acknowledgement and time-out based congestion control mechanism, performance is inherently related to the delay–bandwidth product of the connection. Moreover, round trip time (RTT) measurements are sensitive to delay variations that may cause false time-outs and retransmissions for acknowledgement-based data services. As a result, the congestion control issues for broadband satellite networks are somewhat different from those of low-latency terrestrial networks. Both interoperability as well as performance issues between satellite and terrestrial networks must be addressed before data, voice and video services can be provided over satellite networks.

5.2.2 Attenuation and constraints

The attenuation of free space (called free-space loss, L_{FS}) represents the ratio of received and transmitted power in a link between two isotropic antennas:

$$L_{FS} = (4\pi R/\lambda)^2$$

Where R is propagation distance and λ is wavelength. A GEO satellite and a station situated exactly under the satellite is 35 786 km between the satellite and the station (equal to the altitude of the satellite). Therefore the L_{FS} is of the order of 200 dB at C band and 207 dB at Ku band. Attenuation is also affected by other effects such as rain, clouds, snow, ice and gas in the atmosphere.

Satellite communication bandwidth being a limited resource will continue to be a precious asset. Achieving availability rates of 99.95% at very low bit error rate (BER) is costly. Lowering required availability rates by even 0.05% dramatically lowers satellite link costs. An optimum availability level must be a compromise between cost and performance.

There are constraints in general in choosing the satellite link parameters due to regulations, operational constraints and propagation conditions. The regulations are administered by the ITU-R, ITU-T and ITU-D. They define space radio-communication services in terms of transmission and/or reception of radio waves for specific telecommunication applications. The concept of a radio communication service is applied to the allocation of frequency bands and analysis of conditions for sharing a given band among compatible services. The operational constraints relate to realisation of a C/N_0 ratio, provision of an adequate satellite antenna beam for coverage of a service area with a specified value of satellite antenna gain, level of interference between satellite systems, orbital separation between satellites operating in identical frequency bands and minimum of total cost.

Therefore the design of high-speed transmission faces great challenges to achieve error performance objectives.

5.3 The GEO satellite ATM networking architecture

In this section, the discussion on GEO satellite ATM networking architecture is based on the design of the CATALYST project. The CATALYST project was funded within the European Framework Programme Research in Advanced Communication in Europe phase II (RACE II) to develop an experimental broadband satellite ATM network for interconnection to geographically dispersed broadband networks called 'broadband islands'. The CATALYST demonstration took place in 1992–3 and involved the first transmission of ATM cells over satellite in Europe.

A modular approach was used in the design to interface different networks and the satellite, converting network packets to and from ATM cells. The functions of the main building blocks of the demonstrator are described here.

5.3.1 The ground segment

To make use of the existing satellite systems, development has been mainly on the ground segment. Many modules were developed, where each module had buffer(s) for packet/cell conversion and/or traffic multiplexing. The buffers are also used for absorbing high-speed burst traffic.

Therefore, the satellite ATM system can be designed to be capable of interconnecting different networks with capacities in the range of 10 to 150 Mbit/s (10 Mbit/s for Ethernet, 34 Mbit/s for DQDB, 100 Mbit/s for FDDI and 150 Mbit/s for ATM networks). Figure 5.3 illustrates the model of the ground equipment. A brief description of these modules is also given in the following.

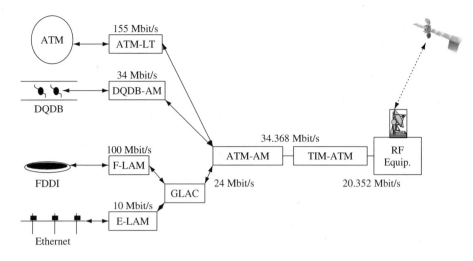

Figure 5.3 Ground-segment modules

- The ATM-LT provides an interface with a speed of 155 Mbit/s between the ATM network and the ground-station ATM equipment. It is also the termination point of the ATM network and passes the ATM cells to the ATM adaptation module (ATM-AM).
- The Ethernet LAN adaptation module (E-LAM) provides an interface to the Ethernet local area network.
- The FDDI LAN adaptation module (F-LAM) provides an interface to the FDDI network.
- The generic LAN ATM converter (GLAC) module converts the FDDI and Ethernet packets into ATM cells, then passes the cells to the ATM-AM module.
- The DQDB adaptation module (DQDB-AM) provides an interface to the DQDB network with a small buffer. It converts DQDB packets into ATM cells and then passes them to the ATM-AM.
- The ATM-AM is an ATM adapter. It multiplexes the ATM cell streams from the two ports into one ATM cell stream. This module passes the cells to the terrestrial interface module for ATM (TIM-ATM) and provides an interface between the terrestrial network and the satellite ground-station.
- The TIM-ATM has two buffers with a 'ping-pong' configuration. Each buffer can store up to 960 cells. The cells are transmitted from one buffer while the ATM-AM feeds the cells into the other buffer. Transmissions of the buffers are switched every 20 ms.

5.3.2 The space segment

In the demonstrator system, the EUTELSAT II satellite was used in good weather conditions making use of 36 MHz bandwidth of a transponder. It achieved transmission capacity of approximately 20 Mbit/s. The capacity has to be shared by a number of earth stations when multiple broadband islands are interconnected. It was a trade-off to provide good required QoS and efficient utilisation of the satellite resources (bandwidth and transmission power).

Compared to the propagation delay, the delay within the ground segment was insignificant. Buffering in the ground-segment modules could cause variation of delay, which was affected by the traffic load on the buffer. Most of the variation was caused in the TIM-ATM buffer. It caused an estimated average delay of 10 ms and worst-case delay of 20 ms. Cell loss occurred when the buffer overflowed. The effects of delay, delay variation and cell loss in the system could be controlled to the minimum by controlling the number of applications, the amount of traffic load and allocating adequate bandwidth for each application.

5.3.3 Satellite bandwidth resource management

The TDMA system was used with frame length of 20 ms which was shared by the earth stations. Each earth station was limited to the time slots corresponding to the allocated transmission capacity up to maximum 960 cells (equivalent to 20.352 Mbit/s). The general TDMA format is shown in Figure 5.4.

There are three levels of resource management (RM) mechanisms. The first level is controlled by the network control centre (NCC) and allocates the bandwidth capacity to each earth station. The allocation is in the form of burst time plans (BTP). Within each BTP, burst times are specified for the earth station, which limit the number of cells in bursts the earth stations can transmit. In the CATALYST demonstrator, the limit is that each BTP is less than or equal to 960 ATM cells and the sum of the total burst times is less than or equal to 1104 cells.

The second level is the management of the virtual paths (VPs) within each BTP. The bandwidth capacity that can be allocated to the VP is restricted by the BTP. The third level is the management of the virtual channels (VCs). It is subject to the available bandwidth resource of the VP. Figure 5.5 illustrates the resource management mechanisms of the bandwidth capacity. Each station is allocated a time slot within the burst time plan. Each time slot is further divided to be allocated according to the requirements of VPI and VCI. The allocation of the satellite bandwidth is done when the connections are established. Dynamic changing, allocation, sharing, or re-negotiation of the bandwidth during the connection is for further study.

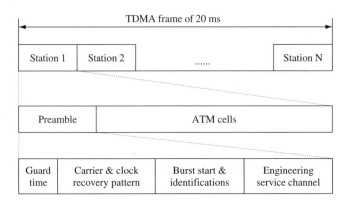

Figure 5.4   TDMA frame format (earth station to satellite)

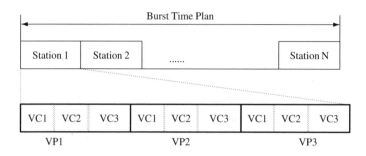

Figure 5.5   Satellite resource management

To effectively implement resource management, the allocation of the satellite link band-width can be mapped into the VP architecture in the ATM networks and each connection mapped into the VC architecture. The BTP can be a continuous burst or a combination of a number of sub-burst times from the TDMA frame.

The burst-time plan, data arrival rate and buffer size of the ground station have an important impact on the system performance. To avoid buffer overflow the system needs to control the traffic arrival rate, burst size or allocation of the burst-time plan. The maximum traffic rate allowed, to prevent the buffer overflow, is a function of the burst-time plan and burst size for a given buffer size, and the cell loss ratio is a function of traffic arrival rate and allocated burst-time plan for a given buffer size.

5.3.4 Connection admission control (CAC)

CAC is defined as the set of actions taken by the network at the call set-up phase in order to establish if sufficient resources are available to establish the call through the whole network at its required QoS and maintain the agreed QoS of existing calls. This also applies to re-negotiation of connection parameters within a given call. In a B-ISDN environment, a call can require more than one connection for multimedia or multiparty services such as video-telephony or videoconference.

A connection may be required by an on-demand service, or by permanent or reserved services. The information about the traffic descriptor and QoS is required by the CAC mechanism to determine whether the connection can be accepted or not. The CAC in the satellite has to be the integrated part of the whole-network CAC mechanisms.

5.3.5 Network policing functions

Networking policing functions make use of usage parameter control (UPC) and network parameter control (NPC) mechanisms. UPC and NPC monitor and control traffic to protect the network (particularly the satellite link) and enforce the negotiated traffic contract during the call. The peak cell rate has to be controlled for all types of connections. Other traffic parameters may be subject to control such as average cell rate, burstiness and peak duration.

At cell level, cells are allowed to pass through the connection if they comply with the negotiated traffic contract. If violations are detected, actions such as cell tagging or discarding are taken to protect the network.

Apart from UPC/NPC tagging users may also generate different priority traffic flows by using the cell loss priority bit. This is called priority control (PC). Thus, a user's low-priority traffic may not be distinguished by a tagged cell, since both user and network use the same CLP bit in the ATM header. Traffic shaping can also be implemented in the satellite equipment to achieve a desired modification of the traffic characteristics. For example, it can be used to reduce peak cell rate, limit burst length and reduce delay variation by suitably spacing cells in time.

5.3.6 Reactive congestion control

Although preventive control tries to prevent congestion before it actually occurs, the satellite system may experience congestion due to the earth-station multiplexing buffer or switch output buffer overflow. In this case, where the network relies only on the UPC and no feedback information is exchanged between the network and the source, no action can be taken once congestion has occurred. Congestion is defined as the state where the network is unable to meet the negotiated QoS objectives for the connections already established. Congestion control (CC) is the set of actions taken by the network to minimise the intensity, spread and duration of congestion. Reactive CC becomes active when there is indication of any network congestion.

Many applications, mainly those handling data transfer, have the ability to reduce their sending rate if the network requires them to do so. Likewise, they may wish to increase their sending rate if there is extra bandwidth available within the network. These kinds of applications are supported by the ABR service class. The bandwidth allocated for such applications is dependent on the congestion state of the network. Rate-based control is recommended for ABR services, where information about the state of the network is conveyed to the source through special control cells called resource management (RM) cells. Rate information can be conveyed back to the source in two forms:

- Binary congestion notification (BCN) using a single bit for marking the congested and not congested states. BCN is particularly attractive for satellites due to their broadcast capability.
- Explicit rate (ER) indication is used by the network to notify the source the exact bandwidth it should use to avoid congestion.

The earth stations may determine congestion status either by measuring the traffic arrival rate or by monitoring the buffer status.

5.4 Advanced satellite ATM networks

Until the launch of the first regenerative INTELSAT satellite in January 1991, all satellites were transparent satellites. Although the regenerative, multibeam and on-board ATM switch satellites have potential advantages, they increased the complexity on reliability, the effect on flexibility of use, the ability to cope with unexpected changes in traffic demand (both volume and nature) and new operation procedures. Advanced satellite ATM networks tried to explore the benefit of on-board processing and switching, multibeam satellite and LEO/MEO constellation, although complexity is still the main concern for satellite payloads.

5.4.1 Radio access layer

The radio access layer (RAL) for satellite access must take into account the performance requirements for GEO satellites. A frequency-independent specification is preferred. Parameters to be specified include range, bit rates, transmit power, modulation/coding, framing formats and encryption. Techniques for dynamically adjusting to varying link conditions and coding techniques for achieving maximum bandwidth efficiencies need to be considered.

The medium access control (MAC) protocol is required to support the shared use of the satellite channels by multiple switching nodes. A primary requirement for the MAC protocol is to ensure bandwidth provisioning for all the traffic classes, as identified in UNI. The protocol should satisfy both the fairness and efficiency criteria.

The data link control (DLC) layer is responsible for the reliable delivery of ATM cells across the GEO satellite link. Since higher layer performance is extremely sensitive to cell loss, error control procedures need to be implemented. Special cases for operation over simplex (or highly bandwidth asymmetric) links need to be developed. DLC algorithms tailored to special specific QoS classes also need to be considered.

Wireless control is needed for support of control plane functions related to resource control and management of the physical, MAC and DLC layers specific to establishing a wireless link over GEO satellites. This also includes meta-signalling for mobility support.

5.4.2 On-board processing (OBP) characteristics

OBP is in itself a vast domain that is the subject of much activity in the USA, Japan and Europe. All commercial civil satellites to date have used transparent transponders, which consist of nothing more than amplifiers, frequency changers and filters. These satellites adapt to changing demands, but at the cost of high space segment tariffs and high-cost, complex earth terminals. OBP aims to put the complexity in the satellite and to reduce the cost of the use of the space segment and the cost of the earth terminals. There are varying degrees of processing on board satellites:

- regenerative transponder (modulation and coding);
- on-board switching;
- access format conversion (e.g. FDMA-TDM); and
- flexible routing.

They may not all be present in one payload and the exact mix will depend on applications. The advantages rendered by the use of OBP are as summarised:

- Regenerative transponders: the advantage of the regenerative scheme is that the uplinks and downlinks are now separated and can be designed independently of each other. With conventional satellites $(C/N)_U$ and $(C/N)_D$ is additive; with regenerative transponders they are separated. This can be translated into an improved BER performance as reduced degradation is now present. Regenerative transponders can withstand much higher levels of interference for the same overall $(C/N)_T$.

- Multirate communications: with OBP it is possible to convert on the satellite between low- and high-rate terminals. This allows ground terminals operating at various rates to communicate with each other via a single hop. Transparent transponders require rate conversion terrestrially and hence necessitate two hops. Multirate communications implies both multicarrier demodulators and baseband switches.

These add up to much reduced complexity and cheaper ground terminals.

5.4.3 The ATM on-board switch

There are potential advantages in performance and flexibility for the support of services by placing switching functions on board satellites. It is particularly important for satellite constellations with spot beam coverage and/or inter-satellite communications, as it allows building networks upon constellation satellites therefore relying less on ground infrastructure. Figure 5.6 illustrates the protocol stack on board satellite and on the ground.

In the case of ATM on-board switch satellites, the satellite acts as a switching point within the network (as illustrated by Figure 5.6) and is interconnected with more than two terrestrial network end points. The on-board switch routes ATM cells according to the VPI/VCI of the header and the routing table when connections are set up. It also needs to support the signalling protocols used for UNI as access links and for NNI as transit links.

On-board switching (OBS) satellites with high-gain multiple spot beams have been considered as key elements of advanced satellite communications systems. These satellites support small, cost-effective terminals and provide the required flexibility and increased utilisation of resources in a burst multimedia traffic environment.

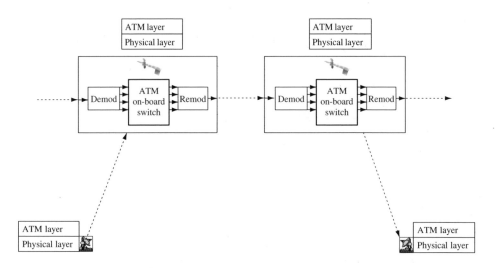

Figure 5.6   Satellite with ATM on-board switch

Although employing an on-board switch function results in more complexity on board the satellite, the following are the advantages of on-board switches.

- Lowering the ground-station costs.
- Providing bandwidth on demand with half the delay.
- Improving interconnectivity.
- Offering added flexibility and improvement in ground-link performance, i.e. this allows earth stations in any uplink beam to communicate with earth stations in any downlink beam while transmitting and receiving only a single carrier.

One of the most critical design issues for on-board processing satellites is the selection of an on-board baseband switching architecture. Four typical types of on-board switches are proposed:

- circuit switch;
- fast packet switch (can be variable packet length);
- hybrid switch;
- ATM cell switch (fixed packet length).

These have some advantages and disadvantages, depending on the services to be carried, which are summarised in Table 5.1.

From a bandwidth efficiency point of view, circuit switching is advantageous under the condition that the major portion of the network traffic is circuit switched. However, for burst traffic, circuit switching results in a lot of wasted bandwidth capacity.

Fast packet switching may be an attractive option for a satellite network carrying both packet-switched traffic and circuit-switched traffic. The bandwidth efficiency for circuit-switched traffic will be slightly less due to packet overheads.

In some situations, a mixed-switch configuration, called a hybrid switch consisting of both circuit and packet switches, may provide optimal on-board processor architecture. However, the distribution of circuit- and packet-switched traffic is unknown, which makes the implementation of such a switch a risk.

For satellite networking, fixed-size fast packet switching, such as ATM cell switching, is an attractive solution for both circuit- and packet-switched traffic. Using statistical multiplexing of cells, it could achieve the highest bandwidth efficiency despite a relatively large header overhead per cell.

In addition, due to on-board mass and power-consumption limitations, packet switching is especially well suited to satellite switching because of the sole use of digital communications. It is important that satellite networking follows the trends of terrestrial technologies for seamless integration.

5.4.4 Multibeam satellites

A multibeam satellite features several antenna beams which provide coverage of different service zones as illustrated by Figure 5.7. As received on board the satellite, the signals appear at the output of one or more receiving antennas. The signals at the repeater outputs must be fed to various transmitting antennas.

Table 5.1 Comparison of various switching techniques

Switching architecture	Circuit switching	Fast packet switching	Hybrid switching	Cell switching (ATM switching)
Advantages	• Efficient bandwidth utilisation for circuit-switched traffic • Efficient if network does not require frequent traffic reconfiguration • Easy to control congestion by limiting access into the network	• Self-routing • Does not require control memory for routing • Transmission without reconfiguring of the on-board switch connection • Easy to implement autonomous private networks • Provides flexibility and efficient bandwidth utilisation for packet-switched traffic • Can accommodate circuit-switched traffic	• Handles a much more diverse range of traffic • Optimisation between circuit switching and packet switching • Lower complexity on board than fast packet switch • Can provide dedicated hardware for each traffic type	• Self-routing with a small VC/VP • Does not require control memory for routing • Transmission without reconfiguring on-board switch connection • Easy to implement autonomous private networks • Provides flexibility and efficient bandwidth utilisation for all traffic sources • Can accommodate circuit-switched traffic • Speed comparable to Fast packet switching
Disadvantages	• Reconfiguration of earth station time/frequency plans for each circuit set up • Fixed bandwidth assignment (not flexible) • Very inefficient bandwidth utilisation when supporting packet-switched traffic • Difficult to implement autonomous private networks	• For circuit-switched traffic higher overheads than circuit switching due to packet headers. • Contention and congestion may occur	• Cannot maintain maximum flexibility for future services because the future distribution of satellite circuit and packet traffic is unknown • Waste of satellite resources in order to be designed to handle the full capacity of satellite traffic	• For circuit-switched traffic somewhat higher overheads than packet switching due to 5 byte ATM header. • Contention and congestion may occur

The spot-beam satellites provide advantages to the earth-station segment by improving the figure of merit G/T on the satellite. It is also possible to reuse the same frequency band several times in different spot beams to increase the total capacity of the network without increasing the allocated bandwidth. However, there is interference between the beams.

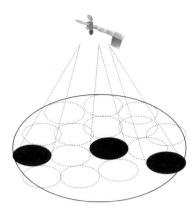

Figure 5.7   Multibeam satellite

One of the current techniques for interconnections between coverage areas is on-board switching-satellite-switched TDMA (SS/TDMA). It is also possible to have packet-switching on-board multibeam satellites.

5.4.5 LEO/MEO satellite constellations

One of the major disadvantages of GEO satellites is caused by the distance between the satellites and the earth stations. They have traditionally mainly been used to offer fixed telecommunication and broadcast services. In recent years, satellite constellations of low/medium earth orbit (LEO/MEO) for global communication have been developed with small terminals to support mobility. The distance is greatly reduced. A typical MEO satellite constellation such as ICO has 10 satellites plus two spares, and an LEO such as SKYBRIDGE has 64 satellites plus spares.

Compared to GEO networks, LEO/MEO networks are much more complicated, but provide a lower end-to-end delay, less free-space loss and higher overall capacity. However, due to the relatively fast movement of satellites in LEO/MEO orbit relative to user terminals, satellite handover is an important issue.

Constellations of LEO/MEO satellites can also be an efficient solution to offer highly interactive services with a very short round-trip propagation time over the space segment (typically 20/100 ms for LEO/MEO as compared to 500 ms for geostationary systems). The systems can offer similar performances to terrestrial networks, thus allowing the use of common communication protocols and applications and standards.

5.4.6 Inter-satellite links (ISL)

The use of ISL for traffic routing has to be considered. It must be justified that this technology will bring a benefit, which would make its inclusion worthwhile or to what extent on-board switching, or some other form of packet switching, can be incorporated into its use.

The issues that need to be discussed when deciding on the use of ISL include:

- networking considerations (coverage, delay, handover);
- the feasibility of the physical link (inter-satellite dynamics);
- the mass, power and cost restrictions (link budget).

The mass and power consumption of ISL payloads are factors in the choice of whether to include them in the system, in addition to the possible benefits and drawbacks. Also the choice between RF and optical payloads is now possible because optical payloads have become more reliable and offer higher link capacity. The tracking capability of the payloads must also be considered, especially if the inter-satellite dynamics are high. This may be an advantage for RF ISL payloads.

Advantages of ISLs can be summarised as the following:

- Calls may be grounded at the optimal ground station through another satellite for call termination, reducing the length of the terrestrial 'tail' required.
- A reduction in ground-based control may be achieved with on-board baseband switching – reducing delay (autonomous operation).
- Increased global coverage – oceans and areas without ground stations.
- Single network control centre and earth station.

Disadvantages of ISLs can be summarised as the following:

- Complexity and cost of the satellites will be increased.
- Power available for the satellite/user link may be reduced.
- Handover between satellites due to inter-satellite dynamics will have to be incorporated.
- Replenishment strategy.
- Frequency coordination.
- Cross-link dimensioning.

5.4.7 Mobile ATM

Hand-off control is a basic mobile network capability that allows for the migration of terminals across the network backbone without dropping an ongoing call. Because of the geographical distances involved, hand-off for access over GEO satellite is expected not to be an issue in most applications. In some instances, for example intercontinental flights, a slow hand-off between GEO satellites with overlapping coverage areas will be required.

Location management refers to the capability of one-to-one mapping between mobile node 'name' and current 'routing-id.' Location management primarily applies to the scenario involving switching on board the satellite.

5.4.8 Use of higher frequency spectrum

Satellite constellations can use the Ku band (11/14 GHz) for connections between user terminals and gateways. High-speed transit links between gateways will be established using either the Ku or the Ka band (20/30 GHz).

According to the ITU radio regulation, GEO satellite networks have to be protected from any harmful interference from non-geostationary systems. This protection is achieved through angular separation using a predetermined hand-over procedure based on the fact that the positions of geostationary and constellation satellites are permanently known and predictable. When the angle between a gateway, the LEO/MEO satellite in use by the gateway and the geostationary satellite is smaller than one degree, the LEO/MEO transmissions are stopped and handed over to another LEO/MEO satellite, which is not in similar interference conditions.

The constellations provide a cost-effective solution offering a global access to broadband services. The architectures are capable of: supporting a large variety of services; reducing costs and technical risks related to the implementation of the system; ensuring a seamless compatibility and complement with terrestrial networks; providing flexibility to accommodate service evolution with time as well as differences in service requirements across regions; and optimising the use of the frequency spectrum.

5.5 ATM performance

ITU (ITUT-I356) defines parameters for quantifying the ATM cell transfer performance of a broadband ISDN connection. This ITU recommendation includes provisional performance objectives for cell transfer, some of which depend on the user's selection of QoS class.

5.5.1 Layered model of performance for B-ISDN

ITU (ITUT-I356) defines a layered model of performance for B-ISDN, as shown in Figure 5.8.

It can be seen that the network performance (NP) provided to B-ISDN users depends on the performance of three layers:

- The physical layer, which may be based on plesiochronous digital hierarchy (PDH), synchronous digital hierarchy (SDH) or cell-based transmission systems. This layer is terminated at points where the connection is switched or cross-connected by equipment using the ATM technique, and thus the physical layer has no end-to-end significance when such switching occurs.
- The ATM layer, which is cell-based. The ATM layer is physical media and application independent and is divided into two types of sublayer: the ATM-VP layer and the ATM-VC layer. The ATM-VC layer always has end-to-end significance. The ATM-VP layer has no user-to-user significance when VC switching occurs. ITUT-I356 specifies network performance at the ATM layer, including the ATM-VC layer and ATM-VP layer.
- The ATM adaptation layer (AAL), which may enhance the performance provided by the ATM layer to meet the needs of higher layers. The AAL supports multiple protocol types, each providing different functions and different performance.

5.5.2 ATM performance parameters

ITUT-I356 also defines a set of ATM cell transfer performance parameters using the cell transfer outcomes. All parameters may be estimated on the basis of observations

Satellite Networking: principles and protocols

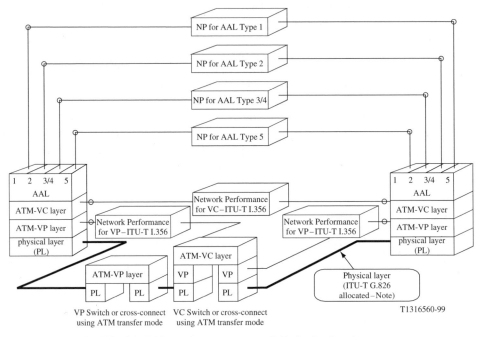

NOTE – The need for additional physical layer performance parameters and objectives is under study.

Figure 5.8 Layered model of performance for B-ISDN (ITUT-1356) (Reproduced with the kind permission of ITU.)

at the measurement points (MPs). Following is a summary of ATM performance parameters:

- Cell error ratio (CER) is the ratio of total errored cells to the total of successfully trans-ferred cells, plus tagged cells, plus errored cells in a population of interest. Successfully transferred cells, tagged cells and errored cells contained in severely errored cell blocks are excluded from the calculation of the cell error ratio.
- Cell loss ratio (CLR) is the ratio of total lost cells to total transmitted cells in a population of interest. Lost cells and transmitted cells in severely errored cell blocks are excluded from the calculation of the cell loss ratio. Three special cases are of interest, CLR0, CLR0 + 1 and CLR1, considering the CLR tag in the ATM cell header.
- Cell misinsertion rate (CMR) is the total number of misinserted cells observed during a specified time interval divided by the time interval duration (equivalently, the number of misinserted cells per connection second). Misinserted cells and time intervals associated with severely errored cell blocks are excluded from the calculation of the cell misinsertion rate.

- Severely errored cell block ratio (SECBR) is the ratio of total severely errored cell blocks to total cell blocks in a population of interest.
- The definition for cell transfer delay can only be applied to successfully transferred, errored and tagged cell outcomes. Cell transfer delay (CTD) is the time between the occurrences of two corresponding cell transfer events.
- Mean cell transfer delay is the arithmetic average of a specified number of cell transfer delays.
- Two cell transfer performance parameters associated with cell delay variation (CDV) are defined as illustrated in Figure 5.9. The first parameter, one-point cell delay variation, is defined based on the observation of a sequence of consecutive cell arrivals at a single MP. The second parameter, two-point cell delay variation, is defined based on the observations of corresponding cell arrivals at two MPs that delimit a virtual connection portion. The two-point CDV gives the measurement of end-to-end performance (see Figure 5.9).

The two-point CDV (v_k) for cell k between MP_1 and MP_2 is the difference between the absolute cell transfer delay (x_k) of cell k between the two MPs and a defined reference cell transfer delay $(d_{1,2})$ between those MPs: $v_k = x_k - d_{1,2}$.

The absolute cell transfer delay (x_k) of cell k between MP_1 and MP_2 is the difference between the cell's actual arrival time at $MP_2(a_{2k})$ and the cell's actual arrival time at $MP_1(a_{1k})$: $x_k = a_{2k} - a_{1k}$. The reference cell transfer delay $(d_{1,2})$ between MP_1 and MP_2 is the absolute cell transfer delay experienced by cell 0 between the two MPs.

5.5.3 Impact of satellite burst errors on the ATM layer

ATM was designed for transmission on a physical medium with excellent error characteristics, such as optical fibre, which has improved dramatically in performance since the 1970s. Therefore, many of the features included in protocols that cope with an unreliable channel were removed from ATM. While this results in considerable protocol simplification in the optical fixed networks ATM was designed for, it also causes severe problems when ATM is transmitted over an error-prone channel, such as the satellite, wireless and mobile networks.

The most important impact of burst errors on the functioning of the ATM layer is the dramatic increase in the cell loss ratio (CLR). The eight-bit ATM header error control (HEC) field in the ATM cell header can correct only single-bit errors in the header. However, in a burst error environment, if a burst of errors hits a cell header, it is likely that it will corrupt more than a single bit. Thus the HEC field becomes ineffective for burst errors and the CLR rises dramatically.

It has been shown by a simplified analysis and confirmed by actual experiments that for random errors, CLR is proportional to the square of the bit error rate (BER); and for burst errors, CLR is linearly related to BER. Hence, for the same BER, in the case of burst errors, the CLR value (proportional to BER) is orders of magnitude higher than the CLR value for random errors (proportional to the square of BER). Also, since for burst errors, CLR is linearly related to BER, the reduction in CLR with reduction in BER is not as steep as in the case of channels with random errors (where CLR is proportional to the square of BER). Finally, for burst errors, the CLR increases with decreasing average burst length. This

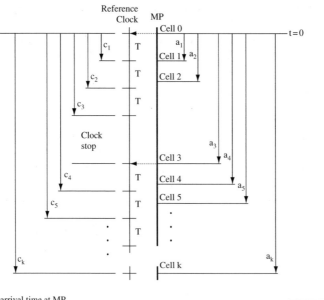

Variables:

a_k Cell k actual arrival time at MP

c_k Cell k reference arrival time at MP

y_k 1-point CDV

$$y_k = c_k - a_k$$

a) Cell delay variation – 1-point definition

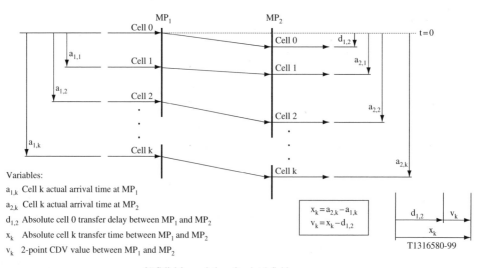

Variables:

$a_{1,k}$ Cell k actual arrival time at MP_1

$a_{2,k}$ Cell k actual arrival time at MP_2

$d_{1,2}$ Absolute cell 0 transfer delay between MP_1 and MP_2

x_k Absolute cell k transfer time between MP_1 and MP_2

v_k 2-point CDV value between MP_1 and MP_2

$$x_k = a_{2,k} - a_{1,k}$$
$$v_k = x_k - d_{1,2}$$

T1316580-99

b) Cell delay variation – 2-point definition

Figure 5.9 Cell delay variation parameter definitions (ITUT-1356) (Reproduced with the kind permission of ITU.)

is because for the same number of total bit errors, shorter error bursts mean that a larger number of cells are affected.

Another negligible but interesting problem is that of misinserted cells. Since eight HEC bits in the ATM cell header are determined by 32 other bits in the header, there are only 2^{32} valid ATM header patterns out of 2^{40} possibilities (for 40 ATM header bits). Thus for a cell header, hit by a burst of errors, there is a $2^{32}/2^{40}$ chance that corrupted header is a valid one. Moreover, if the corrupted header differs from a valid header by only a single bit, HEC will 'correct' that bit and accept the header as a valid one. Thus for every valid header bit pattern (out of 2^{32} possibilities), there are 40 other patterns (obtained by inverting one bit out of 40) that can be 'corrected'. The possibility that the 'error burst' hit the header in one of these patterns is $40 \times 2^{32}/2^{40}$. Thus overall, there is a $41 \times 2^{32}/2^{40}(= 41/256 \approx 1/6)$ chance that a random bit pattern, emerging after an ATM cell header is hit by a burst of errors, will be taken as a valid header. In that case a cell, that should have been discarded, is accepted as a valid cell. (Errors in the payload must be detected by the transport protocol at the end points.) Such a cell is called a 'misinserted' cell. Also, the probability P_{mi} that a cell will be misinserted in a channel with burst errors is around 1/6th of the cell loss ratio on the channel, i.e.,

$$P_{mi} \approx (1/6) \times CLR$$

Since CLR can be written as a constant times BER, the misinserted cell probability is also a constant times BER, i.e.,

$$P_{mi} = k \times BER$$

The cell insertion rate, C_{ir}, the rate at which cells are inserted in a connection, is obtained by multiplying this probability by the number of ATM cells transmitted per second (r), divided by total possible number of ATM connections (2^{24}), i.e.,

$$C_{ir} = (k \times BER \times r)/2^{24}$$

Because of the very large number of total possible ATM connections, the cell insertion rate is negligible (about one inserted cell per month) even for high BER ($\approx 10^{-4}$) and data rates ($\approx 34\,\text{Mbit/s}$). Therefore, transition from random errors to burst errors causes the ATM CLR metric to rise significantly.

5.5.4 Impact of burst errors on AAL protocols

The cyclic error detection codes employed by AAL protocols type 1, 3/4 and 5 are susceptible to error bursts in the same way as the ATM HEC code. A burst of errors that passes undetected through these codes may cause failure of the protocol's mechanism or corruption in data. AAL type 1's segmentation and reassembly (SAR) header consists of four bits of sequence number (SN) protected by a three-bit CRC code and a single-bit parity check.

There is a $15/255 = 1/17$ chance that an error burst on the header will not be detected by the CRC code and parity check. Such an undetected error at the SAR layer may lead to synchronisation failure at the receiver's convergence sublayer. AAL 3/4 uses a 10-bit CRC at the SAR level.

Here, burst errors and scrambling on the satellite channel increase the probability of undetected error. However, full byte interleaving of the ATM cell payload can reduce undetected error rate by several orders of magnitude by distributing the burst error into two AAL 3/4 payloads. The price to be paid for distributing burst error into two AAL payloads is doubling of the detected error rate and AAL 3/4 payload discard rate. AAL type 5 uses a 32-bit CRC code that detects all burst errors of length 32 or less. For longer bursts, the error detection capability of this code is much stronger than that of AAL 3/4 CRC. Moreover, it uses a length check field, which finds out loss or gain of cells in an AAL 5 payload, even when CRC code fails to detect it. Hence it is unlikely that a burst error in AAL 5 payload would go undetected.

It can be seen that ATM AAL 1 and 3/4 are susceptible to burst errors, as there are less redundant bits used for protections. AAL 5 is more robust against burst errors by using more redundant bits.

5.5.5 Error control mechanisms

There are three types of error control mechanisms: re-transmission mechanism, forward error control (FEC) and interleaving techniques to improve quality for ATM traffic over satellite.

Satellite ATM networks try to maintain BER below 10^{-8} in clear sky operation 99% of the time. The burst error characteristics of FEC-coded satellite channels adversely affect the performance of physical, ATM and AAL protocols. The interleaving mechanism reduces the burst error effect of the satellite links.

A typical example of FEC is to use an outer Reed–Solomon (RS) coding/decoding in concatenation with 'inner' convolutional coding/Viterbi decoding. Outer RS coding/decoding will perform the function of correcting error bursts resulting from inner coding/decoding. RS codes consume little extra bandwidth (e.g. 9% at 2 Mbit/s).

HEC codes used in ATM and AAL layer headers are able to correct single bit errors in the header. Thus, if the bits of N headers are interleaved before encoding and de-interleaved after decoding, the burst of errors will get spread over N headers such that two consecutive headers emerging after de-interleaving will most probably never have more than a single bit in error. Now the HEC code will be able to correct single bit errors and by dual mode of operation, no cell/AAL PDU will be discarded. Interleaving involves reshuffling of bits on the channel and there is no overhead involved. However, the process of interleaving and de-interleaving requires additional memory and introduces delay at both sender and receiver.

Burst errors can be mitigated by using FEC and 'interleaving' techniques. The performance of these schemes is directly related to the code rate (bandwidth efficiency) and/or the coding gains (power efficiency), provided the delay involved is acceptable to any ATM-based application.

5.5.6 Enhancement techniques for satellite ATM networks

In satellite ATM networks, we have to exploit the FEC coding and interleaving, and trade off between transmission quality in terms of bit error performance and satellite resources such as bandwidth and power:

- ATM was designed for transmission on a physical medium with excellent error characteristics, such as optical fibre. It has less overhead, by reducing error controls, but it also causes severe problems when ATM is transmitted over an error-prone channel, such as the satellite link.
- Satellite systems are usually power or bandwidth limited and in order to achieve reliable transmission FEC codes are often used in satellite modems. With such codes (typically convolutional codes), the incoming data stream is no longer reconstructed on a symbol-by-symbol basis. Rather some redundancy in the data stream is used.
- On average, coding reduces the BER or alternatively decreases transmission power needed to achieve a certain QoS for a given S/N ratio, at the expense of coding overhead. However, when decoding makes mistakes, in general a large number of bits is affected, resulting in burst errors. Because ATM was designed to be robust with respect to random single bit errors, burst errors can degrade the performance of ATM considerably.

Hence some enhancement techniques can be developed to make the transmission of ATM cells over the satellite link more robust. The performance of these techniques is directly related to the code rate (bandwidth efficiency) and/or the coding gain (power efficiency), provided the processing delay involved is acceptable to any ATM-based application.

For large earth stations operating at high data rates, the enhancement techniques try to deal with burst errors.

- By interleaving the ATM cell headers (not the payload) of several cells the performance of ATM in a random single bit error channel (e.g. AWGN channel) can be achieved. Note that interleaving merely reshuffles the bits on the channel (to spread the bit errors among ATM cell headers) and does not produce additional overhead which might decrease the overall bit rate. However, interleaving requires memory at the transmitter and the receiver, and it introduces additional delay. Assuming an average number of 30 bit errors in an error burst, interleaving over 100 cell headers seems to be sufficient. This requires a memory of only about 10 kbytes and introduces a delay of 840 μs at 50 Mbit/s and a delay of 21 ms at 2 Mbit/s. Since the above interleaving scheme requires a continuous data stream, there are problems using it for portable terminals where single ATM cells may be transmitted.
- Another way of correcting the burst errors due to FEC techniques applied to satellite links are Reed–Solomon (RS) codes. This type of block codes, which are based on symbols, have been identified as performing particularly well in concatenation with convolutional FEC codes, because of their ability to correct bursts of errors.
- Moreover, error bursts longer than what the RS code can correct should be spread over several blocks to take advantage of the error correction capabilities of the block code. This can be done by interleaving between the two codes.

For broadband small and portable terminals, rapid deployment and relocation are important requirements. The transmission bit rates can be up to but normally below 2.048 Mbit/s. Since inter-cell interleaving is not feasible because only a few cells may be transmitted from the terminal, mechanisms which protect single cells have to be found. Interleaving within an entire ATM cell (not only the header), so-called intra-cell interleaving, leads to a performance gain which is too small to be effective.

It can be improved by using additional coding to protect the ATM cells. Note that this introduces additional overheads and therefore reduces the useful data bit rate. There are several reasons why FEC or concatenated FEC may not be suitable for enhancing ATM performance over wideband satellite links. First, if only FEC coding is used, than symbol interleaving is usually used to spread the burst errors over several ATM cell headers. The resulting interleaving delay (which is inversely proportional to the data rate) may be too large at a low rate for certain applications. Second if RS codes are used to correct burst of errors in concatenation with FEC either additional bandwidth has to be provided or the data rate has to be reduced.

It is also possible to improve ATM performance by enhancing equipment which optimises the ATM protocols over a satellite link. This allows the data link layer to be optimised using a combination of protocol conversions and error control techniques. At the transmitter, standard ATM cells are modified to suit the satellite link. At the receiver, error recovery techniques are performed and the modified ATM cells (S-ATM cells) are converted into standard ATM cells.

The main aim of modifying standard ATM cell is to minimise the rather large ATM header overhead which is 5 bytes per 48 byte payload. Of the ATM header information, the address field (which is divided into the VPI and VCI) occupies 24 bits. This allows up to 16 million VC to be set up. Considering that in particular CBR connection cells all carry the same address information in the header, there may be methods not to duplicate the same information. The use of 24 bits for address space may be considered a waste of bandwidth for this scenario.

One method to protect the ATM cell header is, when interleaving is not possible, to compress the 24 bits address space to eight bits so that the saved bits can be used to store the duplicate header information (except the HEC field) of the previous cell. The HEC is still computed over the first four bytes of the header and inserted into the fifth byte of the header. Therefore if a cell header contains errors, the receiver can store the payload in a buffer and recover the header information from the next cell provided that its header does not also contain errors. This method does not intend to protect payload. Studies show that this method provides considerable improvements in CLR compared to standard ATM transmission and even compared to interleaving.

Another alternative is to use three-byte HEC instead of one-byte HEC, which is inadequate for the satellite environment.

5.6 Evolution of ATM satellite systems

While fibre optics is rapidly becoming the preferred carrier for high bandwidth communication services, satellite systems can still play an important role in the B-ISDN. The satellite network configuration and capacity can be increased gradually to match the increasing B-ISDN traffic during the evolution toward broadband communications.

The role of satellites in high-speed networking will evolve according to the evolution of the terrestrial ATM based networks. However, two main roles can be identified in two scenarios of the broadband network development:

- The initial phase when satellites will compensate the lack of sufficient terrestrial high bit rate links mainly by interconnecting a few regional or national distributed broadband networks, usually called 'broadband islands'.
- The maturation phase when the terrestrial broadband infrastructure will have reached some degree of maturity. In this phase, satellites are expected to provide broadcast service and also cost-effective links to rural areas complementing the terrestrial network. At this phase satellite networks will provide broadband links to a large number of end users through a UNI for accessing broadband networks. This allows high flexibility concerning topology, reconfiguration and network expansion. Satellites are also ideal for interconnecting mobile sites and provide a back-up solution in case of failure of the terrestrial systems.

In the first scenario, satellite links provide high bit rate links between broadband nodes or broadband islands. The CATALYST demonstrator provided an example for this scenario and considerations for compatibility between satellite and terrestrial networks. The interfaces with satellite links in this mode are of the NNI type. This scenario is characterised by a relatively small number of large earth stations, which have a relatively large average bit rate.

In the second scenario the satellite can also be located at the border of broadband networks to provide access links to a large number of users. This scenario is characterised by a large number of earth stations whose average and peak bit rates are limited. The traffic at the earth station is expected to show large fluctuations. Dynamic bandwidth allocation mechanisms are used for flexible multiple access.

The problem for efficient use of satellite resources is due to the unpredictable nature of burst traffic and the long delay of the satellite link to reallocate and manage satellite resources. More research has to be carried out on efficient multiple access schemes for satellite systems. The use of OBP satellites with cell-switching capabilities and spot beams would half this delay and bring several advantages for interconnecting a high number of users. By using on-board cell switching the utilisation of the satellite bandwidth can be maximised by statistically multiplexing the traffic in the sky.

The use of GEO satellites to deliver ATM services has proven feasible. However, delivery of high bit rate ATM services to transportable or mobile terminals via satellite requires low delays, low terminal power requirements and high minimum elevation angles. It is a natural evolution path to exploit satellites at much lower altitudes such as MEO and LEO orbit heights. Satellites at these lower altitudes have much smaller delays and lower terminal power requirements than satellites in GEO orbit. Research is still going on to find the most suitable orbit and multiple access schemes to deliver broadband services to small portable and mobile terminals.

The major factor affecting the direction of satellite broadband networking comes from terrestrial networks where networks are evolving towards all-IP solutions. Therefore, it is a logical step to investigate IP routers on board satellites.

Further reading

[1] ITU-T Recommendation I.150, *B-ISDN ATM Functional Characteristics*, November 1995.

[2] ITU-T Recommendation I.211, *B-ISDN Service Aspects*, March 1993.

[3] ITU-T Recommendation I.356, *On B-ISDN ATM Layer Cell Transfer Performance*, October 1996.

[4] ITU-T Recommendation I.361, *ITU-T 'B-ISDN ATM Layer Specification*, November 1995.

[5] ITU-T Recommendation I.371, *Traffic Control and Congestion Control in B-ISDN*, May 1996.

[6] ITU-T Recommendation G826, 'Error performance parameters and objectives for international constant bit rate digital paths at or above the primary rate', 02/1999.

[7] Ors, T., 'Traffic and congestion control for ATM over satellite to provide QoS', PhD thesis, University of Surrey, 1998.

[8] RACE CFS, Satellites in the B-ISDN, general aspects, *RACE Common Functional Specifications D751*, Issue D, December 1993.

[9] Sun, Z., T. Ors and B.G. Evans, Satellite ATM for broadband ISDN, *Telecommunication Systems*, 4:119–31, 1995.

[10] Sun, Z., T. Ors and B.G. Evans, ATM-over-satellite demonstration of broadband network interconnection, *Computer Communications, Special Issue on Transport Protocols for High Speed Broadband Networks*, **21**(12), 1998.

Exercises

1. Explain the design issues and concepts concerning ATM over satellites.
2. Explain the CATALYST GEO satellite ATM networking and advanced satellite networking with LEO/MEO constellations.
3. Use a sketch to explain the major roles of satellites in broadband networks with ATM over satellite networking and also the protocol stacks of the broadband network interconnection and terminal access configurations.
4. Explain the differences between satellites with transparent and on-board switching payload for ATM networks, and discuss advantages and disadvantages.
5. Explain ATM performance issues and enhancement techniques for satellite ATM networks.
6. Explain different on-board processing and on-board switching techniques, and discuss their advantages and disadvantages.
7. Discuss the advantages and disadvantages of ATM networks based on GEO, MEO and LEO satellites.

6

Internet Protocol (IP) over Satellite Networks

This chapter aims to provide an introduction to the Internet protocol (IP) over satellite networks. It explains satellite networking from different viewpoints: protocol centric, network centric and satellite centric. It also explains: how to encapsulate IP packet into different frames of different network technologies; IP extensions including IP multicast, IP security and IP QoS; the concepts of DVB over satellite (DVB-S and DVB-RCS); and IP QoS architectures. When you have completed this chapter, you should be able to:

- Understand the concepts of satellite IP networking.
- Understand IP packet encapsulation concepts.
- Describe different views of satellite networks.
- Describe IP multicast over satellite.
- Explain DVB and related protocol stack.
- Explain DVB over satellite including DVB-S and DVB-RCS.
- Explain IP over DVB-S and DVB-RCS security mechanisms.
- Knows IP QoS performance objectives and parameters and QoS architectures of Intserv and Diffserv.

6.1 Different viewpoints of satellite networking

Like terrestrial networks, satellite networks are increasingly carrying more and more Internet traffic, which now exceeds telephony traffic. Currently, Internet traffic is mainly due to classical Internet services and applications, such as WWW, FTP and emails. Satellite networks only need to support the classical Internet network applications in order to provide traditional best-effort services.

Satellite Networking: Principles and Protocols   Zhili Sun
© 2005 John Wiley & Sons, Ltd

The convergence of the Internet and telecommunications led to the development of voice over IP (VoIP), video conference over IP and broadcasting services over IP. Therefore, IP packets are expected to carry additional classes of services and applications over satellite networks, requiring quality of service (QoS) from IP networks. Much research and development have been carried out in satellite networking to support the new real-time multimedia and multicast applications requiring QoS.

IP has been designed to be independent of any network technology so that it can be adapted to all available networking technologies. For satellite networks, there are three of the satellite networking technologies concerning IP over satellites:

- *Satellite telecommunication networks* – these have provided traditional satellite services (telephony, fax, data, etc.) for many years, and also provide Internet access and Internet subnet interconnections by using point-to-point links.
- *Satellite shared medium packet networks based on the very small aperture terminal (VSAT) concept* – these have supported transaction types of data services for many years, and are also suitable for supporting IP.
- *Digital video broadcasting (DVB)* – IP over DVB via satellite has the potential to provide broadband access for global coverage. DVB-S provides one-way broadcasting services. User terminals can only receive data via satellite. For Internet services, the return links are provided using dial-up links over telecommunication networks. DVB-RCS provides return links via satellite so that user terminals can access the Internet via satellite. This removes all constraints due to return links over terrestrial telecommunications networks, hence allowing great flexibility and mobility for the user terminals.

6.1.1 Protocol-centric viewpoint of satellite IP network

The protocol-centric viewpoint of satellite IP networks emphasises the protocol stack and protocol functions in the context of the reference model. Figure 6.1 illustrates the relationship between IP and different network technologies. IP provides a uniform network hiding away all differences between different technologies; different networks may transport IP packets in different ways.

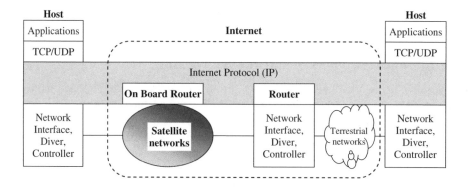

Figure 6.1   Relationship between IP and different network technologies

Satellite networks include connection-oriented networks, shared medium point-to-multipoint connectionless networks, broadcasting networks for point-to-point communications and point-to-multipoint communications. Terrestrial networks include LAN, MAN, WAN, dial-up, circuit networks and packet networks. LAN is often based on a shared medium and WAN point-to-point connections.

6.1.2 Satellite-centric viewpoint of global networks and the Internet

The satellite-centric viewpoint emphasises the satellite network itself, i.e. the satellite (GEO or non-GEO) is viewed as a fixed infrastructure and all ground infrastructures are viewed in relation to the satellite. Figure 6.2 illustrates a satellite-centric viewpoint of global networks. Figure 6.3 shows mapping from the earth-centric viewpoint to a GEO satellite-centric viewpoint of earth and LEO satellites ($\vec{O_G} = \overline{OO}_G$ is vector from O to O_G the location of the GEO satellite, and $\left|\vec{r}\right| = R_G$ is the GEO orbit with radius of R_G) that the earth surface and satellite orbits can be expressed as:

$$\vec{\gamma}^2 = \left(\frac{\left(\vec{r} - \vec{O_G}\right)^2}{2R_G} - 1 \right) (R_G - R_E)$$

where R_E is the radius of earth, and

$$\vec{\gamma}^2 = \left(\frac{\left(\vec{r} - \vec{O_G}\right)^2}{2R_G} - 1 \right) (R_G - R_L)$$

where R_L is the radius of the LEO satellite orbit.

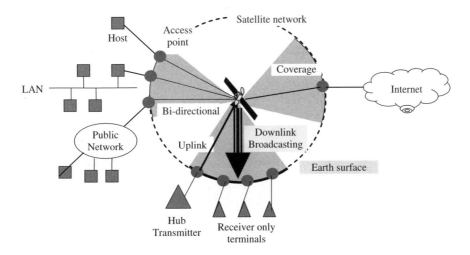

Figure 6.2   Satellite-centric viewpoint of global networks

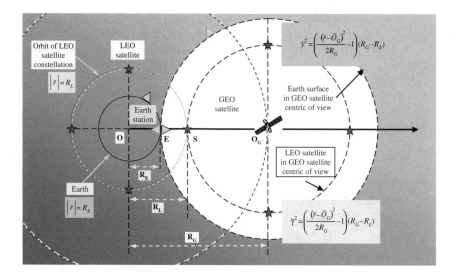

Figure 6.3 Mapping from earth-centric view to GEO-centric view

To support IP, the satellite network must support data frames to carry IP packets across the network technology. The router takes the IP packet from frames of one type of network and repackages the IP packet into frames of another type of network to make them suitable for transmission in the network technologies.

6.1.3 Network-centric viewpoint of satellite networks

Satellite systems and technologies concern two aspects: the ground segment and the space segment. In the space segment (satellite communication payload), various types of technology can be used including transparent (bent-pipe) transponder, on-board processor, on-board circuit switch, on-board packet switch (also possible ATM switch), on-board DVB-S or DVB-RCS switch or IP router. The network-centric view of satellite networks emphasises networking functions rather than satellite technologies. However, users see different types of networks and logical connections rather than satellite technologies and physical implementations. Figure 6.4 shows a network-centric view of satellite networks.

Figure 6.4 Network-centric view of satellite networks

All these additional functions increase the complexity of the satellite payload capable of supporting multiple spot-beam 'star' (point-to-multipoint centred on a gateway earth station) and 'mesh' (multipoint-to-multipoint) topologies, hence the risk of failure, but they also provide great benefit of optimised use of bandwidth and power resources. Future satellites with on-board DVB switching will be able to integrate broadcast and interactive services by combining DVB-S and DVB-RCS standards. A DVB-S regenerative payload can multiplex information from diverse sources into a standard downlink DVB-S stream. Another example of the use of DVB on-board switching is to interconnect LANs using IP over MPEG-2 encapsulation, via a regenerative satellite payload.

Implementation of these functions depend on the demands of network operators and secure manufacturing to produce reliable and cost-effective satellites.

6.2 IP packet encapsulation

IP packet encapsulation is an aspect for IP over any network technology. It is a technique used to encapsulate an IP packet into the data frame, so that it is suitable for transmission over the network technology. Different network technologies may also use different frame formats, frame sizes or bit rate for transporting IP packets. IP packet encapsulation puts the packet into the payload of a data link layer frame for transmission over the network. For example, Ethernet, token ring and wireless LANs have their own standard frame formats to encapsulate IP packet.

6.2.1 Basic concepts

Due to different framing formats, different encapsulation techniques may be used. Sometimes, an IP packet may be too large to fit into the frame payload. In such a case, the IP packet has to be broken up into smaller segments (fragmented) so that the IP packet can be carried over several frames. In this case, additional overhead is added to each of the segments so that on arriving at the destination, the original IP packet can be reassembled from the segments. It can be seen that the encapsulation process may have a significant impact on network performance due to the additional processing and overhead. Figure 6.5 illustrates the concept of encapsulation of IP packets.

6.2.2 High-level data link control (HDLC) protocol

HDLC is an international standard of layer 2 (link layer) protocols. It is an important and also widely used layer 2 protocol. It defines three types of stations (standard, secondary and

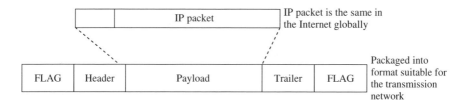
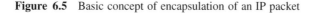

Figure 6.5   Basic concept of encapsulation of an IP packet

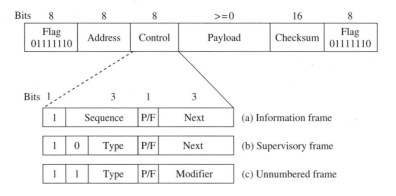

Figure 6.6 HDLC frame structure

combined), two link configurations (balanced and unbalanced) and three data transfer modes (normal response, asynchronous response, asynchronous balanced response). Figure 6.6 shows the HDLC frame structure.

It is bit oriented based on a bit-stuffing technique, and consists of two flags of the eight-bit pattern 01111110 to identify the start and end of the frame, an eight-bit address field to identify multiple terminals, an eight-bit control field to be used for three types of frames (information, supervision and unnumbered), a payload field to carry data (network layer data including IP packet) and 16 bits for CRC error check.

6.2.3 Point-to-point protocol (PPP)

The HDLC frame is adapted for the point-to-Point protocol (PPP), which is the Internet standard widely used for dial-up connections. The PPP handles error detection, supports multiple protocols in addition to IP, allows addresses be negotiated at connection time and permits authentication. Figure 6.7 shows the frame structure of the PPP.

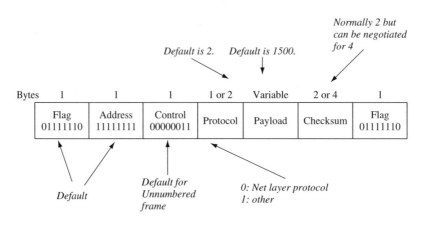

Figure 6.7 Frame structure of the point-to-point protocol (PPP)

6.2.4 Media access control

HDLC and PPP were designed for transmission over point-to-point connection media. For a network of shared media, an additional layer, known as the media access control (MAC) sublayer of the link layer, is used to connect a large number of stations into the network. Without giving full details, Figure 6.8 illustrates a format of MAC frame.

6.2.5 IP over satellite

To support IP over satellites, the satellite networks need to provide a frame structure so that the IP datagram can be encapsulated into the frame and transported via satellite from one access point to other access points. In a satellite environment, the frame can be based on standard data link layer protocols. Encapsulation of IP is also defined in the existing networks, such as dial-up link, ATM, DVB-S and DVB-RCS, which support Internet protocols or interwork with the Internet. ATM networks use ATM adaptation layer type 5 (AAL5) to encapsulate IP packets for transmission over the ATM network, and in DVB-S, IP packets including multicast are encapsulated in an Ethernet-style header using a standard called multi-protocol encapsulation (MPE).

It is also possible to encapsulate an IP packet into another IP packet, i.e., to create a tunnel to transport the IP packets of one Internet across another Internet network.

6.3 Satellite IP networking

The particular benefits provided by satellites include their geographically extended global coverage (including land, sea and air), their efficient delivery to a large number of users on a large scale, and the low marginal cost of adding additional users. A satellite can play several different roles in the Internet:

- Last mile connections (as shown in Figure 6.9): user terminals directly access the satellite, which provides direct forward and return links. Traffic sources connect to the satellite feeder or hub stations through the Internet, tunnelling or dial-up links. It is the last mile to reach user terminals.
- Transit connections (as shown in Figure 6.11): the satellite provides connections between Internet gateways or ISP gateways. The traffic is routed through the satellite links according to specified routing protocols and defined link metrics in the networks so as to minimise connection costs and to meet required QoS constraints for the given traffic sources.
- First mile connections (as shown in Figure 6.10): the satellite provides forward and return link connections directly to a large number of ISPs. IP packets start from the servers as the first mile of their journey to user terminals. As with the last mile connections, the server

MAC control	Destination MAC address	Sources MAC address	Payload for encapsulated IP packet	Checksum

Figure 6.8 Format of a MAC frame

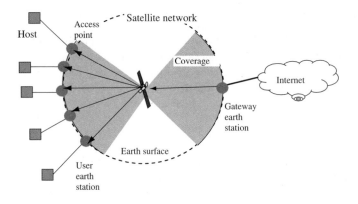

Figure 6.9   Satellite-centric view of last mile connections to the Internet

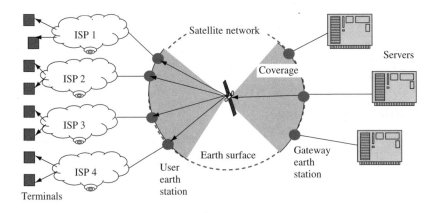

Figure 6.10   Satellite-centric view of first mile connections to the Internet

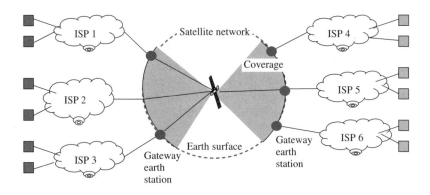

Figure 6.11   Satellite-centric view of transit connections to the Internet

can connect to the satellite feeder or hub stations directly or through the Internet tunnelling or dial-up links.

6.3.1 Routing on board satellites

The benefit of an IP router in the sky is that it allows satellite networks to be integrated into the global Internet using the standard routing algorithms. The Internet consists of a collection of subnetworks, also known as autonomous systems (AS) or domains.

In GEO satellite networks, there is normally only one satellite covering a large area to form a subnet; there is no routing within the satellite network. With a constellation, there are many satellites forming a subnet to cover the earth. Therefore, routing within the constellation satellite network is required. The link relationships between the satellites within the same orbit plane are fixed, but change dynamically in different planes. Since the locations of all the satellites in the orbits are predictable, it is possible to make use of the predictions to dynamically update the routing tables on board satellites and enhance the routing algorithm.

6.3.2 IP mobility in satellite networks

Due to the wide coverage of GEO satellites, we can consider that terrestrial networks are connected to the same satellite subnet permanently, and to user terminals during communication sessions. However, for a network with a constellation of LEO satellites the relationships between the satellite network and user terminals and terrestrial networks are changing continuously. Therefore, there are several issues concerned with mobility:

- Re-establishing the physical connections with the satellite networks.
- Timely updating the routing table so that IP packets can be routed to the right destination.
- Mobility within the satellite networks.
- Mobility between terrestrial networks and satellite networks.

Here, discussions are based on the Internet standard protocol for mobile IP (RFC 2002). In the standard solution, it allows the mobile node to use two IP addresses: a fixed home address and a care-of address, which changes at each new point of attachment. We take the satellite-centric view that the satellite network is fixed but everything on the earth is moving including user terminals and the terrestrial network as shown in Figure 6.12.

In the IP mobility standard, existing transport layer connections are maintained as the mobile node moves from one place to another; IP addresses remain the same. Most of the Internet applications used today are based on TCP. A TCP connection is indexed by a quadruplet of *source IP addresses*, *destination IP addresses*, *source port number* and *destination port number*. Changing any of these four numbers will cause the connection to be disrupted and lost. On the other hand, correct delivery of packets to the mobile node's current point of attachment depends on the network number contained within the mobile node's IP address, which changes at new points of attachment.

In mobile IP, the home address is static and is used, for instance, to identify TCP connections. The care-of address changes at each new point of attachment and can be thought

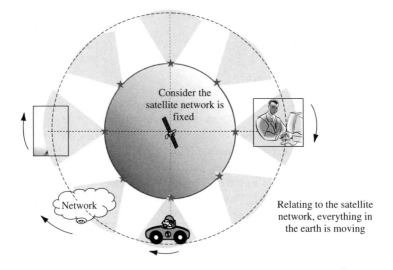

Figure 6.12   Satellite-centric view of fixed satellites with earth moving

of as the mobile node's topologically significant address; it indicates the network number and thus identifies the mobile node's point of attachment with respect to the network topology. The home address makes it appear that the mobile node is continually able to receive data on its home network, where mobile IP requires the existence of a network node known as the home agent. Whenever the mobile node is not attached to its home network (and is therefore attached to what is termed a foreign network), the home agent gets all the packets destined for the mobile node and arranges to deliver them to the mobile node's current point of attachment.

When the mobile node moves to a new place, it registers its new care-of address with its home agent. To get a packet to a mobile node from its home network, the home agent delivers the packet from the home network to the care-of address. The further delivery requires to the care-of address to transform or redirect the IP packet. When the packet arrives at the care-of address, the reverse transformation is applied so that the packet once again appears to have the mobile node's home address as the destination IP address.

When the packet arrives at the mobile node, addressed to the home address, it will be processed properly by TCP.

In mobile IP the home agent redirects packets from the home network to the care-of address by constructing a new IP header that contains the mobile node's care-of address as the destination IP address. This new header then shields or encapsulates the original packet, causing the mobile node's home address to have no effect on the encapsulated packet's routing until it arrives at the care-of address. Such encapsulation is also called tunnelling, which bypasses the usual effects of IP routing.

Mobile IP, then, is best understood as the cooperation of three separate mechanisms:

- Discovering the care-of address: agent advertisement and agent solicitation (RFC1256).
- Registering the care-of address: the registration process begins when the mobile node enters coverage of a foreign agent, sends a registration request with the care-of address

information. When the home agent receives this request, it (typically) adds the necessary information to its routing table, approves the request, and sends a registration reply back to the mobile node. The registration is authenticated by using Message Digest 5 (RFC1321).

- Tunnelling to the care-of address: the default encapsulation mechanism that must be supported by all mobility agents is IP-within-IP (tunnelling). Minimal encapsulation is slightly more complicated than tunnelling, because some of the information from the tunnel header is combined with the information in the inner minimal encapsulation header to reconstitute the original IP header. On the other hand, header overhead is reduced.

6.3.3 Address resolution

Address resolution is also called address mapping and configuration. Different network technologies may use different addressing schemes for assigning addresses, also called physical addresses, to devices. For example, an IEEE 802 LAN uses a 48-bit address for each attached device, an ATM network may use 15-digit decimal address and ISDN uses the ITU-T E.164 address scheme. Similarly, in a satellite network each ground earth station or gateway station has a physical address for circuit connections or packet transmissions. However, the routers that are interconnected by the satellite network know only the IP addresses of the other routers. Therefore, address mapping between each IP address and its related physical address is required, so that packet exchanges between the routers can be carried out through the satellite network using the physical addresses. The precise details of this mapping depend on the underlying data link layer protocols used over the satellite.

6.4 IP multicast over satellite

The success of satellite digital broadcast services (for TV and radio) and the asymmetric nature of IP traffic flow have been combined resulting in satellite systems that support high-speed Internet access. From here, it is a natural step to consider further exploiting satellites' broadcast capability by investigating IP multicast over satellites. Satellite networks can be part of an IP multicast routing tree at the source, trunk or end branch forwarding IP packets towards their destination. Figure 6.13 illustrates an example of star and mesh topologies used in the GEOCAST project on IP multicast over GEO satellite funded within the EU 5th framework programme.

6.4.1 IP multicast

We now proceed to review IP multicast technology. Multicast allows a communications network source to send data to multiple destinations simultaneously while transmitting only a single copy of the data to the network. The network then replicates the data and fans it out to recipients as necessary. Multicast can be considered as part of a spectrum of three types of communications:

- Unicast: transmitting data from a single source to a single destination (for example, downloading a web page from a server to a user's browser, or copying a file from one server to another).

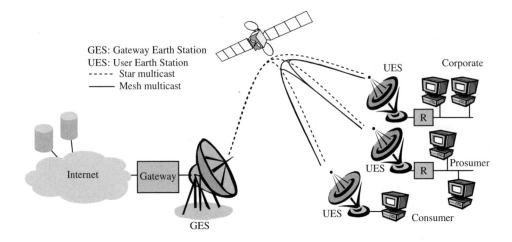

Figure 6.13 The GEOCAST system as an example of star and mesh topologies

- Multicast: transmitting data from a single source to multiple destinations. The definition also encompasses communications where there may be more than one source (i.e. multipoint-to-multipoint). Videoconferences provide an example of the latter, where each participant can be regarded as a single source multicasting to the other participants in the videoconference.
- Broadcast: transmitting data from a single source to all receivers within a domain (e.g. within a LAN, or from a satellite to all receivers within a satellite spot beam).

The advantages of multicast are as follows:

- Reduced network bandwidth usage: for example, if data packets are being multicast to 100 recipients the source only sends a single copy of each packet. The network forwards this to the destinations, only making multiple copies of the packet when it needs to send packets on different network links to reach all destinations. Thus only a single copy of each packet is transmitted over any link in the network, and the total network load is reduced compared to 100 separate unicast connections. This is particularly beneficial on satellite systems where resources are limited and expensive.
- Reduced source processing load: the source host does not need to maintain state information about the communications link to each individual recipient.

Multicast can be either best effort or reliable. 'Best effort' means that there is no mechanism to guarantee that the data sent by any multicast source is received by all or any receivers, and is usually implemented by a source transmitting UDP packets on a multicast address. 'Reliable' means that mechanisms are implemented to ensure that all receivers of a multicast transmission receive all the data that is sent by a source: this requires a reliable multicast protocol.

6.4.2 IP multicast addressing

Each terminal or host in the Internet is uniquely identified by its IP address. In IP Version 4, an IP address has 32 bits, divided into a network number and a host number, which respectively identify a network and the terminal attached to the network. A normal unicast IP datagram includes a source address and destination address in the IP packet header; routers use the destination address to route the packet from the source to the destination. Such a mechanism cannot be used for multicast purposes, since the source terminal may not know when, where and which terminals will try to receive the packet.

Consequently, a range of addresses is set aside for multicast purposes only. The range of addresses, called Class D addresses, is from 224.0.0.0 to 239.255.255.255. Unlike Classes A, B and C, these addresses are not associated with any physical network number or host number, but instead are associated with a multicast group that is like a radio channel; members of the group receive multicast packets sent to this address, and the address is used by multicast routers to route IP multicast packets to users that register for a multicast group. The mechanism by which a terminal registers for a group, IGMP, is described below.

6.4.3 Multicast group management

In order to make efficient use of network resources, the network sends multicast packets only to those networks and subnets that have users belonging to the multicast group. The Internet group membership protocol (IGMP) allows hosts or terminals to declare an interest in receiving a multicast transmission. IGMP supports three main types of message: report, query and leave.

A terminal wishing to receive a multicast transmission issues an IGMP join report, which is received by the nearest router. This report specifies the IP multicast Class D address of the group being joined. The router then uses a multicast routing protocol (described below) to determine a path to the source. To confirm the state of terminals receiving multicast, a router occasionally issues an IGMP query to terminals on its network/subnetwork. When a terminal receives such a query, it sets a separate timer for each of its (potentially many) group memberships. When each timer expires, the terminal issues an IGMP report to confirm that it still wishes to receive the multicast transmission. However, in order to suppress duplicate reports for the same Class D group address, if a terminal has already heard a report for that group from another terminal it stops its timer and does not send a report. This has the benefit of avoiding flooding the subnetwork with IGMP reports.

When a terminal wishes to finish receiving the multicast transmission it issues an IGMP leave request. The leave message is supported in IGMP Version 2. In Version 1, a host or terminal quietly changes its state to non-member, and no message is sent to the router. If all the members of a group in a subnet have left, the router does not forward any more multicast packets to that subnet.

6.4.4 IP multicast routing

In a normal IP router used for unicast, the routing table contains information that specifies paths that lead to a given IP destination addresses. However, this routing table is not useful for IP multicast since multicast packets do not contain information about the location of the packet's destinations. Therefore different routing protocols and routing tables have to be used. Multicast routing protocols address the issue of identifying a route for data to be transmitted across a network from a source to all its destinations, while minimising the total network resources required for this.

In IP multicast, the routing table is effectively built from destinations to the sources rather than from sources to destinations, since only the source address in the IP datagram corresponds to a single physical location. Tunnelling techniques may also be used to support multicast over routers that do not have multicast capabilities.

A number of multicast routing protocols have been developed by the IETF. These include multicast extensions to OSPF (M-OSPF), distance vector multicast routing protocol (DVMRP), protocol-independent multicast-sparse mode (PIM-SM) and PIM dense mode (PIM-DM), and core-based tree (CBT).

Here we briefly review the underlying principle of operation of two protocols. DVMRP and PIM-DM are 'flood and prune' algorithms: in these protocols, when a source starts sending data, the protocols flood the network with the data. All routers that have no multicast recipients attached send a prune message back towards the source (they know they have no receivers because they have received no IGMP join reports). These protocols have the disadvantage that a 'prune' state is required in all routers (i.e. 'I have pruned on this multicast address'), including those routers with no multicast recipients downstream.

Flood and prune protocols use reverse path forwarding (RPF) to forward multicast packets from a source to the recipients: the RPF interface for any packet is the interface that the router would use to send unicast packets to the packet source (Figure 6.14 illustrates this principle in a terrestrial network). If a packet arrives on the RPF interface it is flooded to all other interfaces (unless they have been pruned), but if the packet arrives on any other interface it is silently discarded. This ensures efficient flooding and prevents packet looping.

DVMRP uses its own routing table to compute the best path to the source, whereas PIM-DM uses an underlying unicast routing protocol.

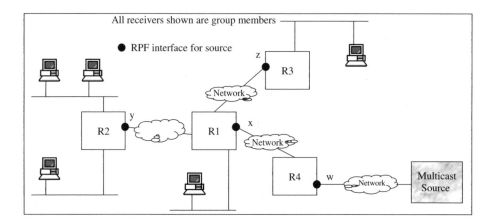

Figure 6.14   RPF terrestrial example

6.4.5 IP multicast scope

Scoping is the mechanism that controls the geographical scale of a multicast transmission, by making use of the time to live (TTL) field in the IP header. It tells the network how far (in terms of router hops) any IP packet is allowed to propagate, allowing IP multicast sources to specify whether packets should be sent only to the local subnetwork, or to larger domains or the whole Internet. This is achieved by each router reducing the TTL by 1 when forwarding the packet to the next hop, and discarding the packet if the TTL is 0. Each subnet may additionally have filters or a firewall to discard some packets according to its security policy, which may be beyond the control of the multicast source.

It can be seen in a satellite network that even with a small TTL value, IP multicast packets can reach a very large number of members of a multicast group scattered over a very large geographical area.

6.4.6 IGMP behaviour in satellite environments

In a satellite environment, multicast group management together with the scoping mechanism may provide an efficient solution to support IP multicast with large numbers of users distributed over a large area. However, IGMP over satellites raises interoperability issues, as we now describe.

In a conventional terrestrial LAN, an IGMP report is heard by other multicast receivers on the LAN, and this prevents flooding of the LAN with multiple reports. In a satellite system, individual ground stations cannot hear each other; given the large number of multicast receivers that are expected in satellite systems (potentially of the order of 10^5 or 10^6) multiple IGMP reports could cause significant flooding of the satellite network with IGMP traffic. One of a number of adaptations of IGMP and multicast must therefore be implemented. Two options are as follows, illustrated with an example of multicast from an uplink gateway earth station out to multiple end-user terminals each with a router as shown in Figure 6.15:

- Multicast channels can be statically configured to be transmitted across the satellite link to each downlink router, with IGMP traffic only operating between a router and the end-user terminal as shown in Figure 6.15(a). There is no transmission of IGMP traffic across the air interface in this case. This is a simple option, but potentially wastes scarce satellite channel capacity if there are no listeners on a particular multicast channel within any spot beam.
- Multicast channels are (as in conventional terrestrial networks) only transmitted across the satellite link if there is one or more listening end user. IGMP messages are transmitted across the air interface. When the uplink router receives an IGMP report from a terminal following an IGMP query, either the router must retransmit the IGMP report via the satellite to all ground stations to avoid flooding, or else other receivers will also transmit IGMP reports resulting in flooding as shown in Figure 6.15(b).

In architectures that have no router on the downlink side, IGMP 'snooping' can be used to forward multicast traffic to group members while avoiding transmission of IGMP traffic over the air interface.

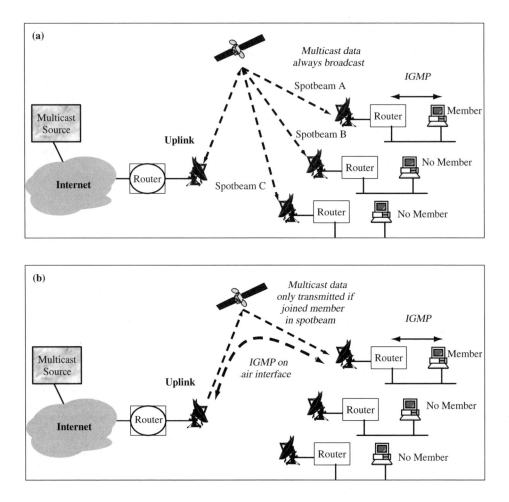

Figure 6.15   IGMP over satellite: (a) static and (b) dynamic multicast

A satellite system that dynamically allows multicast from any user becomes still more complex. For example, for a satellite with an on-board ATM switch, with retransmission of IGMP messages enabled, separate point-to-multipoint virtual circuits (VCs) would need to be established, sourced at each ground station within a satellite spot beam.

6.4.7 Multicast routing protocols in a satellite environment

We illustrate the issues in transmitting multicast routing protocols across a satellite with two examples based on multicast interior gateway routing protocols.

In the first example, we consider a flood and prune algorithm (such as is used in DVMRP or PIM-DM). When a source starts to transmit, the data is flooded across the network, as shown in Figure 6.14 for a terrestrial network. In Figure 6.16(a) the underlying data link layer supports a point-to-multipoint connection (for example, ATM), and the data from the source is correctly flooded out from router R4 to routers R1, R2 and R3. This requires a point-to-multipoint circuit from every such source on the multicast group; this could be expensive in the case of a large multicast group dynamically configured so that every satellite terminal can potentially transmit data from a data source. On the other hand, in Figure 6.16(b) the source transmits through router R4 to the uplink gateway router R1. This router then has to flood the data back out through its RPF interface in order to multicast to routers R2 and

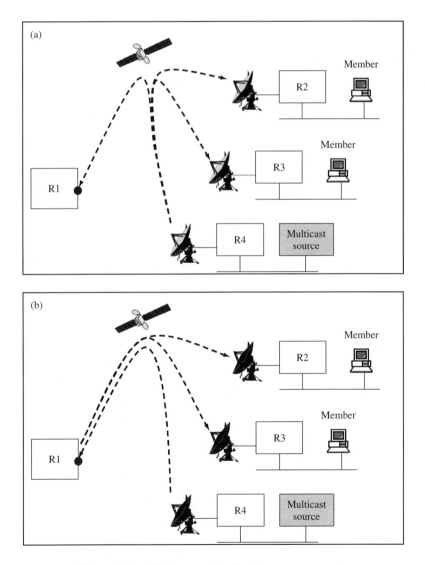

Figure 6.16 Multicast routing flooding: two approaches

R3. This is in contravention of the normal RPF algorithm, and requires modification of the routing algorithm.

In the second example, we consider the CBT multicast routing protocol. This protocol creates a tree that joins receiving members of the multicast group. When a source transmits to the group, the data is forwarded by all network routers until it reaches either the core of the tree or a router on the multicast tree. The tree then propagates the data both out to its downstream leaves and back up into the core. In general therefore, the tree carries multicast traffic in both directions, depending on where the data from the source first reaches the tree. However, satellite links with terrestrial return paths have different forward and return path routes, and so are not suitable for such bi-directional multicast routing protocols.

6.4.8 Reliable multicast protocols over satellites

Reliable multicast protocols address the issue of ensuring that data is multicast from a source to all the multicast recipients and that each packet sent by the source is successfully received by all recipients. Reliable multicast protocols usually also ensure ordered and non-duplicated delivery of packets. Since they provide an end-to-end service they are conventionally regarded as transport layer protocols in the context of the OSI reference model.

A wide range of reliable multicast protocols has been developed and described in the literature. One reason for this is that efficient multicast is a much more complex problem than efficient unicast, and consequently many multicast protocols have been developed for specific classes of application. Two examples of different application classes are real-time applications (requiring low delay with moderate packet loss acceptable) and multicast file transfer (requiring zero packet loss, but delay-insensitive), each of which has its own specific multicast requirements. These are different examples of the conversational services.

Two of the principal drawbacks associated with satellite links are their error characteristics and the round-trip delay, particularly in geostationary links. Historically, satellite links have had high bit error rates, and in addition the channel coding typically used on satellites to maintain a low bit error rate means that errors tend to occur in bursts. The consequent corruption of data means that when there are a large number of multicast end-users there is a significant probability that one or more recipients will not receive the data; this has implications for the design of reliable multicast network protocols.

The high round-trip delay times, especially of satellites in geostationary orbits, is well known to have an adverse impact on two-way real-time communications (for example, telephone conversations or videoconferences), and also affects the behaviour of network protocols such as TCP. In addition, it also needs to deal with transmission errors, acknowledgement and security. A number of mitigating techniques for TCP traffic have been developed for satellite networking. However, no corresponding standard mechanisms have yet been developed for reliable multicast protocols.

In summary, developing reliable multicast protocols and optimising them, particularly for scalability, throughput, flow control and congestion control, is an ongoing research issue both for terrestrial networks and for networks that include satellite links.

6.5 Basic network security mechanisms

Security in general is intended to protect the end-user identity (including their exact location), data traffic to and from the user, signalling traffic and also to protect the network operator against use of the network without appropriate authority and subscription.

The basic mechanics in the Internet at network layer used to provide security include authentication using public key systems, privacy using public and secret key systems and access control using firewalls and passwords.

Internet security is a very important and also very difficult problem particularly in satellite networking, as the Internet covers the world across political and organisational boundaries. It also involves how and when communicating parties (such as users, computer, services and network) can trust each another, as well as understand the network hardware and protocols.

6.5.1 Security approaches

Security coding can proceed by two approaches:

- *Layer-to-layer approach*: in this case, a computer layer (usually, layer 3 – IP layer or layer 4 – TCP and UDP layer) receives an uncoded file from the above layers, encapsulates the file in a protocol data unit (PDU), and codes the whole frame before sending it to the other end. There, the corresponding layer of the peer entity will decode the PDU before sending the file to the higher layers. This requires, however, that those routers on the network are able to deal with completely coded frames.
- *End-to-end approach*: in this case, the files are coded directly at the application layer by the user, and a coded file is handed out to the lower layers for delivery. This means that only the data payload of the frames is coded (contrary to the previous case, where all the frames were coded).

In the second case, cryptography has only indirect consequences on network traffic, and this only if the coding algorithm has an effect on the size of the data to be transmitted. This is the case for hashing functions or algorithms like RSA.

In the first case, this kind of coding implies an overhead in the frames, thus decreasing the useful load of data carried. This kind of mechanism is implemented in IPv4 and IPv6, with different mechanisms.

In IPv4, cryptography is an option that is activated in the 'Options' field of the header (6th: 32-bit row), in IPv6, it is included as an 'extra header' (since the Options field is used in IPv6) of 64 bits.

Another possible consequence, besides the added headers and the variation of the size of data, is the apparition of messages for exchange of session keys, which never happens under normal (i.e. without cryptography) circumstances.

6.5.2 Single-direction hashing functions

A single-direction hashing function $H(M)$ operates on a message M of arbitrary length. It gives as output a fixed length hashing code $h = H(M)$.

Numerous functions take a variable-length input and give back a fixed-length output, but single-direction hashing functions have additional properties that make them useful:

- given M, it is easy to calculate h
- given h, it is difficult to find M
- given M, it is difficult to find another message M' such as $H(M) = H(M')$.

'Difficulty' depends on the level of security specific to each situation, but the majority of existing applications define 'difficulty' as 'needing 2^{64} or more operations to solve'. Current functions of this type include the MD4, MD5 and secure hash algorithm (SHA). From a network point of view, those algorithms are frequently used for authentication purposes.

6.5.3 Symmetrical codes (with secret keys)

An algorithm of coding with a secret key transforms a message M of arbitrary length into a coded message $E_k(M) = C$ of same length using a key k; and the reverse transformation $(D_k(M))$ uses the same key (Figure 6.17). Those algorithms verify the following characteristics:

- $D_k(E_k(M)) = M$
- given M and k, it is easy to calculate C
- given C and k, it is easy to calculate M
- given M and C, it is difficult to find k

Of course, in this case, difficulty is directly linked to the length of $k : 2^{56}$ for the data encryption standard (DES) algorithm and 2^{128} for the international data encryption algorithm (IDEA). Those algorithms are used in networks for 'encapsulating security payload' purposes (i.e. coding data), commonly used in the area of electronic commerce.

6.5.4 Asymmetrical codes (with public/private keys)

Contrary to the preceding case, those algorithms use two different keys (Figure 6.18): one key e to encrypt (called the public key) and one key d to decrypt (called the private key).

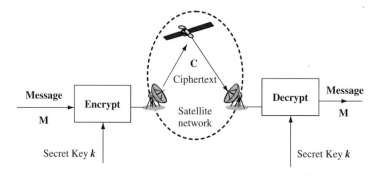

Figure 6.17   Secret key system

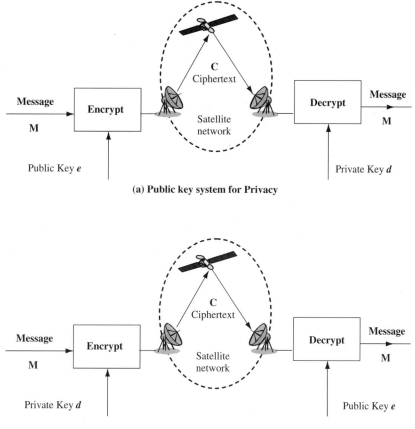

(a) Public key system for Privacy

(b) Public key system for Authentication

Figure 6.18   Public key system for privacy and authentication

Let's define $C = E_e(M)$ and $M = D_d(C)$. We have the following properties:

- $D_d(E_e(M)) = M$
- given M and e, it is easy to calculate C
- given C and d it is easy to calculate M
- given M and C, it is difficult to find e or d
- given e, it is difficult to find d
- given d, it is difficult to find e

The two keys being 'independent', the coding key can be widely known, this is why it has been christened the public key. The private key, in contrast, is only known to the entity decoding the message. The most common algorithm of this type is RSA (for the names of its authors: Rivest, Shamir and Adleman). In networks, those algorithms are used mostly for coding transmissions (Figure 6.18(a)) or authentication (Figure 6.19(b)) between two or more people wishing to communicate in a secure way.

6.6 Satellite networking security

The challenge of security in satellite environments is considered to be one of the main obstacles to the widespread deployment of satellite IP multicast and satellite multimedia applications in general. The main problem is that eavesdropping and active intrusion are much easier than in terrestrial fixed or mobile networks because of the broadcast nature of satellites. In addition, the long delays and high bit error rates experienced on satellite systems may cause loss of security synchronisation. This demands a careful evaluation of encryption systems to prevent QoS degradation because of security processing. A further issue, specific to multicast, is that the number of members in a multicast group can be very large and can change very dynamically.

6.6.1 IP security (IPsec)

Here we only give a brief discussion of the topics relating to IP security (IPsec).

The IPsec protocol suite is used to provide interoperable cryptographically based security services (i.e. confidentiality, authentication and integrity) at the IP layer. It is composed of an authentication protocol: authentication header (AH), a confidentiality protocol: encapsulated security payload (ESP) and it also includes an Internet security association establishment and key management protocol (ISAKMP).

IP AH and ESP may be applied alone or in combination with each other. Each protocol can operate in one of two modes: transport mode or tunnel mode. In transport mode (see Figure 6.19), the security mechanisms of the protocol are applied only to the upper layer data and the information pertaining to IP layer operation as contained in the IP header is left unprotected. In tunnel mode (see Figure 6.20), both the upper layer protocol data and the IP header of the IP packet are protected or 'tunnelled' through encapsulation. The transport mode is intended for end-to-end protection and can be implemented only by the source and destination hosts of the original IP datagram. Tunnel mode can be used between firewalls.

IPsec allows us to consider security as an end-to-end issue, managed by the entities that own the data; this compares with the data link layer security, which is provided by the satellite operator or network operator.

Filters can also be set up in the firewalls to block some IP packets from entering the network based on the IP addresses and port numbers. It is also possible to have security mechanisms at the transport layer such as secure socket layer (SSL), at the link layer or physical layer.

Original IP Header	Authentication Header (AH)	TCP Header	Data

Figure 6.19 Transport mode in IPv4

Encapsulation IP Header	Authentication Header (AH)	Original IP Header	TCP Header	Data

Figure 6.20 Tunnelling mode (the same for both IPv4 and IPv6)

6.6.2 Satellite VPN

A firewall consists of two routers performing IP packet filtering and an application gateway for higher layer checking shown in Figure 6.21. The inside one checks outgoing packets; the outside one checks incoming packets. An application gateway, between the routers, performs further examination of higher layer protocol data including TCP, UDP, email, WWW and other application data. This configuration is to make sure that no packets get in or out without having to pass through the application gateway. Packet filters are table driven and check the raw packets. The application gateway checks contents, message sizes and headers. IPsec is used to provide secure delivery between the corporate network sites across public Internet.

6.6.3 IP multicast security

In secure IP multicast, one of the principal issues is that of ensuring that the key used to encrypt traffic is known to all the member of the group, and only to those members: this is the issue of key management and distribution. The size and dynamics of the multicast group have a great impact on the key management distribution system, especially for large groups. There are several architectures for key management that are currently the subject of research. Another area of significant research effort is that of ensuring that key management is scalable to the large groups that are expected in satellite multicast; one of the most promising such mechanisms is the logical key hierarchy and its derivatives. These keys could then be used in security architecture such as IPsec. This research is being conducted independently of any satellite considerations, but the results are expected to be applicable to secure IP multicast satellite systems.

To deal with the complexity of updating keys (re-key) at a very large scale, the concept of logical key hierarchy (LKH) can be used as shown in Figure 6.22. Keys are organised into a tree structure. Each of the users is allocated a chain of keys allowing some overlaps from leaves to root. Users can be grouped based on the tree so that they share some common keys, therefore a single message can be broadcasted to update keys of the group of users.

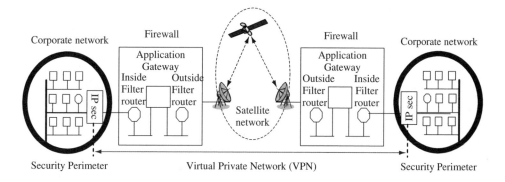

Figure 6.21 Firewall consisting of two routers and one gateway

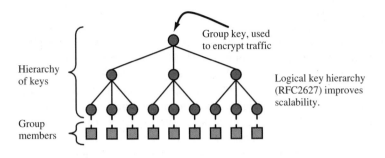

Figure 6.22 Illustration of logical key hierarchy (LKH)

6.7 DVB over satellite

Satellite technology is well known to many people due to satellite broadcasting. The number of antennas outside many homes indicate how many families are receiving TV programmes through satellite broadcasting. The DVB Project (digital video broadcasting, DVB) started the development of a system for digital television broadcasting via satellite (DVB-S) in 1992 and finalised the specification in 1993.

The DVB-S system has been designed with a modular structure, based on independent subsystems, so that the other DVB systems, which were defined later (DVB-C: cable, DVB-T: terrestrial), could maintain a high level of commonality with it. The MPEG-2 source coding and multiplexing subsystem are common to all the broadcasting systems, and only the 'channel adapters', providing channel coding and modulation, are specifically designed to optimise the performance on each media (satellite, cable, terrestrial). To support Internet services for DVB-S, the return channel uses terrestrial networks (Figure 6.23).

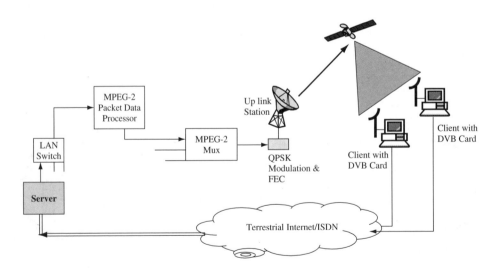

Figure 6.23 DVB-S with return channel via terrestrial networks

6.7.1 MPEG-2 source coding and multiplexing DVB-S streams

The Motion Picture Expert Group (MPEG) has developed MPEG-2 which specifies coding formats for multiplexing and de-multiplexing of streams of audio, video and other data into a form suitable for transmission or storage (Figure 6.24).

Each elementary stream (ES) output by an MPEG audio, video and (some) data encoders contains a single type of (usually compressed) signal.

Each ES is input to an MPEG-2 processor, which accumulates the data into a stream of packetised elementary stream (PES) packets (see Figure 6.25). Each PES has a size up to maximum of 65 536 bytes.

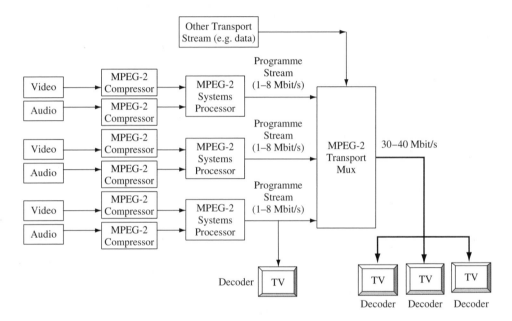

Figure 6.24   MPEG-2 source coding and multiplexing DVB-S streams

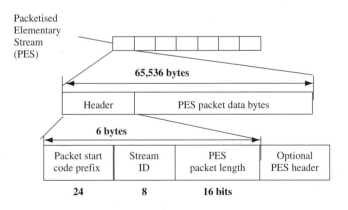

Figure 6.25   MPEG-2 packetised elementary stream (PES)

Each PES packet contains information such as the packet length, PES priority, packet transmission rate and presentation and decoding timestamp information to identify the stream and for layered coding.

6.7.2 *DVB over satellite (DVB-S)*

The DVB system extends MPEG-2 transport facilities by adding programme guides (both teletext style and magazine style formats), specifications for conditional access (CA), and an optional return channel for interactive services with various types of packet. DVB transmission via satellite (often known as DVB-S) defines a series of options for sending MPEG-TS packets over satellite links (Figure 6.26). The size of each MPEG-TS packet is 188 bytes.

Using DVB, a single 38 Mbit/s satellite DVB transponder (DVB-S) may be used to provide one of a variety of services (Figures 6.27 and 6.28):

- four to eight standard TV channels (depending on programme style and quality);
- two high definition TV (HDTV) channels;
- 150 radio programmes;
- 550 ISDN-style data channels at 64 kbit/s;
- a variety of other high and low rate data services.

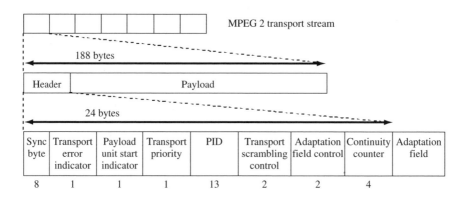

Sync byte	Transport error indicator	Payload unit start indicator	Transport priority	PID	Transport scrambling control	Adaptation field control	Continuity counter	Adaptation field
8	1	1	1	13	2	2	4	

Figure 6.26 MPEG-2 transport stream (MPEG-TS)

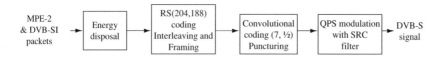

Figure 6.27 DVB-S and DVB-RCS transmission

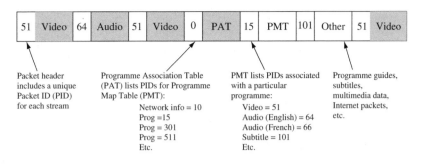

Figure 6.28 DVB service information (DVB-SI) and MPEG signalling

The signalling information includes:

- Program association table (PAT) lists the PIDs of tables describing each programme. The PAT is sent with the well-known PID value of 0x000.
- Conditional access table (CAT) defines the type of scrambling used and PID values of transport streams, which contain the conditional access to entitlement management message (EMM). The CAT is sent with the well-known PID value of 0x001.
- Program map table (PMT) defines the set of PIDs associated with a programme, e.g. audio, video, etc.
- Network information table (NIT) with PID = 10 contains details of the bearer network used to transmit the MPEG multiplex, including the carrier frequency.
- Digital storage media command and control (DSM-CC) contains messages to the receivers.

The service information includes:

- Bouquet association table (BAT) groups services into logical groups.
- Service description table (SDT) describes the name and other details of services.
- Time and date table (TDT) with PID = 14 provides the present time and date.
- Running status table (RST) with PID = 13 provides the status of a programmed transmission and allows for automatic event switching.
- Event information table (EIT) with PID = 12 provides details of a programmed transmission.
- Time offset table (TOT) with PID = 11 gives information relating to the present time and date and local time offset.

6.7.3 DVB security

The DVB system by contrast only provides link layer security. IPsec ESP tunnel mode provides the best security; however, the cost of this is the addition of a new IP header of 20 bytes, which is a large overhead to add to a satellite system.

In DVB, two levels of security can be applied:

- DVB common scrambling; and
- individual user scrambling in the forward and return link.

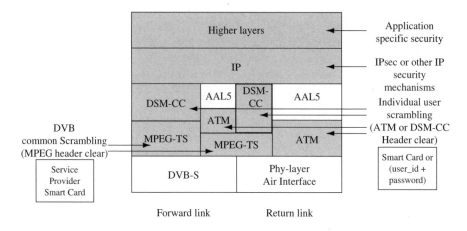

Figure 6.29 IP stack and security in DVB-S and DVB-RCS (© ETSI 2003. © EBU 2003. Further use, modification, redistribution is strictly prohibited. ETSI standards are available from http://www.etsi.org/services_products/freestandard/home.htm and http://pda.etsi.org/pda/.)

Although the user/service provider could use their own security systems above the data link layer, it is usually desirable to provide a security system at the data link layer so that the satellite link is secure without recourse to additional measures. Link level security is particularly desired by satellite access network operators in order to secure satellite links and provide their clients (such as ISPs) with data confidentiality. For DVB, the satellite interactive network is based on the DVB/MPEG-TS standard. The security concept is shown in Figure 6.29.

6.7.4 Conditional access in DVB-S

Conditional access (CA) is a service that allows broadcasters to restrict certain programming products to certain viewers, by encrypting the broadcast programmes. Consequently, the programmes must be decrypted at the receiving end before they can be decoded for viewing. CA offers capabilities such as pay TV (PTV), interactive features such as video-on-demand (VOD) and games, the ability to restrict access to certain material (such as movies) and the ability to direct messages to specific set-top boxes (perhaps based on geographic region).

DVB conditional access originated as a broadcast security mechanism that allows a source to determine which individual receivers are able to receive particular broadcast programmes. CA requires two principal functions: (a) the ability to encode (or 'scramble') a transmission and decode it (or 'descramble') at the receiver; and (b) the ability to specify which receivers are capable of descrambling the transmission.

As Figure 6.30 shows, the transmission from a source to all receivers comprises a set of scrambled MPEG components (video, audio and data), entitlement control messages (ECM) and entitlement management messages (EMM). The ECM identify the CA services, and for each CA service carry the control word (CW), in an encrypted form, and any other parameters required to access the service. The EMM are a set of messages that identify the entitlements (permissions) of any individual user.

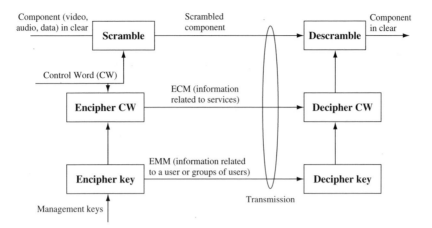

Figure 6.30 DVB conditional access

In addition, the subscriber management system (SMS) maintains and stores commercial aspects of customer relationships (registration, granting of entitlements, invoicing and accounting), and the subscriber authorisation system (SAS) encrypts code words and delivers them to the descrambler.

At the receiving end, it is the job of the set-top box (STB) to descramble the CA encryption and decode the MPEG-2 streams for viewing. Each packet has associated with it (in its header) a program identifier (PID). The conditional access table (CAT) has a well-known PID value = 1. This table can be used to identify the PID values of the transport packets containing the EMM. The de-multiplexer processor also constructs the program map table (PMT) from non-encrypted packets; this gives the PID values of all the transport streams associated with a particular programme. Private data associated with the programme can also be included in this table, for example, the PID value of the packets that contain ECM. All these tables (signalling messages) are transmitted in the clear, which is an inherent security weakness in DVB-S systems.

6.7.5 DVB-RCS interactive service and IP over DVB

The interactive satellite architecture consists of a ground station (hub), one or more satellites in the forward direction, a satellite interactive terminal (return channel satellite terminal, RCST) at the user's location and a satellite in the return direction.

The forward path carries traffic from the ISP to the individual user, and it is multiplexed into a conventional DVB/MPEG-2 broadcast stream at a broadcast centre (the hub) and relayed to the RCST. Figure 6.31 shows the protocol stack and Figure 6.32 shows multi-protocol encapsulation (MPE) for IP over DVB.

The return channel path operates as part of a digital network, with the hub station providing the gateway to other (satellite and terrestrial) networks. The satellite terminal employs a scheduled MF-TDMA scheme to access the network and participate in bi-directional communications. MF-TDMA allows a group of terminals to communicate with a hub using a set of carrier frequencies, each of which is divided into time slots. There are four types of

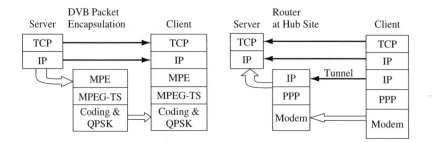

Figure 6.31   DVB-S and DVB-RCS protocol stack

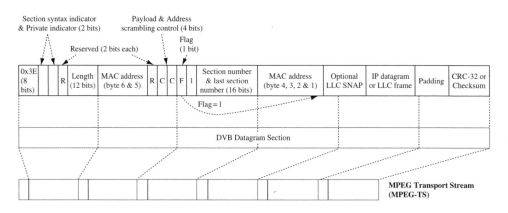

Figure 6.32   IP over DVB: multi protocol encapsulation (MPE)

bursts: traffic (TRF), acquisition (ACQ), synchronisation (SYNC) and common signalling channel (CSC).

There is a new development on DVB-RCS with satellite on-board processors for DVB streams de-multiplexing and re-multiplexing. In the future, the Ka band will be explored for higher capacity and smaller antenna sizes; there will be tighter integration with IP technology, protocols and architecture including network management and IP security over the satellite link; and there will also be more integration between DVB and UMTS, where the two systems can complement each other.

6.7.6 DVB-RCS security

The DVB-RCS standard provides much more advanced security procedures (in comparison to DVB-S CA), which enable satellite terminal authentication and key exchanges with a network control centre (NCC).

DVB-RCS security can be divided into two phases: phase 1 is the authentication during the logon procedure. During this phase a security session key is agreed between the satellite terminal and the NCC. In phase 2, the session key is used for the encryption of all subsequent messages between UES and NCC. The authentication is based on a long-term secret shared between NCC and UES, called a cookie, which is 160 bits long and stored in non-volatile storage (such as a smart card). The NCC maintains a database of the cookie values of the UES on its network. Cookie values can be updated occasionally as dictated by security policy, but they are less vulnerable than session keys. Anti-cloning measures can also be implemented using message sequence numbering.

A separate consideration is security of the space segment. In satellite systems with DVB on-board switching, message integrity between the NCC and the OBP is important to make sure that configuration messages originate from the NCC. The major constraint in the OBP is its limited memory and computational power, since the computational cost of message integrity can be high. This cost depends on the type of algorithms used: for example, message integrity can be provided using public-key digital signatures, which are computationally heavy, or using MAC (message authentication code) with secret keys, which are computationally lighter. The use of secret keys implies the need for a key agreement, where keys can be stored in the OBP at installation time or agreed using the DVB-RCS key exchange mechanisms.

6.7.7 DVB security and IP multicast security

DVB-S conditional access is used today for digital broadcasting over satellite and can also be used to secure multicast communications over satellites at the MPEG-TS level. Descrambling in DVB-S is programme-based, where a whole programme will be scrambled with the same CW. In a TV broadcast, the programme may contain video, audio and data, each with a specific PID; for IP transmission, the IP datagrams are encapsulated using MPE and transmitted on a specific PID. The main drawback is that the DVB-S scrambling system favours a centralised ECM and EMM, and its use for dynamically changing the IP of a multicast group is limited.

The number of PIDs is limited to 8192, and if there is one PID per multicast group this could easily constrain the total number of IP multicast groups that the satellite supports: the alternative is to support several multicast groups per PID, or all groups on a single PID. On the other hand, the DVB-RCS standard provides more advanced security procedures for satellite terminal authentication and key exchanges with the satellite network operator. However, it does not provide security procedures for terminal-to-terminal communications (the 'mesh' scenario of Figure 6.13). DVB-RCS only allows a single key per terminal, and therefore does not allow different multicast groups to be encrypted with different keys.

6.8 Internet quality of service (IP QoS)

The original Internet protocol (IP) was design for connectionless networks with best effort to deliver IP packets across the Internet. Best effort means no QoS requirement. In the next generation Internet, best effort is not good enough. It needs to provide new services and applications with different classes of QoS including guaranteed QoS and controlled load QoS, in addition to the best-effort services. These presented great challenges to the new generation network to provide IP-related QoS. Important network QoS parameters include end-to-end delay, delay variation and packet loss. These have to be measured in an end-to-end reference path, where the propagation delay of satellite links should be taken into account properly.

There are many issues on IP-based networks and services defined by the ITU-T (G.1000), which take into account:

- Dynamic allocations of resources (like packet loss and delay) among network segments.
- Assuring that required end-to-end network performance objectives are achieved.
- Seamless signalling of desired end-to-end QoS across both network and end-user interfaces.
- Performance monitoring of IP-based networks and services.
- Rapid and complete restoration of IP layer connectivity following severe outages (or attacks) of heavily loaded networks.

The ITU-T (Y.1540) defines parameters that may be used in specifying and assessing the performance of speed, accuracy, dependability and availability of IP packet transfer of international IP data communication services. The defined parameters apply to end-to-end, point-to-point IP service and to the network portions that provide such service. Connectionless transport is a distinguishing aspect of the IP service that is considered in this recommendation.

The end-to-end IP service refers to the transfer of user-generated IP datagrams (i.e. IP packets) between two end hosts as specified by their complete IP addresses.

6.8.1 Layered model of performance for IP service

Figure 6.33 shows the layered model of performance for IP service. It illustrates the layered nature of the performance of IP service. The performance provided to IP service users depends on the performance of other layers:

- Lower layers that provide (via 'links') connection-oriented or connectionless transport supporting the IP layer. Links are terminated at points where IP packets are forwarded (i.e., routers or switches) and thus have no end-to-end significance. Links may involve different types of technologies, for example, ATM, SDH, PDH, mobile and wireless etc.

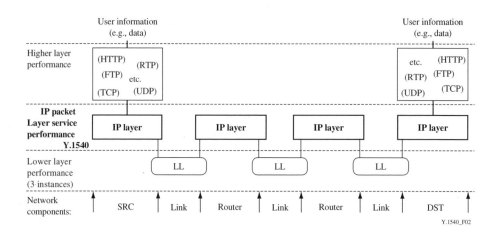
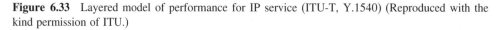

Figure 6.33 Layered model of performance for IP service (ITU-T, Y.1540) (Reproduced with the kind permission of ITU.)

- The IP layer that provides connectionless transport of IP datagrams (i.e., IP packets). The IP layer has end-to-end significance for a given pair of source and destination IP addresses. Certain elements in the IP packet headers may be modified by networks, but the IP user data may not be modified at or below the IP layer.
- Higher layers, supported by IP, that further enable end-to-end communications. Upper layers may include, for example, TCP, UDP, FTP, RTP, RTCP, SMTP and HTTP. The higher layers will modify and may enhance the end-to-end performance provided at the IP layer.

6.8.2 IP packet transfer performance parameters

IP packet transfer delay (IPTD) is defined by the ITU-T (Y.1540) for all successful and errored packet outcomes across a basic section or a network section ensemble (NSE). IPTD is the time $(t_2 - t_1)$ between the occurrence of two corresponding IP packet reference events, ingress event $IPRE_1$ at time t_1 and egress event $IPRE_2$ at time t_2, where $(t_2 > t_1)$ and $(t_2 - t_1) \leq T_{max}$. If the packet is fragmented within the NSE, t_2 is the time of the final corresponding egress event. The end-to-end IP packet transfer delay is the one-way delay between the MP at the SRC and DST as illustrated in Figure 6.34.

Mean IP packet transfer delay is the arithmetic average of IP packet transfer delays used as an indicator of overall performance.

End-to-end two-point IP packet delay variation (IPDV) is the variation in IP packet transfer delay. Streaming applications might use information about the total range of IP delay variation to avoid buffer underflow and overflow. Variations in IP delay will cause

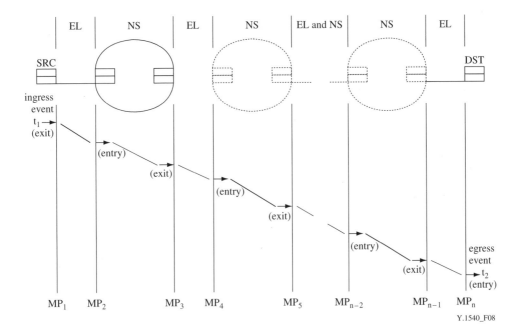

Figure 6.34 IP packet transfer delay events [ITU-Y.1540] (illustrated for the end-to-end transfer of a single IP packet) (Reproduced with the kind permission of ITU.)

TCP retransmission timer thresholds to grow and may also cause packet retransmissions to be delayed or packets to be retransmitted unnecessarily.

IP packet error ratio (IPER) is the ratio of total errored IP packet outcomes to the total of successful IP packet transfer outcomes plus errored IP packet outcomes in a population of interest.

IP packet loss ratio (IPLR) is the ratio of total lost IP packet outcomes to total transmitted IP packets in a population of interest. Metrics for describing one-way loss patterns is stated in IETF RFC3357. Consecutive packet loss is of particular interest to certain non-elastic real-time applications, such as voice and video.

Spurious IP packet rate at an egress MP is the total number of spurious IP packets observed at that egress MP during a specified time interval divided by the time interval duration (equivalently, the number of spurious IP packets per service-second).

IP packet severe loss block ratio (IPSLBR) is the ratio of the IP packet severe loss block outcomes to total blocks in a population of interest. This parameter can identify multiple IP path changes due to routing updates, also known as route flapping, which cause significant degradation to most user applications.

6.8.3 IP network performance objectives for QoS classes

ITU-T recommendation Y.1540 addresses the topic of network transfer capacity (the effective bit rate delivered to a flow over a time interval), its relationship to the packet transfer QoS parameters and objectives specified for each QoS class. The IP network performance objectives are shown in Table 6.1.

Table 6.1 Provisional IP network QoS class definitions and network performance objectives (Y.1540)

Network performance parameter	Nature of network performance objective	QoS classes					Class 5 Unspecified (U)
		Class 0	Class 1	Class 2	Class 3	Class 4	
IPTD	Upper bound on the mean IPTD	100 ms	400 ms	100 ms	400 ms	1 s	U
IPDV	Upper bound on the $1 - 10^{-3}$ quantile of IPTD minus the minimum IPTD	50 ms	50 ms	U	U	U	U
IPLR	Upper bound on the packet loss probability	10^{-3}	10^{-3}	10^{-3}	10^{-3}	10^{-3}	U
IPER	Upper bound	1×10^{-4}	1×10^{-4}	1×10^{-4}	1×10^{-4}	1×10^{-4}	U

(Reproduced with the kind permission of ITU).

Table 6.2 Guidance for IP QoS classes (Y.1541)

QoS class	Applications (examples)	Node mechanisms	Network techniques
0	Real-time, jitter sensitive, high interaction (VoIP, VTC)	Separate queue with preferential servicing, traffic grooming	Constrained routing and distance
1	Real-time, jitter sensitive, interactive (VoIP, VTC).		Less constrained routing and distances
2	Transaction data, highly interactive (signalling)	Separate queue, drop priority	Constrained routing and distance
3	Transaction data, interactive		Less constrained routing and distances
4	Low loss only (short transactions, bulk data, video streaming)	Long queue, drop priority	Any route/path
5	Traditional applications of default IP networks	Separate queue (lowest priority)	Any route/path

(Reproduced with the kind permission of ITU).

Transfer capacity is a fundamental QoS parameter having primary influence on the performance perceived by end users. Many user applications have minimum capacity requirements; these requirements should be considered when entering into service agreements. Y.1540 does not define a parameter for capacity; however, it does define the packet loss parameter. Lost bits or octets can be subtracted from the total sent in order to provisionally determine network capacity.

Theoretically, IP over satellite networks is not able to provide Class 0 or 2 services (refer to Table 6.2), due to their real-time characteristics, but the advantage factor of satellite should be taken into consideration.

6.8.4 Guidance on IP QoS class usage

Table 6.2 gives some guidance for the applicability and engineering of the network QoS classes (Y.1541). To support QoS there are two architectures that are defined by the IETF: integrated services (commonly known as Intserv) and differentiated services (commonly known as Diffserv).

6.9 Integrated services (Intserv) architectures for QoS

Within the Internet, each node (switch or router) deals with protocols up to the IP layer packet. The Internet provides only best-effort IP datagram transmission. IP packets are sent from a source to a destination without any guarantee that the packet will reach its destination. It is only suitable for elastic applications that tolerate packet delays and packet losses; the best-effort model at the network layer is compensated by the TCP at the transport layer introduced in the end systems (clients or servers) to provide reliability by acknowledgements and retransmission mechanisms.

However, the emerging real-time applications have very different characteristics and requirements to data applications. They are inelastic, hence are less tolerant of delay variation and need specific network conditions in order to perform well. To support the range of QoS, the IP architecture has to be extended to provide support for real-time services.

6.9.1 Integrated services architecture (ISA) principles

The primary goal of the integrated services architecture (ISA) and QoS model is to provide IP applications with end-to-end 'hard' QoS guarantees, where the application may explicitly specify its QoS requirements and these will be guaranteed by the network.

The Intserv architecture is a framework developed within the IETF to provide individualised QoS guarantees to individual application sessions (RFC1633). It is a reservation-based QoS architecture, designed to guarantee fair sharing of resources (both link bandwidth and router buffers) among users by dynamically controlling and managing the bandwidth via resource reservation and admission control. It uses the resource reservation protocol (RSVP) (RFC2475) as the signalling mechanism for specifying an application's QoS requirements and identifying the packets to which these requirements apply. The two key features of the Intserv architecture are:

- Reserved resources: routers need to know the amounts of resources (link bandwidth and buffers) already reserved for ongoing sessions, and available for allocations.
- Session set up: a session must reserve sufficient resources at each network router from source-to-destination path to ensure that its end-to-end QoS requirement is met. This call set up (also known as call admission) process requires the participation of each router on the path. Each router must determine the local resources required by the session, consider the amount of its resources that are already committed to other ongoing sessions, and determine whether it has sufficient resources to satisfy the QoS requirement without violating local QoS guarantees.

The building blocks relevant to the Intserv approach are resource reservation, admission control, traffic classification, traffic policing, queuing and scheduling.

There are two types of routers: edge router (ER) and core router (CR). The functions of ER are to control flows into the network domain, including explicit per-flow admission control, per-flow classification, per-flow signalling and per-flow resource reservation. The functions of CR are to forward the IP packets as fast as possible, based on information in the IP packets set by the ER.

In order for a router to determine whether or not its resources are sufficient to meet the QoS requirements of a session, that session must first declare its QoS requirement, as well as characteristics of the traffic. The signalling entity request specification (R_Spec) defines the specific QoS being requested by a connection; traffic specification (T_Spec) on the other hand characterises the traffic. The RSVP protocol is currently the signalling protocol of for this purpose.

A session (application) is only allowed to send its data once its request for resources is granted. It is also important that granting a request must not be at the expense of other commitments already in place. A successful reservation request results in installation of states at RSVP-aware nodes. As long as the application honours its traffic profile, the network meets its service commitments by maintaining per-flow states and using queuing and scheduling disciplines.

6.9.2 The resource reservation protocol (RSVP)

RSVP is the signalling protocol used in the Intserv model to reserve network resources (bandwidth and buffer space) for their data flows. RSVP requests are carried through the network, visiting each node along the routed path used to the destination. At each node (router), RSVP attempts to reserve resources for the particular flow. Hence, RSVP software must run in the hosts (senders and receivers) and the routers. It is also a flow-based protocol, i.e. classification is done on each and every flow. Resources reserved need to be refreshed within a specified time limit – otherwise the resources are released upon the expiry of this time interval. This is also known as a 'soft-state' reservation. The two key characteristics of RSVP are:

- It provides reservation for bandwidth in multicast applications such as audio/video conferencing and broadcasting. It is also used for unicast traffic, however, unicast requests are handled as a special case.
- It is receiver-oriented, i.e. the receiver of the data flow initiates and maintains the resource reservation used for that flow.

There are two main components of RSVP – the packet classifier and the packet scheduler installed on the host to make QoS decisions about the packets sent in by applications. The communications among various components existing in an RSVP-enabled host and router is as shown in Figure 6.35. RSVP reserves bandwidth and advises the network on the correct queue management and packet discard policies. RSVP-enabled routers will then invoke their admission control and packet-scheduling mechanisms based on the QoS requirements. The admission control module decides whether or not there are enough resources locally to grant the reservation without violating resources already committed to existing connections. The packet-scheduling module is a key component because this is the module that manifests the different services to different flows.

RSVP first queries the local decision modules to find out whether the desired QoS can be provided (this may involve resource-based decisions as well as policy-based decisions). It then sets up the required parameters in the packet classifier and the packet scheduler. The packet classifier implements the process of associating each packet with the appropriate reservation so that it can be handled correctly. This classification is done by examining the packet header. The packet classifier also determines the route of the packet based on these parameters. The scheduler makes the forwarding decisions to achieve the desired QoS. When the link layer at the host has its own QoS management capability, then the packet scheduler negotiates with it to obtain the QoS requested by RSVP. In the other case, for example, when the host is using a leased line, the scheduler itself allocates packet transmission capacity. It may also allocate other system resources like CPU time, buffers, etc.

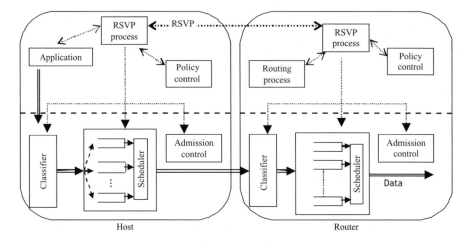

Figure 6.35 Interaction between the different RSVP components

Two basic messages used in RSVP are the PATH and RESV messages. A PATH message is initiated by the sender and is addressed directly to the destination. It sets up states along the path to be followed by the application packets from the sender to the specified destination. This path is determined by the underlying routing protocol. A PATH message includes information such as the previous hop (the previous RSVP-aware entity on the path), the sender's T_Spec and ADSPEC (the advertising specification used to capture the path characteristics). At each router along this path, a local RSVP entity updates these parameters in its memory and amends some of the information carried by the PATH message.

Upon receiving the PATH message, the receiver will decide whether or not to actually receive the data from the sender. Should it wish to continue with the session the receiver constructs a RESV message based on the advertisement information carried by the PATH message and sends this message back towards the sender along the path already set up. Routers along the path will then invoke their RSVP processes and reserve the required resources extracted from the receiver's R_Spec information contained in the RESV message. When the receiver has successfully reserved resources over the entire path, a success message is returned. The same RESV message is sent about once every 30 s should the receiver wish to retain the reservation. If any one router rejects the reservation, the request is denied and an error message is generated. Resources already reserved at intermediate nodes will then be released.

RSVP is not a routing protocol and it does not perform its own routing. Like any other IP traffic, it relies on the underlying IP routing protocols to determine the path for both its data and control traffic. As the routing information adapts to network topology changes (due to link or router failure), RSVP reservations are carried over to the new path calculated by the routing protocols. This flexibility helps RSVP to function effectively with current and future unicast or multicast routing protocols. It is specially suited for multicast applications – RSVP scales to very large multicast groups because it uses receiver-oriented reservation requests that merge as they progress up the multicast tree. If the RESV message arrives at a router where the desired QoS reservation (or one greater) is already in place for another receiver in the same multicast group, then the RESV message need not travel any further. The two (or more) receivers can share the reservation.

6.9.3 Intserv service classes

In terms of QoS support, Intserv defines two classes of service, in addition to the existing best-effort service (BES): guaranteed services (GS) and controlled load services (CLS). Typically, the total capacity is divided and allocated in proportions to accommodate the three different service classes.

- *Guaranteed service (GS)*. It guarantees firm bounds (mathematically provable) on the maximum end-to-end packet delay by reserving a rate at each router. It guarantees that packets will arrive within the requested delivery time and will not be discarded due to queue overflows (provided the flow's traffic conforms to the specified traffic parameters). This service is designed for applications requiring a fixed amount of delay.
  GS traffic must be policed at the network access points to ensure conformance to the T_Spec. Non-conforming packets are usually forwarded as BES traffic. GS also requires traffic shaping and any packets failing this process will be forwarded as BE traffic.
- *Controlled load service (CLS)*. It allocates resources such that a high proportion of traffic using this service will experience conditions very close to an uncongested network. CLS aims to emulate a lightly loaded network although the network as a whole may in fact be heavily loaded. In other words, the session may assume that a 'very high percentage' of its packets will successfully pass through the router without being dropped and will experience a queuing delay in the router under a light load condition.
- *Best-effort service (BES)*. This is the service provided by the current Internet.

An important difference between CLS and BES is that CLS does not noticeably deteriorate as the network load increases and regardless of the level of load increase. BES on the other hand will experience progressively worse service as the network load increases. However, CLS makes no quantitative guarantees about performance – it does not specify what constitutes a 'very high percentage' of packets or what QoS closely approximates that of an unloaded network element.

CLS also requires traffic policing. Non-conforming CLS flows must not be allowed to affect the QoS offered to conforming CLS flows or to unfairly affect the handling of BES traffic.

6.10 Differentiated services (Diffserv) for QoS

Diffserv allows IP traffic to be classified into a finite number of priority and/or delay classes. Traffic classified as having a higher priority and/or delay class receives some form of preferential treatment over traffic classified into a lower class. The differentiated services architecture (DSA) does not attempt to give explicit 'hard' end-to-end guarantees. Instead, at congested routers, the aggregate of traffic flows with a higher class of priority has a higher probability of getting through.

6.10.1 DSA principles

The DSA approach is intended to provide scalable and flexible service discrimination without the signalling overhead or significant changes to the Internet infrastructure as required by the Intserv/RSVP architecture. This approach aims to provide the ability to handle different 'classes' of traffic in different ways within the Internet. The need for scalability arises from the fact that hundreds of thousands of simultaneous source–destination traffic flows may be present at backbone networks. The need for flexibility arises from the fact that new service classes may appear and old service classes may become obsolete. The Diffserv architecture is flexible in the sense that it does not define specific services or service classes (e.g., as is the case with Intserv). Instead, the Diffserv architecture provides the functional components, within the network architecture, to allow such services to be built.

There is no reservation in the Diffserv QoS architecture. It provides differential treatments to a consolidation of flows where traffic is aggregated into groups or classes throughout the network. This approach consists of marking packets by setting bits in the packet header, specifically the type of service (TOS) field in the IPv4 packets and the traffic class field in IPv6 packets (Figure 6.36). In the TOS byte structure as shown below, the first three bits represents the precedence bits and can be used to indicate need for a low delay or high throughput or low loss rate service. MBZ is the must be zero bit. This byte is renamed to DS field in Diffserv and has the following structure (Figure 6.37).

The first six bits of the DS field is known as the differentiated services code point (DSCP); the last two bits are currently unused (CU). By setting these bits appropriately, different services requiring different treatment will be 'tagged' with different priority levels. This differentiation allows the network (routers) to recognise the type of service required and handle the packets accordingly, usually by some form of priority queuing management and packet scheduling schemes. Note that the key approach here is the use of packet headers to carry information required by these schemes, hence eliminating the need for signalling protocols to control the mechanisms that are used to select different treatments for the individual packets. As a result, the requirement for maintaining state information at every node is reduced substantially – the amount of information needed is now proportional to the number of services instead of the number of application flows, as is the case with Intserv.

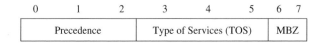

Figure 6.36   Type of service (TOS) field

Figure 6.37   Differentiated service (DS) field

The architectural framework of the Diffserv approach consists of two sets of functional elements:

- Edge functions: packet marking (classification) and traffic conditioning. These functions are implemented at the incoming edge of the network (ingress) i.e. either at a Diffserv-capable host that generates traffic or at the first Diffserv-capable router that the traffic passes through. Packets entering the network will be marked, i.e. the first six bits of the DS field of the packet's header is set to some value. The mark that the packet receives depends on the measured temporal properties of the flow the packet belongs to and compared against a predefined traffic profile. The mark identifies the class of traffic or more specifically the behaviour aggregate (BA) the packet belongs to. Different behaviour aggregates will then receive different treatments or service within the core network (the Diffserv domain). After being marked, the packet may be allowed entry into the network immediately, delayed for some time before being forwarded or discarded altogether. This is performed by the traffic conditioning function to ensure compliance to the predefined profile.
- Core function: forwarding. When a DS-marked packet arrives at a Diffserv-capable router, the packet is forwarded onto its next hop according to the so-called per-hop behaviour (PHB) associated with that packet's BA. The per-hop behaviour influences how a router's buffers and link bandwidth are shared among the competing classes of traffic. A crucial tenet of the Diffserv architecture is that a router's PHB will be based only on packet markings, i.e. the class of traffic to which a packet belongs. It will not distinguish packets based on source–destination address. The implication of this approach is that the core routers will no longer need to keep state information for source–destination pairs – an important consideration when meeting the scalability requirement.

In essence, the Diffserv architecture defines three main components: the traffic classifiers, which select packets and assigns their DSCP values; the traffic conditioners, which mark and enforce rate limitations; and the PHB, which enforces differentiated packet treatments.

Before differentiated services can be extended across a Diffserv network domain, a service level agreement (SLA) is first established between the subscriber and the network/service provider. The SLA basically establishes the policy criteria and defines traffic profiles. Among others, an SLA contains policies such as monitoring provisions, billing and accounting agreements and availability levels. However, one key subset of the SLA is the traffic conditioning agreement (TCA). The TCA defines traffic profiles, performance metrics (e.g. throughput, latency and drop probability) and instructions on how both in- and out-of-profile packets (with respect to the agreed traffic profile) are to be handled. The contents of the SLA, especially the TCA, will be used by both the subscriber when submitting traffic to the network, and the network/service provider when handling the submitted traffic.

6.10.2 Traffic classification

Traffic classification is an important function to be undertaken at the Diffserv network point of entry (ingress). The purpose of this function is to identify packets belonging to a certain class that may receive differentiated services. From the classification result, traffic profile and the corresponding policing, marking and shaping rules of the incoming packets can be derived. Packet classification is done by the packet classifier. The classifier selects packets based either on the DSCP only or a combination of one or more header fields. The first of these classifiers is known as the BA classifier and the second the multi-field (MF) classifier. Once the packets are classified, they are steered to the appropriate marking function where the DS field value of the packets is set accordingly.

6.10.3 Traffic conditioning

A traffic conditioner is an entity that applies some traffic control function to incoming packets to ensure the traffic flow adheres to the TCA rules. These functions include:

- Marking i.e. setting the DSCP in a packet that has already been classified, based on well-defined rules.
- Metering, which compares the incoming packets with the negotiated traffic profile and determines whether the packet is within the negotiated traffic profile or not. It will then decide whether to re-mark, forward, delay or drop a packet even though the actual decision on what to do to a packet is not defined in the Diffserv architecture. The aim is to make the Diffserv components flexible enough to accommodate a wide and constantly evolving set of services.
- Shaping, which delays packets within a traffic stream to cause the stream to conform to the negotiated traffic profile.
- Shaper or dropper, which discards packets based on specified rules e.g. when the traffic stream violates the negotiated traffic profile.

A logical view of these components is shown in Figure 6.38.

6.10.4 Diffserv per hop behaviour (PHB)

The third set of Diffserv functional element is the packet forwarding function performed by the core Diffserv-capable routers. This forwarding function known as the per hop behaviour

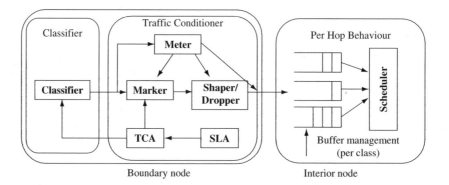

Figure 6.38 Logical view of Diffserv components

(PHB) is defined as 'a description of the externally observable forwarding behaviour of a Diffserv node applied to a particular Diffserv behaviour aggregate'. There are several important considerations embedded within this definition:

- A PHB can result in different classes of traffic (i.e., traffic with different DS field values) receiving different performance (i.e., different externally observable forwarding behaviour).
- While a PHB defines differences in performance (behaviour) among classes, it does not mandate any particular mechanism for achieving these behaviours. As long as the externally observable performance criteria are met, any implementation mechanism and any buffer/bandwidth allocation policy can be used. For example, a PHB would not require that a particular packet queuing discipline, e.g., a priority queue versus a weighted-fair-queuing queue versus a first-come-first-served queue, be used to achieve a particular behaviour.
- Differences in performance must be observable, and hence measurable.

An example of a simple PHB is one that guarantees that a given class of marked packets receives at least a certain percentage of the outgoing link bandwidth over some interval of time. Another PHB might specify that one class of traffic will always receive strict priority over another class of traffic, i.e. if a high priority packet and low priority packet are present in a router's queue at the same time, the high priority packet will always leave first.

Diffserv defines a base set of PHB. These PHB are in turn defined by a set of forwarding behaviour that each router along the path adheres to, i.e. each PHB would correspond to a particular forwarding treatment given to the packets, implemented by means of buffer management and packet scheduling mechanisms. There are currently three proposed PHBs:

- The default (DE) PHB is equivalent to the best-effort forwarding already existing in today's IP networks. Packets marked with this service are sent into the network without adhering to any particular rules and the network will deliver as many of these packets as possible as soon as possible without any performance guarantees.
- The expedited forwarding (EF) PHB specifies that the departure rate of a class of traffic from a router must equal or exceed a configured rate. That is, during any interval of time, the class of traffic can be guaranteed to receive enough bandwidth so that the output rate of the traffic equals or exceeds this minimum configured rate. Note that the EF PHB implies some form of isolation among traffic classes, as this guarantee is made independently of the traffic intensity of any other classes that are arriving to a router. Thus, even if the other classes of traffic are overwhelming router and link resources, enough of those resources must still be made available to the EF class to ensure that it receives its minimum rate guarantee. It assures bandwidth availability regardless of the number of flows sharing the link. EF PHB thus provides a class with the simple abstraction of a link with a minimum guaranteed link bandwidth. It can be used to build an end-to-end service that requires low loss, low delay, low jitter and assured bandwidth service (also known as premium service). It essentially emulates a virtual leased line.
- The assured forwarding (AF) PHB is more complex. AF PHB divides traffic into four classes, where each AF class is guaranteed to be provided with some minimum amount of bandwidth and buffering. Within each class, packets are further partitioned into one of three

'drop preference' categories. When congestion occurs within an AF class, a router can then discard (drop) packets based on their drop preference values. Low drop precedence packets are protected from loss by preferentially discarding higher drop precedence packets. By varying the amount of resources allocated to each class, an ISP can provide different levels of performance to the different AF traffic classes.

Because there are only three PHBs or traffic classes, only the first three bits of the DSCP are needed to denote the traffic class a packet belongs to; the remaining three bits are set to zero. Out of the first three bits, the first two are actually used to denote the traffic class. This is then used to select the appropriate queue (each traffic class is allocated its own queue at the output port). The third bit is used to indicate the drop preference inside each queue/class.

As mentioned previously, the Diffserv architecture only defines the DS and PHB fields of a packet header. It does not mandate any specific implementation mechanisms in order to achieve the service differentiation. The service provider will have the responsibility and flexibility to implement appropriate traffic handling mechanisms that best fit the specific service differentiation they wish to offer. These traffic handling mechanisms are basically traffic filtering (classification), queue management and packet scheduling mechanisms. Hence careful design of these mechanisms is needed to ensure the desired service(s) is achievable while keeping the design as simple as possible.

6.10.5 Supporting Intserv across the satellite network Diffserv domain

Both Intserv and Diffserv approaches go beyond the best-effort service model by defining some kind of agreement between the users and the network/service providers. From this agreement, a 'service profile' can then be built and classified according to a specific service's spatial and temporal requirements. In terms of the spatial requirements, Intserv/RSVP provides the maximum detail – the flow to which the agreement applies is fully specified from the source to the destination and along the path taken. Diffserv, on the other hand, provides a coarser approach – a user may require all or a fraction of his traffic to be given a better service than best effort. In terms of the temporal requirements, again Intserv/RSVP provides a more flexible approach – dynamic agreements can be set up and released on demand depending on the need of the user. Diffserv supports a static agreement where the duration of the agreement is defined on a contractual basis between the user and the service provider (in the form of the SLA).

Although it is clear the approaches taken by both Intserv and Diffserv are contradictory to each other, the two architectures can be complementary to each other. Existing alone, Intserv would definitely suffer from the scalability problems – although it promises tightly controlled QoS on an end-to-end basis, its processing overhead is just too much for an Internet of a decent coverage. Diffserv on the other hand, only guarantees QoS on an aggregate basis (per class basis) – there is no guarantee to the individual flows making up the class.

However, by combining the advantages from the two models, it is possible to build a scalable QoS architecture capable of delivering predictable service guarantees. Diffserv, with its focus on the needs of large networks, can be deployed in high-speed transit networks. A hybrid architecture can consist of peripheral domains (access networks) that are Intserv/RSVP-aware interconnected by a Diffserv core. A typical configuration is shown in Figure 6.39 where the satellite acts a core network to bridge the access networks.

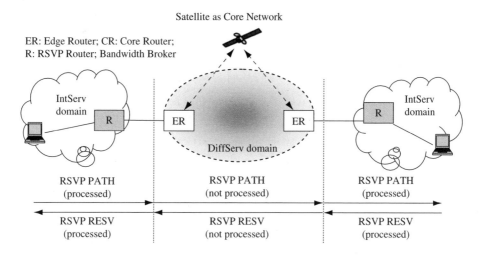

Figure 6.39 Architectural for Intserv networks via satellite Diffserv network

The edge routers at the boundaries of the different network regions would normally be dual function – a standard RSVP function, which interfaces the stub networks and a Diffserv function, which interfaces the transit network. The RSVP half is capable of processing all the RSVP signalling elements. The Diffserv half functions also work as the admission controller into the Diffserv domain. In the simplest scenario, the admission control has information regarding how much bandwidth has been used and how much is left available. Using this information as parameters in the RESV messages, the edge router is able to determine whether or not to permit a new connection. If the request is accepted, the traffic is then mapped to an appropriate PHB and its associated DSCP is marked in the packet header. It can be assumed that insert's GS is mapped to EF PHB and CLS to the highest priority AF PHB.

The signalling process for end-to-end QoS is triggered by the sending host generating a PATH message. On its way towards the receiver, this PATH message is only processed in the Intserv domains that it passes through. At the edge router straddling the boundaries, the PATH state is installed in the router and the PATH message is carried through the Diffserv transit network unprocessed.

Upon receiving the PATH message, the receiver generates a RESV message and this is sent back to the sender via the same path the PATH message took. The request may of course be rejected while it is still in the stub Intserv domain, according to standard Intserv admission control. At the boundary, the RESV triggers admission control at the edge router. The router compares the requested resources to the corresponding Diffserv service level. The request is approved when there are enough resources and the request fits the SLA. The RESV message is admitted into the Diffserv transit and continues upstream towards the sender unprocessed. Upon entering the Intserv stub again, normal Intserv processes resume until the RESV message reaches the intended sender. The RESV message terminates at the sender with information regarding the specified traffic flow and its corresponding Diffserv service level.

Further reading

[1] Akyildiz, I.F. *et al.*, Satellite ATM networks: a survey, *IEEE Communications*, **35**(7): 30–43, 1997.

[2] Allman, M., Glover, D. and Sanchez, L. *Enhancing TCP over Satellite Channels using Standard Mechanisms*, IETF RFC2488, January 1999.

[3] Ballardie, A., *Core-based Trees (CBT version 2) Multicast Routing*, IETF RFC2189, September 1997.

[4] Bem, D.J., Wieckowski, T.W. *et al.*, Broadband satellite systems, *IEEE Communications*, **3**(1): 2–14, 2000.

[5] Blake, S. *et al.*, *An Architecture for Differentiated Services*, IETF RFC2475, December 1998.

[6] Braden, R., Clark, D. and Shenker, S., *Integrated Services in the Internet Architecture: an Overview*, IETF RFC1633, June 1994.

[7] Deering, S., Estrin, D.L. *et al.*, The PIM architecture for wide-area multicast routing, *IEEE Transactions: Networking*, **4**(2): 153–62, 1996.

[8] Estrin, D. *et al.*, *Protocol-independent Multicast – Sparse Mode (PIM-SM): Protocol Specification*, IETF RFC2362, June 1998.

[9] ETSI EN301790. *Digital Video Broadcasting (DVB)* Interaction Channel for Satellite Distribution Systems, 2000.

[10] Fenner, W., *Internet Group Management Protocol, Version 2*, IETF RFC2236, November 1997.

[11] Howarth, M.P., Cruickshank, H. and Sun, Z., Unicast and multicast IP error performance over an ATM satellite link, *IEEE Comms. Letters*, **5**(8): 340–2, 2001.

[12] Howarth, M., Iyngar, S., Sun, Z. and Cruickshank, H., Dynamics of key management in secure satellite multicast, *IEEE Journal on Selected Areas in Communications: Broadband IP Networks via Satellites, Part I*, **22**(2), 2004.

[13] Kent, S. and Atkinson, R., *Security Architecture for the Internet Protocol*, IETF RFC2401, November 1998.

[14] Koyabe, M. and Fairhurst, G., Reliable multicast via satellite: a comparison survey and taxonomy, *International Journal of Satellite Communications*, **19**(1): 3–28, 2001.

[15] Moy, J., *Multicast Extensions to OSPF*, IETF RFC1584, March 1994.

[16] Rosen, E., Viswanathan, A. and Callon, R., *Multiprotocol Label Switching Architecture*, IETF RFC3031, January 2001.

[17] Sahasrabuddhe, L.H. and Mukherjee, B., Multicast routing algorithms and protocols: a tutorial, *IEEE Network*, **14**(1): 90–102, 2000.

[18] Sun, Z., Broadband satellite networking, *Space Communications, Special issue on On-Board Processing*, **17**(1–3): 7–22, 2001.

[19] Sun, Z., He, D., Cruickshank, H., Liang, L., Sánchez, A. and Tocci, C., Scalable architecture and evaluation for multiparty conferencing over satellite links, *IEEE Journal on Selected Areas in Communications: Broadband IP Networks via Satellites, Part II*, **22**(3), 2004.

[20] Thaler, D., *Border Gateway Multicast Protocol (BGMP): Protocol Specification*, IETF Draft, work-in-progress, draft-ietf-bgmp-spec-06.txt, 20 January 2004.

[21] Waitzman, D., Partridge, C. and Deering, S. *Distance Vector Multicast Routing Protocol*, IETF RFC1075, November 1988.

[22] Wallner, D., Harder, E. and Agee, R., Key Management for Multicast: Issues and Architectures, IETF RFC2627, June 1999.

[23] Yegenoglu, F., Alexander, R. and Gokhale, D., An IP transport and routing architecture for next-generation satellite networks, *IEEE Network*, **14**(5): 32–8, 2000.

[24] RFC 2002, IP Mobility Support, C. Perkins, IETF, October 1996.

[25] RFC 1256, ICMP Router Discovery Messages, S. Deering, IETF, September 1991.

[26] RFC 1321, The MD5 Message-Digest Algorithm, R. Rivest, IEFT, April 1992.

[27] RFC 2627, Key Management for Multicast: Issues and Architectures, D. Wallner, E. Harder and R. Agee, IETF, June 1999.

[28] RFC 3357, One-way Loss Pattern Sample Metrics, R. Koodli and R. Ravikanth, IETF, August 2002.

Exercises

1. Explain the concepts of satellite IP networking.
2. Explain IP packet encapsulation concepts of PPP and IP tunnelling.
3. Use a sketch to explain the satellite-centric view of the global network and Internet.

Exercises (*continued*)

4. Explain IP multicast over satellite.
5. Explain DVB and related protocol stack.
6. Explain DVB over satellite including DVB-S and DVB-RCS.
7. Explain IP over DVB-S and DVB-RCS security mechanisms.
8. Discuss IP QoS performance objectives and parameters and QoS architectures of Intserv and Diffserv.

7

Impact of Satellite Networks on Transport Layer Protocols

This chapter aims to discuss the impact of satellite networks on transport layer protocols including the transmission control protocol (TCP) and their applications. TCP is a reliable transport layer protocol of the Internet protocol stack. TCP provides the protocol for end-to-end communications between a client process in one host and a server process in the other host in the Internet. TCP has neither information on applications nor information on Internet traffic conditions and the transmission technologies (such as LAN, WAN, wireless and mobile and satellite networks). TCP relies on mechanisms including flow control, error control and congestion control between the client and server hosts to recover from transmission error and loss and from network congestion and buffer overflows. All these mechanisms affect the performance of TCP over satellite and hence the Internet applications directly. This chapter also explains the major enhancements designed to improve TCP performance over satellite for a 'satellite-friendly TCP', although not all of these enhancements have become IETF standards, since they may cause some side-effects on the normal TCP operations. This chapter also provides an introduction to real-time transport protocols built on top of the UDP including RTP, RTCP, SAP, SIP, etc., and related applications including voice over IP (VoIP) and multimedia conferencing (MMC). When you have completed this chapter, you should be able to:

- Know the impact of satellite networks on the performance of TCP due to flow control, error control and congestion control mechanisms.
- Carry out a performance analysis on the standard TCP slow-start algorithm and congestion avoidance mechanism, and calculate the utilisation of satellite bandwidth.
- Know the typical mechanisms for TCP-enhancement satellite networks.
- Describe TCP enhancement on the slow-start algorithm.

Satellite Networking: Principles and Protocols   Zhili Sun
© 2005 John Wiley & Sons, Ltd

- Describe TCP enhancement on the congestion avoidance mechanism.
- Describe TCP enhancement on acknowledgement.
- Know TCP enhancement on error recovery mechanisms including fast retransmission and fast recovery.
- Learn the interruptive TCP performance acceleration mechanisms including TCP spoofing and cascading TCP (also known as split TCP).
- Understand the impact of satellite networks on different applications.
- Understand the limitation of TCP enhancement mechanisms based on existing TCP mechanisms.
- Understand real-time protocols including RTP, RTCP, SAP, SIP, etc., and their differences from other application layer protocols such as HTTP and SMTP.
- Understand VoIP and MMC based on the real time transport protocols.

7.1 Introduction

TCP is the protocol for end-to-end communications between processes in different hosts across Internet networks. It is implemented within the client host or server in order to provide applications with reliable transmission services. It is transparent to the Internet, i.e. the Internet treats it only as the payload of IP packets (see Figure 7.1).

The most challenging task of TCP is to provide reliable and efficient transmission services without knowing anything about applications above it or anything about the Internet below it. TCP carries out proper actions according to application characteristics, client and server parameters and network parameters and conditions (particularly satellite networks).

7.1.1 Application characteristics

There is a wide range of applications built on TCP, including remote login, file transfer, email and WWW. The amount of data to be transmitted by TCP can range from a few bytes

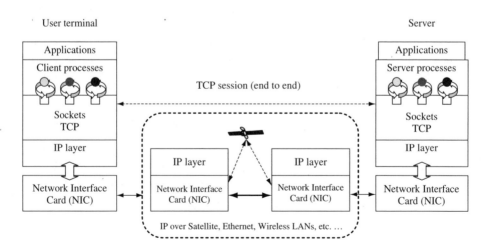

Figure 7.1 The TCP protocol over satellite Internet

to kilobytes, megabytes or even gigabytes. The duration of a TCP session can be as few as a fraction of a second up to many hours. Therefore, the data size of each transaction and total data size of each TCP session are important factors affecting TCP performance.

7.1.2 Client and server host parameters

The current Internet applications built on TCP are elastic, i.e. they can tolerant slow process and transmission of their data. It is this feature together with the TCP that allows us to build the Internet using different types of computers, from PCs to supercomputers, and enables them to communicate with each other.

The main parameters affecting TCP performance include the process power (how fast it can deal with data within the TCP session), buffer sizes (memory space allocated to the TCP session for data buffering) and speeds of network interface cards (how fast the hosts can send data to networks) in both client and server hosts, and round trip delay (RTT) between the client and the server.

7.1.3 Satellite network configurations

Satellite can play many different roles in the Internet. Figure 7.2 shows a typical example of satellite network configurations with the satellite network in the centre connecting two terrestrial access networks.

For ease of discussion, we assume that all constraints are due to the satellite network (long delay, errors, limited bandwidth, etc.). Both access networks and interworking units (routers or switches) are capable of dealing with traffic flows between access networks and the satellite network. The following are some typical satellite network configurations:

- **Asymmetric satellite networks**: DVB-S, DVB-RCS and VSAT satellite networks are configured with bandwidth asymmetry, a larger data rate in the forward direction (from

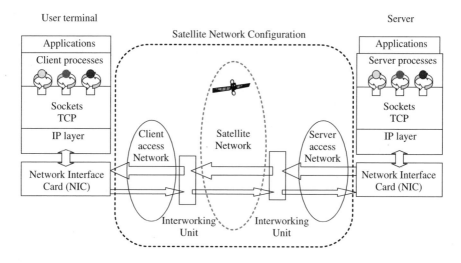

Figure 7.2   Example of satellite network configurations

satellite gateway station to user earth stations) than the return direction (user earth stations to satellite gateway station), because of limits on the transmission power and the antenna size at different satellite earth stations. Receive-only broadcasting satellite systems are unidirectional and can use a non-satellite return path (such as a dialup modem link via telephony networks). The nature of most TCP traffic is asymmetric with data flowing in one direction and acknowledgements in the opposite direction.

- **Satellite link as last hop**: satellite links that provide service directly to end users, as opposed to satellite links located in the middle of a network, may allow for specialised design of protocols used over the last hop. Some satellite providers use the satellite link as a shared high-speed downlink to users with a lower speed, non-shared terrestrial link that is used as a return link for requests and acknowledgements. In this configuration, the client host has direct access to the satellite network.

- **Hybrid satellite networks**: in the more general case, satellite links may be located at any point in the network topology. In this case, the satellite link acts as just another link between two gateways. In this environment, a given connection may be sent over terrestrial links (including terrestrial wireless), as well as satellite links. This is a typical transit network configuration.

- **Point-to-point satellite networks**: in point-to-point satellite networks, the only hop in the network is over the satellite link. This is a pure satellite network configuration.

- **Multiple satellite hops**: in some situations, network traffic may traverse multiple satellite hops between the source and the destination. Such an environment aggravates the satellite characteristics. This is a generic problem with special circumstances or space communications where there are many more constraints due to long delay, error and bandwidth.

- **Constellation satellite networks with and without inter-satellite links (ISL)**: in constellation satellite networks without ISL, multiple satellite hops are used for wide coverage. In constellation satellite networks with ISL, wide coverage is achieved by ISL. The problem is that the route of the network is highly dynamic hence end-to-end delay is variable.

7.1.4 TCP and satellite channel characteristics

The Internet differs from a single network because it consists of different network topologies, bandwidth, delays and packet sizes. TCP is formally defined in RFC793 and updated in RFC1122 and extensions are given in RFC1323 to work in such heterogeneous networks.

TCP is a byte stream, not a message stream and message boundaries are not preserved end to end. All TCP connections are full-duplex connections and point to point. As such TCP does not support multicasting or broadcasting.

The sending and receiving TCP entities exchange data in the form of segments. A segment consists of a fixed 20-byte header (plus an optional part) followed by zero or more data bytes. Two limits restrict the TCP segment size:

- Each segment must fit into the 65 535 byte IP payload (RFC2147 describes adapting TCP and UDP to use IP6 that supports datagrams larger than 65 535 bytes long).
- Each network has a maximum transfer unit (MTU). The segment must fit into the MTU.

In practice, the MTU is a few thousands of bytes and thus defines the upper boundary of the segment size. Satellite channels have several characteristics that differ from most terrestrial channels. These characteristics may degrade the performance of TCP. These characteristics include:

- **Long round trip time (RTT)**: due to the propagation delay of some satellite channels it may take a long time for a TCP sender to determine whether or not a packet has been successfully received at the final destination. This delay affects interactive applications such as telnet, as well as some of the TCP congestion control algorithms.
- **Large delay*bandwidth product**: The delay*bandwidth (DB) product defines the amount of data a protocol should have 'in flight' (data that has been transmitted but not yet acknowledged) at any one time to fully utilise the available channel capacity. The delay is the RTT (end-to-end) and the bandwidth is the capacity of the bottleneck link in the network path.
- **Transmission errors**: satellite channels exhibit a higher bit-error rate (BER) than typical terrestrial networks. TCP assumes that all packet drops are cased by network congestion and reduces its window size in an attempt to alleviate the congestion. In the absence of knowledge about why a packet was dropped (congestion at the network or corruption at the receiver), TCP must assume the drop was due to network congestion to avoid congestion collapse. Therefore, packets dropped due to corruption cause TCP to reduce the size of its sliding window, even though these packet drops do not signal congestion in the network.
- **Asymmetric use**: due to the expense of the equipment used to send data to satellites, asymmetric satellite networks are often constructed. A common situation is that the uplink has less available capacity than the downlink for return channel. This asymmetry may have an impact on TCP performance.
- **Variable round trip times**: in LEO constellations, the propagation delay to and from the satellite varies over time. This may affect retransmission time out (RTO) granularity.
- **Intermittent connectivity**: in non-GSO satellite orbit configurations, TCP connections may be handed over from one satellite to another or from one ground station to another from time to time. This may cause packet loss if not properly performed.

7.1.5 TCP flow control, congestion control and error recovery

As part of implementing a reliable service, TCP is responsible for flow and congestion control: ensuring that data is transmitted at a rate consistent with the capacities of both the receiver and the intermediate links in the network path.

Since there may be multiple TCP connections active in a link, TCP is also responsible for ensuring that a link's capacity is responsibly shared among the connections using it. As a result, most throughput issues are rooted in TCP.

To avoid generating an inappropriate amount of network traffic for the current network conditions, during a connection TCP employs four congestion control mechanisms. These algorithms are:

- slow start;
- congestion avoidance;
- fast retransmit before RTO expires;
- fast recovery to avoid slow start.

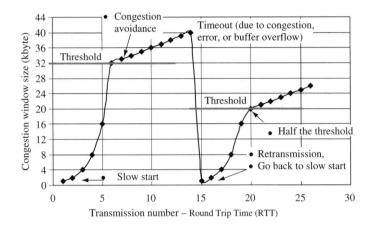

Figure 7.3   An example of TCP operations

These algorithms are described in detail in RFC2581. They are used to adjust the amount of unacknowledged data that can be injected into the network and to retransmit segments dropped by the network.

TCP senders use two state variables to accomplish congestion control. The first variable is the congestion window (cwnd). This is an upper bound on the amount of data the sender can inject into the network before receiving an acknowledgement (ACK). The value of cwnd is limited to the receiver's advertised window. The congestion window is increased or decreased during the transfer based on the inferred amount of congestion present in the network. The second variable is the slow start threshold (ssthresh). This variable determines which algorithm is used to increase the value of cwnd. If cwnd is less than ssthresh the slow-start algorithm is used to increase the value of cwnd. However, if cwnd is greater than or equal to (or just greater than in some TCP implementations) ssthresh the congestion avoidance algorithm is used. The initial value of ssthresh is the receiver's advertised window size. Further more, the value of ssthresh is set when congestion is detected. Figure 7.3 illustrates an example of TCP operations.

The above algorithms have a negative impact on the performance of individual TCP connections' performance because the algorithms slowly probe the network for additional capacity, which in turn wastes bandwidth. This is especially true over long-delay satellite channels because of the large amount of time required for the sender to obtain feedback from the receiver. However, the algorithms are necessary to prevent congestive collapse in a shared network. Therefore, the negative impact on a given connection is more than offset by the benefit to the entire network.

7.2 TCP performance analysis

The key parameter considered here is satellite link utilisation as satellite networks are very expensive and take a long time to build. The performance of TCP over satellite can be calculated as utilisation (U). The TCP transmission may complete before full bandwidth speed has been reached due to the slow-start algorithm, congestion control mechanism, and

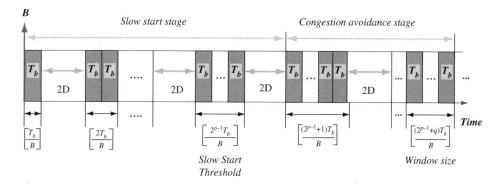

Figure 7.4 TCP segment traffic block bursts

network congestion or network errors. Figure 7.4 illustrates TCP segment traffic block bursts. This section provides analysis and calculation of bandwidth utilisation of TCP connections over a point-to-point satellite network.

7.2.1 First TCP segment transmission

After TCP connection set up, we can calculate the bandwidth utilisation when the TCP has completed the first data segment (T_b).

$$U = \frac{T}{2D+T} = \frac{T_b/B}{2D+T_b/B} = \frac{1}{2(\frac{DB}{T_b})+1} \tag{7.1}$$

Where T is the time to transmit the data T_b, D is propagation delay and B is the bandwidth capacity of the TCP session. It takes $2D$ time to acknowledgement for a successful transmission. It does not take into account the TCP three-way handshake connection set-up delay and connection close-down delay. The TCP transmission can finish when there are no more data for transmission, i.e., the total data size is less than the minimum segment size (MSS). Therefore, the utilisation is shown in Equation (7.1). It can be seen that the delay*bandwidth product is a key parameter affecting TCP performance. For satellite networks, broadband particular, the DB can be very large. It will take the round trip time $(2D)$ and data transmission time (T_b/B) to complete the TCP data transmission: $(2D+T_b/B)$.

7.2.2 TCP transmission in the slow-start stage

The utilisation can be improved if the data size is larger than the MSS traffic block T_b. The transmission will enter the TCP slow-start stage. After successful transmission of the first TCP segment traffic block T_b, two more blocks of T_b are transmitted, then a further two more traffic blocks T_b will be transmitted for each previous successful transmission(s). We can see that the number of traffic blocks T_b increases exponentially as $2^i T_b (i = 1, 2, \ldots, n-1)$

for every round trip time (RTT) if there is no packet loss. The TCP can transmit a data size of $F(n)$ as a sequence of block sizes of $2^i T_b (i = 0, 1, 2, \ldots, n-1)$. Let

$$F(n) = \left(\sum_{i=0}^{n-1} 2^i \right) T_b = (2^n - 1) T_b \tag{7.2}$$

where n is the total number of RTT needed to complete the transmission. We can calculate the utilisation of TCP connection as:

$$U_{F(n)} = \frac{F(n)/B}{2nD + F(n)/B} = \frac{(2^n - 1)T_b/B}{2nD + (2^n - 1)T_b/B} = \frac{1}{1 + \left(\dfrac{2n}{(2^n - 1)T_b} \right)(DB)} \tag{7.3}$$

The time it takes to complete the TCP data transmission is $(2nD + F(n)/B)$.

In a general case, the TCP data transmission may have completed during the slow-start stage. Then we can derive the following general formula for the transmitted data size and link utilisation as the following:

$$F = \left(\left(\sum_{i=0}^{n-1} 2^i \right) + \alpha 2^n \right) T_b = (2^n(1+\alpha) - 1) T_b \quad \text{where } 0 \le \alpha < 1$$

$$U_F = \frac{1}{1 + \left(\dfrac{2(n+1)}{(2^n(1+\alpha) - 1)Tb} \right)(DB)} \tag{7.4}$$

The time it takes to complete the TCP data transmission is $(2(n+1)D + F(n)/B)$.

7.2.3 TCP transmission in congestion avoidance stage

When the transmission data block size reaches the slow-start threshold $2^{p-1}T_b$, the slow-start algorithm stops and the congestion avoidance mechanism starts until it reaches the window size. Then the transmitted data size and link utilisation can be calculated as the following:

$$F_l = \left(\sum_{i=0}^{p-1} 2^i \right) T_b + \left(\sum_{j=1}^{m} (2^{p-1} + j) \right) T_b + \beta 2^{p-1} T_b = \left(2^p + \frac{m^2 + 3m}{2} - 1 \right) T_b$$

$$U_{F_l} = \frac{1}{1 + \dfrac{2(p+m)}{\left(2^p + \dfrac{m^2 + 3m}{2} + \beta - 1 \right) T_b}(DB)} \tag{7.5}$$

where $0 \le \beta < 1$, and $2^m T_b \le W$, the window size. When the transmission reaches the window size, TCP transmits at a constant speed of one window size of data per RTT.

In classical TCP, T_b and W are agreed initially between the client and server for the maximum size. The slow-start threshold and window size change according to the network conditions and rules of TCP. If a packet gets lost, TCP goes back to the slow-start algorithm and the threshold is reduced to a half. The window size depends on how fast the receiver can empty the receive buffer.

The basic assumption is that packet loss is due to network congestion, and such an assumption is true in normal networks but not always true in satellite networks where transmission errors can also be the major cause of packet loss.

7.3 Slow-start enhancement for satellite networks

There are many TCP enhancements to make TCP friendly to satellite. In order to optimise TCP performance, we can adapt some of the parameters and TCP rules to the satellite networking environment:

- Increasing the minimum segment size transmission block size (T_b), but it is limited by the slow-start threshold, congestion window size and receiver buffer size.
- Improving the slow-start algorithm at the start, and when a packet gets lost. It may cause problems such as slowing the receiver and congested networks.
- Improving acknowledgement. This may need additional buffer space.
- Early detecting packet loss due to transmission error rather than network congestion. It may not work if acknowledgements are transmitted over different network paths.
- Improving congestion avoidance mechanisms. This has similar problems to the slow-start algorithm.

One of the major problems is that TCP does not have any knowledge about the total data size and the available bandwidth. If the bandwidth (B) is shared among many TCP connections, the available bandwidth (B) can also be variable. Another is that TCP does not know how the IP layer actually carries the TCP segment across the Internet, because the IP packets may need to be limited in size or split into small packets for the network technologies transporting the IP packets. This makes the TCP a robust protocol that provides reliable services for different applications over different technologies, but is often not very efficient; particularly for satellite networks (see Figure 7.5). The RTT is measured by timing when

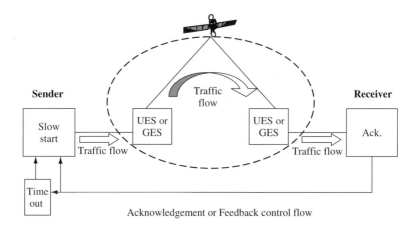

Figure 7.5 Traffic and control flows

the packet was sent out and the acknowledgement returned as M_n, and the average RTT_n calculated with a weight factor α (typically $\alpha = 7/8$, and RTT_0 is set to a default value) as:

$$RTT_n = \alpha RTT_{n-1} + (1 - \alpha)M_n$$

The deviation is calculated with the same weight factor α as:

$$D_n = \alpha D_{n-1} + (1 - \alpha)|M_n - RTT_{n-1}|$$

Then the time out can be calculated as:

$$\text{Timeout} = RTT_n + 4D_n$$

We will now discuss some TCP enhancement techniques. These are optimised to deal with particular conditions in satellite network configurations, but may have side effects or may not be applicable to general network configurations. It is also a great challenge for the enhancement to interwork with existing TCP implementations.

7.3.1 TCP for transactions

In a transaction service, particularly for short data size and TCP session, the utilisation is significantly affected by the connection set-up and connection close-down time. TCP uses a three-way handshake to set-up a connection between two hosts. This connection set-up requires one or 1.5 RTT, depending upon whether the data sender started the connection actively or passively. This start-up time can be eliminated by using TCP extensions for transactions (T/TCP). After the first connection between a pair of hosts is established, T/TCP is able to bypass the three-way handshake, allowing the data sender to begin transmitting data in the first segment sent (along with the SYN – synchronisation number). This is especially helpful for short request/response traffic, as it saves a potentially long set-up phase when no useful data are being transmitted.

As each of the transactions has a small data size, the utilisation of satellite bandwidth can be very low. However, it has the potential for many TCP session hosts to share the same bandwidth to improve bandwidth utilisation. T/TCP requires changes of both the sender and the receiver. While T/TCP is safe to implement in shared networks from a congestion control perspective, several security implications of sending data in the first data segment have been identified.

7.3.2 Slow start and delayed acknowledgement (ACK)

As we have discussed, TCP uses the slow-start algorithm to increase the size of TCP's congestion window (cwnd) at exponential speed. The algorithm is an important safeguard against transmitting an inappropriate amount of data into the network when the connection starts up. However, slow start can also waste available network capacity due to large delay*bandwidth product of the network, especially in satellite networks.

In delayed acknowledgement (ACK) schemes, receivers refrain from acknowledging every incoming data segment. Every second full-sized segment is acknowledged. If a second full-sized segment does not arrive within a given timeout, an ACK must be generated (this timeout cannot exceed 500 ms). Since the sender increases the size of cwnd based on the number of arriving ACKs, reducing the number of ACKs slows the cwnd growth rate. In addition, when TCP starts sending, it sends one segment. When using delayed ACKs a second segment must arrive before an ACK is sent. Therefore, the receiver is always forced to wait for the delayed ACK timer to expire before ACKing the first segment, which also increases the transfer time.

7.3.3 Larger initial window

One method that will reduce the amount of time required by slow start (and therefore, the amount of wasted capacity) is to increase the initial value of cwnd. However, TCP has been extended to support larger windows (RFC1323). The window-scaling options can be used in satellite environments, as well as the companion algorithms PAWS (protection against wrapped wequence space) and RTTM (round-trip time measurements).

By increasing the initial value of cwnd, more packets are sent during the first RTT of data transmission, which will trigger more ACKs, allowing the congestion window to open more rapidly. In addition, by sending at least two segments initially, the first segment does not need to wait for the delayed ACK timer to expire as is the case when the initial size of cwnd is one segment. Therefore, the value of cwnd saves the number of RTT and a delayed ACK timeout. In the standards-track document RFC2581, TCP allows an initial cwnd of up to two segments. It is expected that the use of a large initial window would be beneficial for satellite networks.

The use of a larger initial cwnd value of two segments requires changes to the sender's TCP stack, defined in RFC2581. Using an initial congestion window of three or four segments is not expected to present any danger of congestion collapse, however, it may degrade performance in some networks if the network or terminal cannot cope with such burst traffic.

Using a fixed larger initial congestion window decreases the impact of a long RTT on transfer time (especially for short transfers) at the cost of bursting data into a network with unknown conditions. A mechanism is required to limit the effect of these bursts. Also, using delayed ACKs only after slow start offers an alternative way to immediately ACK the first segment of a transfer and opens the congestion window more rapidly.

7.3.4 Terminating slow start

The initial slow-start phase is used by TCP to determine an appropriate congestion window size for the given network conditions. Slow start is terminated when TCP detects congestion, or when the size of cwnd reaches the size of the receiver's advertised window. Slow start is also terminated if cwnd grows beyond a certain size. TCP ends slow start and begins using the congestion avoidance algorithm when it reaches the slow-start threshold (ssthresh). In most implementations, the initial value for ssthresh is the receiver's advertised window. During slow start, TCP roughly doubles the size of cwnd every RTT and therefore can overwhelm the network with at most twice as many segments as the network can handle.

By setting ssthresh to a value less than the receiver's advertised window initially, the sender may avoid overwhelming the network with twice the appropriate number of segments.

It is possible to use the packet-pair algorithm and the measured RTT to determine a more appropriate value for ssthresh. The algorithm observes the spacing between the first few returning ACKs to determine the bandwidth of the bottleneck link. Together with the measured RTT, the delay*bandwidth product is determined and ssthresh is set to this value. When the cwnd reaches this reduced ssthresh, slow start is terminated and transmission continues using congestion avoidance, which is a more conservative algorithm for increasing the size of the congestion window.

Estimating ssthresh can improve performance and decrease packet loss, but obtaining an accurate estimate of available bandwidth in a dynamic network is very challenging, especially attempting on the sending side of the TCP connection.

Estimating ssthresh requires changes to the data sender's TCP stack. Bandwidth estimates may be more accurate when taken by the TCP receiver, and therefore both sender and receiver changes would be required. It makes TCP more conservative than outlined in RFC2581.

It is expected that this mechanism will work equally well in all symmetric satellite network configurations. However, asymmetric links pose a special problem, as the rate of the returning ACKs may not be the bottleneck bandwidth in the forward direction. This can lead to the sender setting ssthresh too low. Premature termination of slow start can hurt performance, as congestion avoidance opens cwnd more conservatively. Receiver-based bandwidth estimators do not suffer from this problem, but needs changes the TCP in receiver side as well.

Terminating slow start at the right time is useful to avoid overflowing the network, hence avoiding multiple dropped segments. However, using a selective acknowledgement-based loss recovery scheme can drastically improve TCP's ability to quickly recover from multiple lost segments.

7.4 Loss recovery enhancement

Satellite paths have higher error rates than terrestrial lines. Higher error rates matter for two reasons. First, they cause errors in data transmissions, which will have to be retransmitted. Second, as noted above, TCP typically interprets loss as a sign of congestion and goes back into the slow start. Clearly we need to either reduce the error rate to a level acceptable to TCP (i.e., it allows the data transmissions to reach the full window size without suffering any packet loss) or find a way to let TCP know that the datagram loss is due to transmission errors, not congestion (and thus TCP should not reduce its transmission rate).

Loss recovery enhancement is to prevent TCP going to slow start unnecessarily when data segments get lost due to error rather network congestion. Several similar algorithms have been developed and studied that improve TCP's ability to recover from multiple lost segments without relying on the (often long) retransmission timeout. These sender-side algorithms, known as NewReno TCP (one of the TCP implementations) do not depend on the availability of selective acknowledgements (SACK).

7.4.1 Fast retransmission and fast recovery

It is possible during transmission that one or more TCP segments may not reach the other end of the connection, and TCP uses timeout mechanisms to detect those missing segments.

In normal situations, TCP assumes that segments are dropped due to network congestion. This usually results in ssthresh being set to half the current value of the congestion window (cwnd), and the cwnd size is being reduced to the size of one TCP segment. This severely affects TCP throughput. The situation is worse when the loss of TCP segments is not due to network congestion. To avoid the unnecessary process of going back to the slow-start process each time a segment fails to reach the intended destination, the process of fast retransmission was introduced.

The fast retransmission algorithm uses duplicate ACKs to detect the loss of segments. If three duplicate ACKs are received within the timeout period, TCP immediately retransmits the missing segment without waiting for the timeout to occur. Once fast retransmission is used to retransmit the missing data segment, TCP can use its fast recovery algorithm, which will resume the normal transmission process via the congestion avoidance phase instead of slow start as before. However, in this case ssthresh will be reduced to half the value of cwnd, and the value of cwnd is itself halved. This allows faster data transmission than is the case with TCP's normal timeout.

7.4.2 Selective acknowledgement (SACK)

TCP, even with fast retransmission and fast recovery, still performs poorly when multiple segments are lost within a single transmission window. This is due to the fact that TCP can only learn of a missing segment per RTT, due to the lack of cumulative acknowledgements. This limitation reduces TCP throughout.

To improve TCP performance for this situation, selective acknowledgement (SACK) is proposed (RFC2018). The SACK option format allows any missing segments to be identified and typically retransmits them within a single RTT. By adding extra information about all the received segments sequence numbers, the sender is notified about which segments have not been received and therefore need to be retransmitted. This feature is very important in satellite network environments due to occasional high bit-error rates (BER) of the channel, and using larger transmission windows has increased the possibility of multiple segment losses in a single round trip.

7.4.3 SACK based enhancement mechanisms

It is possible to use a conservative extension to the fast recovery algorithm that takes into account information provided by SACKs. The algorithm starts after fast retransmit triggers the resending of a segment. As with fast retransmit, the algorithm reduces cwnd into half of the size when a loss is detected. The algorithm keeps a variable called 'pipe', which is an estimate of the number of outstanding segments in the network. The pipe variable is decremented by one segment for each duplicate ACK that arrives with new SACK information. The pipe variable is incremented by one for each new or retransmitted segment sent. A segment may be sent when the value of pipe is less than cwnd (this segment is either a retransmission per the SACK information or a new segment if the SACK information indicates that no more retransmits are needed).

This algorithm generally allows TCP to recover from multiple segment losses in a window of data within one RTT of loss detection. The SACK information allows the pipe algorithm to decouple the choice of when to send a segment from the choice of what segment to send. It is also consistent with the spirit of the fast recovery algorithm.

Some research has shown that the SACK based algorithm performs better than several non-SACK based recovery algorithms, and that the algorithm improves performance over satellite links. Other research shows that in certain circumstances, the SACK algorithm can hurt performance by generating a large line-rate burst of data at the end of loss recovery, which causes further loss.

This algorithm is implemented in the sender's TCP stack. However, it relies on SACK information generated by the receiver (RFC2581).

7.4.4 ACK congestion control

Acknowledgement enhancement is concerned with the acknowledgement packet flows. In a symmetric network, this is not an issue, as the ACK traffic is much less than the data traffic itself. But for asymmetric networks, the return link has much lower speed than the forward link. There is still the possibility that the ACK traffic overloads the return link, hence restricting the performance of the TCP transmissions.

In highly asymmetric networks, such as VSAT satellite networks, a low-speed return link can restrict the performance of the data flow on a high-speed forward link by limiting the flow of acknowledgements returned to the data sender. If a terrestrial modem link is used as a reverse link, ACK congestion is also likely, especially as the speed of the forward link is increased. Current congestion control mechanisms are aimed at controlling the flow of data segments, but do not affect the flow of ACKs.

The flow of acknowledgements can be restricted on the low-speed link not only by the bandwidth of the link, but also by the queue length of the router. The router may limit its queue length by counting packets, not bytes, and therefore begin discarding ACKs even if there is enough bandwidth to forward them.

7.4.5 ACK filtering

ACK filtering (AF) is designed to address the same ACK congestion effects. Contrary to ACK congestion control (ACC), however, AF is designed to operate without host modifications.

AF takes advantage of the cumulative acknowledgement structure of TCP. The bottleneck router in the reverse direction (the low-speed link) must be modified to implement AF. Upon receipt of a segment, which represents a TCP acknowledgement, the router scans the queue for redundant ACKs for the same connection, i.e. ACKs which acknowledge portions of the window which are included in the most recent ACK. All of these 'earlier' ACKs are removed from the queue and discarded.

The router does not store state information, but does need to implement the additional processing required to find and remove segments from the queue upon receipt of an ACK.

As is the case in ACC, the use of ACK filtering alone would produce significant sender bursts, since the ACKs will be acknowledging more previously unacknowledged data. The sender adaptation (SA) modifications could be used to prevent those bursts, at the cost of requiring host modifications. To prevent the need for modifications in the TCP stack, AF is more likely to be paired with the ACK reconstruction (AR) technique, which can be implemented at the router where segments exit the slow reverse link.

AR inspects ACKs exiting the link, and if it detects large 'gaps' in the ACK sequence, it generates additional ACKs to reconstruct an acknowledgement flow which more closely resembles what the data sender would have seen had ACK filtering not been introduced. AR requires two parameters: one parameter is the desired ACK frequency; while the second controls the spacing, in time, between the releases of consecutive reconstructed ACKs.

7.4.6 Explicit congestion notification

Explicit congestion notification (ECN) allows routers to inform TCP senders about imminent congestion without dropping segments. There are two major forms of ECN:

- The first major form of congestion notification is backward ECN (BECN). A router employing BECN transmits messages directly to the data originator informing it of congestion. IP routers can accomplish this with an ICMP source quench message. The arrival of a BECN signal may or may not mean that a TCP data segment has been dropped, but it is a clear indication that the TCP sender should reduce its sending rate (i.e., the value of cwnd).
- The second major form of congestion notification is forward ECN (FECN). FECN routers mark data segments with a special tag when congestion is imminent, but forward the data segment. The data receiver then echoes the congestion information back to the sender in the ACK packet.

Senders transmit segments with an 'ECN-capable transport' bit set in the IP header of each packet. If a router employing an active queuing strategy, such as random early detection (RED), would otherwise drop this segment, a 'congestion experienced' bit in the IP header is set instead. Upon reception, the information is echoed back to TCP senders using a bit in the TCP header. The TCP sender adjusts the congestion window just as it would if a segment was dropped.

The implementation of ECN requires the deployment of active queue management mechanisms in the affected routers. This allows the routers to signal congestion by sending TCP a small number of 'congestion signals' (segment drops or ECN messages), rather than discarding a large number of segments, as can happen when TCP overwhelms a drop-tail router queue.

Since satellite networks generally have higher bit-error rates than terrestrial networks, determining whether a segment was lost due to congestion or corruption may allow TCP to achieve better performance in high BER environments than currently possible (due to TCP's assumption that all loss is due to congestion). While not a solution to this problem, adding an ECN mechanism to TCP may be a part of a mechanism that will help achieve this goal.

Research shows that ECN is effective in reducing the segment loss rate, which yields better performance especially for short and interactive TCP connections, and that ECN avoids some unnecessary and costly TCP retransmission timeouts.

Deployment of ECN requires changes to the TCP implementation on both sender and receiver. Additionally, deployment of ECN requires deployment of some active queue management infrastructure in routers. RED is assumed in most ECN discussions, because RED is already identifying segments to drop, even before its buffer space is exhausted. ECN simply allows the delivery of 'marked' segments while still notifying the end nodes that congestion is occurring along the path. ECN maintains the same TCP congestion control principles as are used when congestion is detected via segment drops. Due to long propagation delay, the ECN signalling may not reflect the current status of networks accurately.

7.4.7 Detecting corruption loss

Differentiating between congestion (loss of segments due to router buffer overflow or imminent buffer overflow) and corruption (loss of segments due to damaged bits) is a difficult problem for TCP. This differentiation is particularly important because the action that TCP should take in the two cases is entirely different. In the case of corruption, TCP should merely retransmit the damaged segment as soon as its loss is detected; there is no need for TCP to adjust its congestion window. On the other hand, as has been widely discussed above, when the TCP sender detects congestion, it should immediately reduce its congestion window to avoid making the congestion worse.

TCP's defined behaviour in terrestrial wired networks is to assume that all loss is due to congestion and to trigger the congestion control algorithms. The loss may be detected using the fast retransmit algorithm, or in the worst case is detected by the expiration of TCP's retransmission timer. TCP's assumption that loss is due to congestion rather than corruption is a conservative mechanism that prevents congestion collapse.

Over satellite networks, however, as in many wireless environments, loss due to corruption is more common than on terrestrial networks. One common partial solution to this problem is to add forward error correction (FEC) to the data that are sent over the satellite or wireless links. However, given that FEC does not always work or cannot be universally applied, it is important to make TCP able to differentiate between congestion-based and corruption-based loss.

TCP segments that have been corrupted are most often dropped by intervening routers when link-level checksum mechanisms detect that an incoming frame has errors. Occasionally, a TCP segment containing an error may survive without detection until it arrives at the TCP receiving host, at which point it will almost always either fail the IP header checksum or the TCP checksum and be discarded as in the link-level error case. Unfortunately, in either of these cases, it is not generally safe for the node detecting the corruption to return information about the corrupt packet to the TCP sender because the sending address itself might have been corrupted.

Because the probability of link errors on a satellite link is relatively greater than on a hardwired link, it is particularly important that the TCP sender retransmit these lost segments without reducing its congestion window. Because corrupt segments do not indicate congestion, there is no need for the TCP sender to enter a congestion avoidance phase, which may waste available bandwidth. Therefore, it can improve TCP performance if TCP can properly differentiate between corruption and congestion of networks.

7.4.8 Congestion avoidance enhancement

During congestion avoidance, in the absence of loss, the TCP sender adds approximately one segment to its congestion window during each RTT. This policy leads to unfair sharing of bandwidth when multiple connections with different RTTs traverse the same bottleneck link, with the long RTT connections obtaining only a small fraction of their fair share of the bandwidth.

One effective solution to this problem is to deploy fair queuing and TCP-friendly buffer management in network routers. However, in the absence of help from the network, there are two possible changes available to the congestion avoidance policy at the TCP sender:

- The 'constant-rate' increase policy attempts to equalise the rate at which TCP senders increase their sending rate during congestion avoidance. It could correct the bias against long RTT connections, but may be difficult to incrementally deploy in an operational network. Further studies are required on the proper selection of a constant (for the constant rate of increase).
- The 'increase-by-K' policy can be selectively used by long RTT connections in a heterogeneous environment. This policy simply changes the slope of the linear increase, with connections over a given RTT threshold adding 'K' segments to the congestion window every RTT, instead of one. This policy, when used with small values of K, may be successful in reducing the unfairness while keeping the link utilisation high, when a small number of connections share a bottleneck link. Further studies are required on the selection of the constant K, the RTT threshold to invoke this policy, and performance under a large number of flows.

Implementation of either the 'constant-rate' or 'increase-by-K' policies requires a change to the congestion avoidance mechanism at the TCP sender. In the case of 'constant-rate', such a change must be implemented globally. Additionally, the TCP sender must have a reasonably accurate estimate of the RTT of the connection. The algorithms outlined above violate the congestion avoidance algorithm as outlined in RFC2581 and therefore should not be implemented in shared networks at this time.

These solutions are applicable to all satellite networks that are integrated with a terrestrial network, in which satellite connections may be competing with terrestrial connections for the same bottleneck link. But increasing the congestion window by multiple segments per RTT can cause TCP to drop multiple segments and force a retransmission timeout in some versions of TCP. Therefore, the above changes to the congestion avoidance algorithm may need to be accompanied by a SACK-based loss recovery algorithm that can quickly repair multiple dropped segments.

7.5 Enhancements for satellite networks using interruptive mechanisms

According to the principle of protocols, each layer of the protocol should only make use of the services provided by the protocol below it to provide services to the protocol above it. TCP is a transport layer protocol providing end-to-end connection-oriented services. Any function between the TCP connection or Internet protocol below it should not disturb or interrupt the TCP data transmission or acknowledgement flows. As the characteristics of satellite networks are known to networking design, there is potential to benefit performance by making using of such knowledge but in an interruptive manner. Two methods have been widely used: TCP spoofing and TCP cascading (also known as split TCP), but they violate the protocol layering principles for network performance. Figure 7.6 illustrates the concept of interruptive mechanisms of satellite-friendly TCP (TCP-sat).

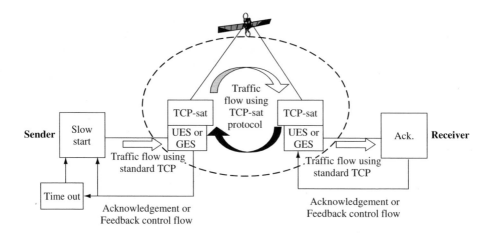

Figure 7.6 The concept of satellite-friendly TCP (TCP-sat)

7.5.1 TCP spoofing

TCP spoofing is an idea for getting around slow start in a practice known for satellite networks particularly GEO satellite links. The idea calls for a router near the satellite link to send back acknowledgements for the TCP data to give the sender the illusion of a short delay path. The router then suppresses acknowledgements returning from the receiver, and takes responsibility for retransmitting any segments lost downstream of the router.

Though TCP spoofing helps to improve TCP performance over satellite, there are a number of problems with this scheme. First, the router must do a considerable amount of work after it sends an acknowledgement. It must buffer the data segment because the original sender is now free to discard its copy (the segment has been acknowledged) and so if the segment gets lost between the router and the receiver, the router has to take full responsibility for retransmitting it. One side effect of this behaviour is that if a queue builds up, it is likely to be a queue of TCP segments that the router is holding for possible retransmission. Unlike an IP datagram, this data cannot be deleted until the router gets the relevant acknowledgements from the receiver.

Second, spoofing requires symmetric paths: the data and acknowledgements must flow along the same path through the router. However, in much of the Internet, asymmetric paths are quite common.

Third, spoofing is vulnerable to unexpected failures. If a path changes or the router crashes, data may be lost. Data may even be lost after the sender has finished sending and, based on the router's acknowledgements, reported data successfully transferred.

Fourth, it does not work if the data in the IP datagram are encrypted because the router will be unable to read the TCP header.

7.5.2 Cascading TCP or split TCP

Cascading TCP, also known as split TCP, is an idea where a TCP connection is divided into multiple TCP connections, with a special TCP connection running over the satellite link.

The thought behind this idea is that the TCP running over the satellite link can be modified, with knowledge of the satellite's properties, to run faster.

Because each TCP connection is terminated, cascading TCP is not vulnerable to asymmetric paths. And in cases where applications actively participate in TCP connection management (such as web caching) it works well. But otherwise cascading TCP has the same problems as TCP spoofing.

7.5.3 The perfect TCP solution for satellite networking

A perfect solution should be able to meet the requirements of user applications, takes into account the characteristics of data traffic and makes full use of network resources (processing power, memory and bandwidth). Current solutions based on the enhancement of existing TCP mechanisms have reached their limits as neither knowledge about applications nor knowledge about networks and hosts (client and server computers) are taken into account. In future networks, with application traffic characteristics and QoS requirements together with knowledge of network resources, it should be possible to achieve a perfect solution for the TCP within the integrated network architecture. It will need new techniques to achieve multi-layer and cross-layer optimisation of protocol architecture. It will have potentially more benefit to satellite networks where efficient utilisation of the expensive bandwidth resources is the main objective.

7.6 Impacts on applications

TCP support a wide range of applications. Different applications have different characteristics; hence they are affected by TCP in different ways. This also tells us that it is impossible to have one perfect solution for all the different applications without knowing the characteristics of these applications. Here we give examples of how different applications may be affected by TCP in satellite networks.

7.6.1 Bulk transfer protocols

The file transfer protocol (FTP) can be found on all TCP/IP installed systems and provides an example for the most commonly executed bulk transfer protocol. FTP allows the user to log onto a remote machine and either download files from or upload files to the machine.

At bandwidths of 64 kbit/s and 9.6 kbit/s, throughput was proportional to the bandwidth available and delay had little effect on the performance. This was due to the 24-kbyte window size, which was large enough to prevent any window exhaustion. At a bandwidth of 1 Mbit/s however, window exhaustion occurred and the delay had a detrimental effect on the throughput of the system. Link utilisation dropped from 98% at 64 kbit/s and 9.6% kbit/s to only 30% for 1 Mbit/s. The throughput, however, was still higher for the 1 Mbit/s case (due to reduced serialisation delay of the data). All transfers were conducted with a 1 Mbyte file, which was large enough to negate the effect of the slow-start algorithm. Other bulk transfer protocols e.g. SMTP and RCP recorded similar performances using a typical application file size.

At 64 kbit/s link capacity the return link could be reduced to 4.8 kbit/s with no effect on the throughput of the system. This was due to the limited bandwidth availability for the outbound connection, which experienced congestion. At 2.4 kbit/s return link bandwidth, transfer showed a 25% decrease in throughput, resulting from ACKs in the return link.

At a 1 Mbit/s outbound link speed, the performance of FTP was affected more by the TCP window size (24 kbytes) than by any variation in the bandwidth of the return link. It was not affected until the return link dropped to 9.6 kbit/s and started to show congestion. A 15% drop in performance was recorded for the return of 9.6 kbit/s. Delay again had a significant effect on the performance at 1 Mbit/s due to the window exhaustion.

The high ratio of outbound to inbound traffic experienced in the FTP session means that it is well suited to links with limited return bandwidth. For a 64 kbit/s outbound link, FTP will perform well with return links down to 4.8 kbit/s.

7.6.2 Semi-interactive protocols

WWW browsers use the HTTP protocol to view graphical pages downloaded from remote machines. The performance of the HTTP protocol is largely dependent on the structure of the HTML files being downloaded.

At bandwidths of 1 Mbit/s and 64 kbit/s the throughput was largely governed by the delay, due to the majority of the sessions being spent in the open/close and slow-start stages of transfer, which are affected by the RTT of the Internet. At 9.6 kbit/s this effect was overshadowed by the serialisation delay caused by the limited bandwidth on the outbound link. With bandwidths of 1 Mbit/s and 64 kbit/s the performance was found as expected. At 9.6 kbit/s the users tended to get frustrated when downloading large files and would abandon the session.

At 1 Mbit/s and 64 kbit/s, the speed of the return link had a far greater effect than any variation in delay. This was due to congestion in the return link, arising because of the low server/client traffic ratio. The lower ratio was a result of the increased number of TCP connections required to download each object. At 9.6 kbit/s the return link was close to the congestion, but still offered throughputs comparable to that at 64 kbit/s. At 4.8 kbit/s the return link became congested and the outbound throughput showed a 50% drop off. A further 50% reduction in the outbound throughput occurred when the return link dropped to 2.4 kbit/s.

For both the 1 Mbit/s and 64 kbit/s inbound, the return link speed was down to 19.2 kbit/s, which was acceptable. Below this rate, users started to become frustrated by the time taken to request a WWW page. A return bandwidth of at least 19.2 kbit/s is therefore recommended for WWW applications.

7.6.3 Interactive protocols

A telnet session allows the user to log onto a remote system, using his computer as a simple terminal. This allows a poor performance computer to make use of the resources of a higher power CPU at a remote site or to access resources not available at a local site.

The telnet sessions were judged subjectively by the user. At 1 Mbit/s and 64 kbit/s, users noticed the changes in delay more than the bandwidth, but at 9.6 kbit/s the delay due to serialisation was the more noticeable effect and became annoying to the user. The performance of interactive sessions was greatly dependent on the type of session. Telnet sessions used to view and move directories/files were performed satisfactorily down to 9.6 kbit/s. Similar performance was observed for other interactive protocols (e.g. rlogin, SNMP, etc.).

During interactive sessions, reducing the bandwidth of the return link increased the serial-isation delay of the routers. This was counterbalanced by the fact that most of the datagrams sent from the remote side consisted of only 1 or 2 bytes of the TCP payload and therefore could be serialised relatively quickly. Reducing the bandwidth was noticeable only to the competent typist where the increased data flow from the remote network resulted in increased serialisation and round trip times.

7.6.4 Distributed methods for providing Internet services and applications

User requests on the Internet are often served by a single machine. Very often and especially when this server exists in a rather distant location, the user experiences reduced throughput and network performance. This low throughput is due to bottlenecks that can be either the server itself or one or more congested Internet routing hops. Furthermore, that server represents a single point of failure – if it is down, access to the information is lost.

To preserve the usability of the information distributed in the Internet such as grid computing and peer-to-peer networks, the following issues need to be addressed at the server level:

- Document retrieval latency times must be decreased.
- Document availability must be increased, perhaps by distributing documents among several servers.
- The amount of data transferred must be reduced – certainly an important issue for anyone paying for network usage.
- Network access must be redistributed to avoid peak hours.
- Improvements in general user perceived performance.

Of course these goals must be implemented to retain transparency for the user as well as backward compatibility with existing standards. A popular and widely accepted approach to address at least some of these problems is the use of caching proxies.

A user may experience high latency when accessing a server that is attached to a network with limited bandwidth. Caching is a standard solution for this type of problem, and it was applied to the Internet (mainly to WWW) early for this reason. Caching has been a well-known solution to increase computer performance since the 1960s. The technique is now applied in nearly every computer's architecture. Caching relies on the principle of locality of reference which assumes that the most recently accessed data have the highest probability of being accessed again in the near future. The idea of Internet caching relies on the same principle.

ICP (Internet caching protocol) is a well-organised, university-based effort that deals with these issues. ICP is currently implemented in the public domain Squid proxy server. ICP is a protocol used for communication among squid caches. ICP is primarily used within a cache hierarchy to locate specific objects in sibling caches. If a squid cache does not have a requested document, it sends an ICP query to its siblings, and the siblings respond with ICP replies indicating a 'HIT' or a 'MISS'. The cache then uses the replies to choose from which cache to resolve its own MISS. ICP also supports multiplexed transmission of multiple object streams over a single TCP connection. ICP is currently implemented on top of UDP. Current versions of Squid also support ICP via multicast.

Another way of reducing the overall bandwidth and the latency, thus increasing the user-perceived throughput, is by using replication. This solution can also provide a more fault-tolerant and evenly balanced system. Replication offers promise towards solving some of the deficiencies of the proxy caching method.

A recent example of replication was the information of NASA's mission to Mars. In that case the information about the mission was replicated in several sites in the USA, Europe, Japan and Australia in order to be able to satisfy the millions of user requests.

7.6.5 Web caching in satellite networks

The concept of web caching is quite popular since many Internet service providers (ISPs) already use central servers to hold popular web pages, thus avoiding the increased traffic and delays created when thousands of subscribers request and download the same page across the network. Caches can be quite efficient but they have several weak points as they are limited by the number of people that are using each cache.

A solution can be provided by using a satellite system to distribute caches among ISPs. This concept can boost Internet performance, since many already fill multiple 1.5 Mbit/s T1 or 2 Mbit/s E1 lines primarily with web traffic. The broadcast satellite link could avoid much of that backhaul, but research is needed for delivering the proof.

Such a satellite system can be useful and becomes significantly exploited in circumstances where bandwidth is expensive and traffic jams and delays are significant, i.e. trans-Atlantic access. For example, a large amount of web content resides in the US and European ISPs face a heavy bandwidth crunch to move data their way. A complete satellite system where caching can be introduced in most of its points (i.e. ISP, Internet, LAN, etc.) is presented in Figure 7.7.

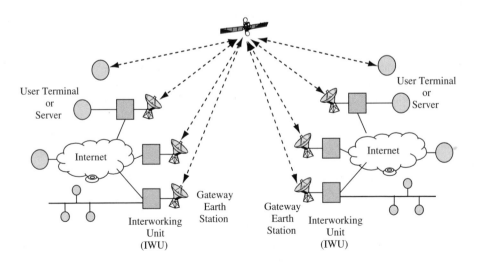

Figure 7.7   Satellite configuration with caches at IWU

7.7 Real-time transport protocol (RTP)

Originally the Internet protocols (e.g. TCP/IP) were primarily specified for the transmission of raw data between computer systems. For a long time the TCP/IP protocol suite was adequate for the transmission of still pictures and other row data-based documents. However, the emergence of modern applications and mainly those based on real-time voice and video present new requirements to the IP protocol suite. Though the former IP is not the ideal protocol for these suite of services, many applications appeared which present real-time (or near real-time) characteristics using IP. Products are available that support streaming audio, streaming video and audio-video conferencing.

7.7.1 Basics of RTP

The real-time transport protocol (RTP) provides end-to-end network transport functions suitable for applications transmitting real-time data, such as audio, video or simulation data, over multicast or unicast network services. RTP does not address resource reservation and does not guarantee QoS for real-time services. The data transport is augmented by a control protocol (RTCP), which allows monitoring of the data delivery in a manner scalable to large multicast networks, and provides minimal control and identification functionality. RTP and RTCP are designed to be independent of the underlying transport and network layers.

Applications typically run RTP on top of UDP to make use of its multiplexing and checksum services. Figure 7.8 illustrates that the RTP is encapsulated into a UDP datagram, which is transported by an IP packet.

Both RTP and RTCP protocols contribute parts of the transport protocol functionality. There are two closely linked parts:

- The real-time transport protocol (RTP), to carry data that has real-time properties.
- The RTP control protocol (RTCP), to monitor the quality of service and to convey information about the participants in an ongoing session.

A defining property of real-time applications is the ability of one party to signal to one or more other parties and initiate a call. Session invitation protocol (SIP) is a client-server protocol that enables peer users to establish a virtual connection (association) between them and then refers to a RTP (real-time transport protocol) (RFC1889) session carrying a single media type. RTP provides end-to-end network transport functions suitable for applications transmitting real-time data, such as audio, video or simulation data, over multicast or unicast network services. RTP does not address resource reservation and does not guarantee QoS for real-time services.

Note that RTP itself does not provide any mechanism to ensure timely delivery or provide other QoS guarantees, but relies on lower layer services to do so. It does not guarantee

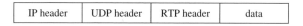

IP header	UDP header	RTP header	data

Figure 7.8 RTP packet encapsulations

delivery or prevent out-of-order delivery, nor does it assume that the underlying network is reliable and delivers packets in sequence.

There are four network components:

- **End system**: an application that generates the content to be sent in RTP packets and/or consumes the content of received RTP packets.
- **Mixer**: an intermediate system that receives RTP packets from one or more sources, possibly changes the data format, combines the packets in some manner and then forwards a new RTP packet.
- **Translator**: an intermediate system that forwards RTP packets with their synchronisation source identifier intact. Examples of translators include devices that convert encodings without mixing, replications from multicast to unicast, and application-level filters in firewalls.
- **Monitor**: an application that receives RTCP packets sent by participants in an RTP session, in particular the reception reports, and estimates the current QoS for distribution monitoring, fault diagnosis and long-term statistics.

Figure 7.9 shows the RTP header format. The first 12 octets are present in every RTP packet, while the list of contribution source (CSRC) identifiers is present only when inserted by a mixer. The fields have the following meaning:

- Version (V): two bits – this field identifies the version of RTP. The current version is two (2). (The value 1 is used by the first draft version of RTP and the value 0 is used by the protocol initially implemented in the 'vat' audio tool.)
- Padding (P): one bit – if the padding bit is set, the packet contains one or more additional padding octets at the end which are not part of the payload. The last octet of the padding

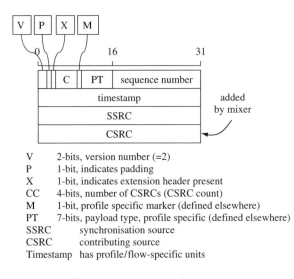

Figure 7.9 RTP header information

contains a count of how many padding octets should be ignored, including last padding octet.

- Extension (X): one bit – if the extension bit is set, the fixed header *must* be followed by exactly one header extension, with a defined format.
- Contribution source (CSRC) count (CC): four bits – the CSRC count contains the number of CSRC identifiers that follow the fixed header.
- Marker (M): one bit – the interpretation of the marker is defined by a profile.
- Payload type (PT): seven bits – this field identifies the format of the RTP payload and determines its interpretation by the application. A set of default mappings for audio and video is specified in the companion RFC1890.
- Sequence number: 16 bits – the sequence number increments by one for each RTP data packet sent, and may be used by the receiver to detect packet loss and to restore packet sequence.
- Timestamp: 32 bits – the timestamp reflects the sampling instant of the first octet in the RTP data packet. The sampling instant *must* be derived from a clock that increments monotonically and linearly in time to allow synchronisation and jitter calculations.
- Synchronisation source (SSRC): 32 bits – the SSRC field identifies the synchronisation source. This identifier *should* be chosen randomly, with the intent that no two synchronisation sources within the same RTP session will have the same SSRC identifier.
- CSRC list: 0 to 15 items, 32 bits each – the CSRC list identifies the contributing sources for the payload contained in this packet. The number of identifiers is given by the CC field. If there are more than 15 contributing sources, only 15 can be identified.

7.7.2 RTP control protocol (RTCP)

The RTP control protocol (RTCP) is based on the periodic transmission of control packets to all participants in the session, using the same distribution mechanism as the data packets. The underlying protocol *must* provide multiplexing of the data and control packets, for example using separate port numbers with UDP. RTCP performs four functions:

- The primary function is to provide feedback on the quality of the data distribution. This is an integral part of the RTP role as a transport protocol and is related to the flow and congestion control functions of other transport protocols. The feedback may be directly useful for control of adaptive encodings, but experiments with IP multicasting have shown that it is also critical to get feedback from the receivers to diagnose faults in the distribution. Sending reception feedback reports to all participants allows whoever is observing problems to evaluate whether those problems are local or global. With a distribution mechanism like IP multicast, it is also possible for an entity such as a network service provider who is not otherwise involved in the session to receive the feedback information and act as a third-party monitor to diagnose network problems. This feedback function is performed by the RTCP sender and receiver reports (RS and RR) – see Figure 7.10.
- RTCP carries a persistent transport-level identifier for an RTP source called the canonical name or CNAME. Since the SSRC identifier may change if a conflict is discovered or a program is restarted, receivers require the CNAME to keep track of each participant. Receivers may also require the CNAME to associate multiple data streams from a given

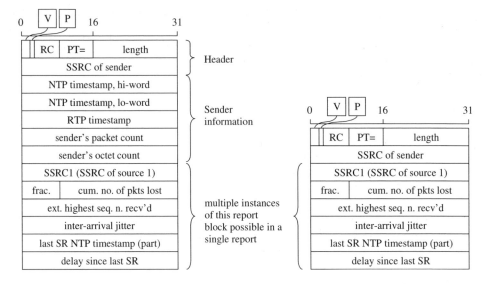

Figure 7.10   Sender report (SR) and receiver report (RR)

participant in a set of related RTP sessions, for example to synchronise audio and video. Inter-media synchronisation also requires the NTP and RTP timestamps included in RTCP packets by data senders.

- The first two functions require that all participants send RTCP packets, therefore the rate must be controlled in order for RTP to scale up to a large number of participants. By having each participant send its control packets to all the others, each can independently observe the number of participants.
- A fourth, *optional* function is to convey minimal session control information, for example participant identification to be displayed in the user interface. This is most likely to be useful in 'loosely controlled' sessions where participants enter and leave without membership control or parameter negotiation.

7.7.3 Sender report (SR) packets

There are three sections. The first section (header) consists of the following fields:

- Version (V): two bits – identifies the version of RTP, which is the same in RTCP packets as in RTP data packets. The current version is two (2).
- Padding (P): one bit – if the padding bit is set, this individual RTCP packet contains some additional padding octets at the end which are not part of the control information but are included in the length field. The last octet of the padding is a count of how many padding octets should be ignored, including itself (it will be a multiple of four).
- Reception report count (RC): five bits – the number of report blocks contained in this packet.
- Packet type (PT): eight bits – contains the constant 200 to identify this as an RTCP SR packet.

- Length: 16 bits – the length of this RTCP packet in 32-bit words minus one, including the header and any padding.
- SSRC: 32 bits – the synchronisation source identifier for the originator of this SR packet.

The second section, the sender information, is 20 octets long and is present in every sender report packet. It summarises the data transmissions from this sender. The fields have the following meaning:

- NTP timestamp: 64 bits – indicates the wall clock time when this report was sent so that it may be used in combination with timestamps returned in reception reports from other receivers to measure round-trip propagation to those receivers.
- RTP timestamp: 32 bits – corresponds to the same time as the NTP timestamp (above), but in the same units and with the same random offset as the RTP timestamps in data packets.
- Sender's packet count: 32 bits – the total number of RTP data packets transmitted by the sender since starting transmission up until the time this SR packet was generated.
- Sender's octet count: 32 bits – the total number of payload octets (i.e., not including header or padding) transmitted in RTP data packets by the sender since starting transmission up until the time this SR packet was generated.

The third section contains zero or more reception report blocks depending on the number of other sources heard by this sender since the last report. Each reception report block conveys statistics on the reception of RTP packets from a single synchronisation source.

SSRC_n (source identifier): 32 bits – the SSRC identifier of the source to which the information in this reception report block pertains, including:

- Fraction lost: eight bits – the fraction of RTP data packets from source SSRC_n lost since the previous SR or RR packet was sent, expressed as a fixed point number with the binary point at the left edge of the field. This fraction is defined to be the number of packets lost divided by the number of packets expected.
- Cumulative number of packets lost: 24 bits – the total number of RTP data packets from source SSRC_n that have been lost since the beginning of reception. This number is defined to be the number of packets expected less the number of packets actually received.
- Extended highest sequence number received: 32 bits – the least significant 16 bits contain the highest sequence number received in an RTP data packet from source SSRC_n, and the most significant 16 bits extend that sequence number with the corresponding count of sequence number cycles.
- Inter-arrival jitter: 32 bits – an estimate of the statistical variance of the RTP data packet inter-arrival time, measured in timestamp units and expressed as an unsigned integer. The inter-arrival jitter J is defined to be the mean deviation (smoothed absolute value) of the difference D in packet spacing at the receiver compared to the sender for a pair of packets.
- Last SR timestamp (LSR): 32 bits – the middle 32 bits out of 64 in the NTP timestamp received as part of the most recent RTCP sender report (SR) packet from source SSRC_n. If no SR has been received yet, the field is set to zero.
- Delay since last SR (DLSR): 32 bits – the delay, expressed in units of 1/65536 seconds, between receiving the last SR packet from source SSRC_n and sending this reception report block. If no SR packet has been received yet from SSRC_n, the DLSR field is set to zero.

7.7.4 Receiver report (RR) packets

The format of the receiver report (RR) packet is the same as that of the SR packet except that the packet type field contains the constant 201 and the five words of sender information are omitted. The remaining fields have the same meaning as for the SR packet.

7.7.5 Source description (SDES) RTCP packet

The SDES packet is a three-level structure composed of a header and zero or more chunks, each of which is composed of items describing the source identified in that chunk. Each chunk consists of an SSRC/CSRC identifier followed by a list of zero or more items, which carry information about the SSRC/CSRC. Each chunk starts on a 32-bit boundary. Each item consists of an eight-bit type field, an eight-bit octet count describing the length of the text (thus, not including this two-octet header), and the text itself. Note that the text can be no longer than 255 octets, but this is consistent with the need to limit RTCP bandwidth consumption.

End systems send one SDES packet containing their own source identifier (the same as the SSRC in the fixed RTP header). A mixer sends one SDES packet containing a chunk for each contributing source from which it is receiving SDES information, or multiple complete SDES packets if there are more than 31 such sources.

The SDES items currently defined include:

- CNAME: canonical identifier (mandatory);
- NAME: name of user;
- EMAIL: address user;
- PHONE: number for user;
- LOC: location of user, application specific;
- TOOL: name of application/tool;
- NOTE: transient messages from user;
- PRIV: application specific/experimental use.

Goodbye RTCP packet (BYE): the BYE packet indicates that one or more sources are no longer active.

Application-defined RTCP packet (APP): the APP packet is intended for experimental use as new applications and new features are developed, without requiring packet type value registration.

7.7.6 SAP and SIP protocols for session initiations

There are several complementary mechanisms for initiating sessions, depending on the purpose of the session, but they essentially can be divided into invitation and announcement mechanisms. A traditional example of an invitation mechanism would be making a telephone call, which is essentially an invitation to participate in a private session. A traditional example of an announcement mechanism is the television guide in a newspaper, which announces the time and channel that each programme is broadcast. In the Internet, in addition to these two extremes, there are also sessions that fall in the middle, such as an invitation to listen to a public session, and announcements of private sessions to restricted groups.

The *session announcement protocol (SAP)* must be one of the simplest protocols around. To announce a multicast session, the session creator merely multicasts packets periodically to a well-known multicast group carrying an SDP description of the session that is going to take place. People that wish to know which sessions are going to be active simply listen to the same well-known multicast group, and receive those announcement packets. Of course, the protocol gets a little more complex when we take security and caching into account, but basically that's it.

The *session initiation protocol (SIP)* works like making a telephone call, e.g. finds the person you're trying to reach and causes their phone to ring. The most important way that SIP differs from making an existing telephone call (apart from that it is an IP-based protocol) is that you may not be dialling a number at all. Although SIP can call traditional telephone numbers, SIP native concept of an address is an SIP URL, which looks very like an email address. Figure 7.11 illustrates a typical SIP call of an initiate and terminate session.

Users may move to a different location. Redirect servers and location servers can be used to assist SIP calls. Figure 7.12 illustrates a typical SIP call using redirect server and location server.

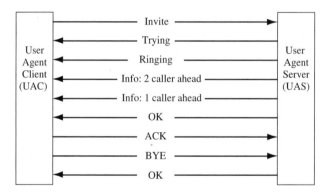

Figure 7.11   A typical SIP call of initiate and terminate session

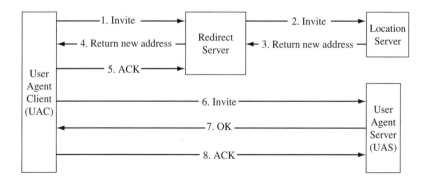

Figure 7.12   A typical SIP call using a redirect server and location server

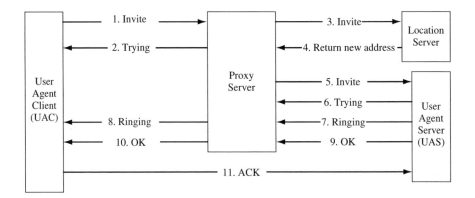

Figure 7.13   A typical SIP call using a proxy server and location server

SIP makes extensive use of proxy servers, each of which looks at the call request, looks up whatever local information it has about the person being called (i.e., the callee), performs any security checks it has been asked by the callee or her organisation to make, and then routes the call onward. Figure 7.13 shows a typical SIP call using a proxy server and location server.

There are two multicast channels per application per session: one for RTCP and the other for RTCP. It allows ad hoc configurations as stand-alone for individual applications; and also allows advertised conference with session directory (SDR) and configuration information.

7.7.7 Session directory service (SDS)

The growth in multicast services and applications has led to some navigation difficulties (just as there are in the WWW). This has led to the creation of a session directory service (SDS). This has several functions:

- A user creating a conference needs to choose a multicast address that is not in use. The session directory system has two ways of doing this: firstly, it allocates addresses using a pseudo-random strategy based on how widespread the conference is going to be according to the user, and where the creator is; secondly, it multicasts the session information out, and if it detects a clash from an existing session announcement, it changes its allocation. This is currently the main mechanism for the management of allocation and listing of dynamic multicast addresses.
- Users need to know what conferences there are on the multicast backbone (Mbone), what multicast addresses they are using, and what media are in use on them. They can use the session directory messages to discover all of this. The latest versions of multicast include a facility for administrative scoping, which allows session creators to designate a logical region of interest outside of which traffic will not (necessarily) flow.
- Furthermore, the session directory tools currently implemented will launch applications for the user.

7.8 Voice over IP

Based on RTP, IP telephony is becoming a mainstream application moving away from proprietary solutions to standards based solutions, providing QoS comparable to the PSTN, and providing transparent interoperability of the IP and PSTN networks.

7.8.1 Gateway decomposition

The *signalling gateway* is responsible for signalling between end users on either network. On the PSTN side, SS7 or ISDN (T1/E1-BRI/PRI) is used, which is then translated to an IP signalling protocol such as SIP or H.323, and transported across the IP network. SAP is used to announce the session. SDP is used to describe the call (or session).

Once a call is set up, the *media gateway* is responsible for transfer of the data, video and audio streams. On the PSTN side, media transport is by PCM-encoded data on TDM streams; on the IP network side, on RTP/UDP. The *media gateway controller* is used to control one or more media gateways.

7.8.2 Protocols

VoIP uses a number of protocols. As far back as 1994, the ITU introduced its H.323 family of protocols, to provide multimedia capability over the Internet. Many vendors have developed and deployed these solutions. In parallel, the IETF introduced many protocols used for IP telephony – RTP, RTSP, RTCP, Megaco, SIP, and SDP. These protocols provide the foundation for standards based IP telephony.

7.8.3 Gatekeepers

Gatekeepers are responsible for addressing, authorisation and authentication of terminal and gateways, bandwidth management, accounting, billing and charging. They may also provide call-routing services. *Terminal* is a PC or stand-alone device running multimedia applications. *Multipoint control units (MCU)* provide support for conferences of three or more terminals.

7.8.4 Multimedia conferencing (MMC)

Multimedia conferencing (MMC) is one of the typical example applications based on IP multicast. It is also well suited for satellite networks with great advantages. It consists of multimedia application with the following components:

- *Voice* provides packet audio in time slices, numerous audio-coding schemes, redundant audio for repair, unicast or multicast, configurable data rates.
- *Video* provides packet video in frames, numerous video-coding schemes, unicast or multicast, configurable data rates.
- *Network Text Editor* can be used for message exchanges
- *Whiteboard* can be used for free-hand drawing.

It should allow local scoped groups, global scope groups and administratively scoped groups, and also unicast traffic gateway (UTG) so that routers routing protocols and multicast domains can be reached by tunnelling, i.e., in a LAN, IP packets are multicasted to all hosts directly; and in a WAN it is a virtual overlay network on top of the Internet. RTP/RTCP is used as the protocol for transmission and control. Overlapping multicast domains can be configured by using different administratively scoped addresses in each of the domains.

7.8.5 Conference control

Conference control provides functions and mechanisms for users to control how to organise, manage and control a conference, with the following control functions:

- Floor control: who speaks? chairman control? distributed control?
- Loose control: one person speaks, grabs channel.
- Strict control: application specific, e.g. lecture.
- Resource reservation: bandwidth requirement and quality of the conference.
- Per-flow reservation: audio only, video only, audio and video.

Further reading

[1] Allman, M., Floyd, S. and C. Partridge, *Increasing TCP's Initial Window*, RFC 2414, September 1998.
[2] Allman, M., Glover, D. and L. Sanchez, *Enhancing TCP over Satellite Channels using Standard Mechanisms*, BCP 28, RFC 2488, January 1999.
[3] Allman, M., Paxson, V. and W. Richard Stevens, *TCP Congestion Control*, RFC 2581, April 1999.
[4] Braden, R., *Transaction TCP – Concepts*, RFC 1379, September 1992.
[5] Braden, R., *T/TCP – TCP Extensions for Transactions: Functional Specification*, RFC 1644, July 1994.
[6] Chotikapong, Y., TCP/IP and ATM over LEO satellite networks, PhD thesis, University of Surrey, 2000.
[7] Chotikapong, Y., Cruickshank, H. and Z. Sun, 'Evaluation of TCP and Internet traffic via low earth orbit satellites, *IEEE Personal Communications over Satellites, Special Issue on Multimedia Communications over Satellites*, **8**(3): 28–34, 2001.
[8] Chotikapong, Y. and Z. Sun, Evaluation of application performance for TCP/IP via satellite links, *IEE Colloquium on Satellite Services and the Internet*, 17 February 2000.
[9] Chotikapong, Y., Sun, Z., Örs, T. and B.G. Evans, Network architecture and performance evaluation of TCP/IP and ATM over satellite, 18th AIAA International Communication Satellite Systems Conference and Exhibit, Oakland, April 2000.
[10] Jacobson, V., *Compressing TCP/IP Headers*, RFC 1144, February 1990.
[11] Mathis, M., Mahdavi, J., Floyd, S. and A. Romanow, *TCP Selective Acknowledgment Options*, RFC 2018, October 1996.
[12] Paxson, V., Allman, M., Dawson, S., Heavens, I. and B. Volz, *Known TCP Implementation Problems*, RFC 2525, March 1999.
[13] Ramakrishnan, K. and S. Floyd, *A Proposal to Add Explicit Congestion Notification (ECN) to IP*, RFC 2481, January 1999.
[14] Stevens, W., *TCP Slow Start, Congestion Avoidance, Fast Retransmit, and Fast Recovery Algorithms*, RFC 2001, January 1997.
[15] Sun, Z., TCP/IP over satellite, in *Service Efficient Network Interconnection via Satellite*, Fun Hu, Y., Maral, G. and Erina Ferro (eds), John Wiley & Son, Inc., pp. 195–212.
[16] Sun, Z., Chotikapong, Y. and C. Chaisompong, Simulation studies of TCP/IP performance over satellite, 18th AIAA International Communication Satellite Systems Conference and Exhibit, Oakland, April 2000.
[17] Sun, Z. and H. Cruickshank, Analysis of IP voice conferencing over geostationary satellite systems, *IEE Colloquium on Satellite Services and the Internet*, 17 February 2000.

[18] RFC 793, Transmission control protocol, Jon Postel, IETF, September 1981.

[19] RFC 1122, Requirements for Internet Hosts – Communication Layers, R. Braden, IETF, October 1989.

[20] RFC 1323, TCP Extensions for High Performance, V. Jacobson, R. Braden and D. Borman, IETF, May 1992.

[21] RFC 2142, Mailbox names for common services, roles and functions, D. Crocker, IETF, May 1997.

[22] RFC 1889, RTP: A Transport Protocol for Real-Time Applications, H. Schulzrinne, S. Casner, R. Frederick and V. Jacobson, IETF, January 1996.

[23] RFC 1890, RTP Profile for Audio and Video Conferences with Minimal Control, H. Schulzrinne, IETF, January 1996.

Exercises

1. Explain how satellite networks affect the performance of TCP due to flow control, error controls and congestion control mechanisms.

2. Discuss typical satellite network configurations for the Internet connections.

3. Explain TCP enhancement for satellite networks based on the slow-start algorithm.

4. Explain TCP enhancement based on the congestion avoidance mechanism.

5. Discuss how to achieve TCP enhancement based on acknowledgement.

6. Calculate the utilisation of satellite bandwidth in the slow-start and congestion avoidance stages.

7. Explain TCP enhancement on error recovery mechanisms including fast retransmission and fast recovery.

8. Explain the pros and cons of TCP spoofing and split TCP (also known as cascading TCP) mechanisms.

9. Explain the limitation of TCP enhancement mechanisms based on existing TCP mechanisms.

10. Discuss the real-time protocols including RTP, RTCP, SAP, SIP, etc., and the HTTP protocol.

11. Compare differences between non-real-time applications, WWW and FTP, and real-time applications, VoIP and MMC.

8

Next Generation Internet (NGI) over Satellite

This chapter aims to introduce next generation Internet (NGI) over satellite. We try not to separate satellites from general networks, as satellites are considered as an integrated part of the Internet. First this chapter introduces new services and applications, modelling and traffic engineering and multi-protocol label switching (MPLS), then it introduces the Internet protocol version 6 (IPv6) including addressing and transitions, and particularly it explains IPv6 over satellite including tunnelling and translation techniques of IPv6 over satellite networks. Finally, as a conclusion, it discusses the future development of satellite networking. When you have completed this chapter, you should be able to:

- Understand the concepts of new services and applications to be supported in future satellite networks.
- Understand the basic principles and techniques for traffic modelling and traffic characterisation.
- Understand the nature of Internet traffic.
- Describe the concepts of traffic engineering in general and Internet traffic engineering in particular.
- Appreciate the principles of MPLS, and interworking with different technologies and traffic engineering concepts.
- Understand IPv6 and its main differences from IPv4.
- Understand IPv6 addressing and transition techniques.
- Understand IPv6 over satellite networks and transition techniques such as tunnelling and translations.
- Appreciate the future development and convergence of satellite networking.

Satellite Networking: Principles and Protocols   Zhili Sun
© 2005 John Wiley & Sons, Ltd

8.1 Introduction

In recent years, we have seen the tremendous spread of mobile networks. The new generations of mobile telephones have become more and more sophisticated, with increasing capabilities of email, WWW access, multimedia messaging, streaming voice and video broadcasting, which go far beyond the original definition of mobile phones.

In terms of software, the mobile phone is more like a computer than a telephone. There are full Internet protocol stacks (TCP/IP), transmission technologies (infrared, wireless, USB, etc.), and various peripheral devices. In computer networks, Ethernet and wireless LANs dominate LANs. In mobile networks, the GSM mobile networks are evolving towards 3G networks. They are converging towards an all-IP solution. The Internet protocol (IP) has also evolved to cope with demands from networking technologies and new services and applications.

Inevitably, satellite networking is also evolving towards an all-IP solution and is following the trends in the terrestrial mobile and fixed networks. In addition to user terminals, services and applications are also converging, i.e., satellite network terminals aim to be the same as terrestrial network terminals providing the same user interface and functionality. As the current satellite networks integrate with terrestrial networks, it is not difficult to see that future satellite terminals will be fully compatible with standard terrestrial network terminals, but with a different air interface in the lower layers of the protocol stack (physical and link layers only).

In traditional computer networks, network designers were not very concerned with QoS, traffic engineering and security. For real-time services, QoS and traffic engineering are important and have been successfully implemented in telephony networks for nearly a century. More and more people own portable computers and mobility is now a new requirement. More and more business transactions, commercial and public uses of the Internet make security a very important Internet issue.

The original design of the Internet did not take all these new requirements and the large scale of today's Internet into consideration. There is also concern that IPv4 may soon run out of IP addresses. Although IPv6 has started to address these issues, we are still a long way from a perfect solution.

So far, we have completed our discussion about the transition from physical layer to transport layer. Now it is time to discuss new services and applications (starting from information processing) and the development of satellite networks and related issues, including traffic modelling and characterisation, MPLS, traffic engineering and IPv6.

8.2 New services and applications

We have discussed various kinds of network services which we expect to support over satellite networks. The services information has to be encoded in proper formats to be suitable for transmission, and decoded at the receiver. The new services and applications should include major components of high-quality digital voice, image and video (and combinations of these) across broadband networks. Here we briefly discuss some of these topics.

8.2.1 Internet integrated services

One of the principal functions of network layer protocols is to offer universal connectivity, and a uniform service interface, to higher layer protocols – in particular, to transport layer protocols – independent of the nature of the underlying physical network. Correspondingly, the function of transport layer protocols is to provide session control services (e.g. reliability) to applications, without being tied to particular networking technologies.

Unless applications run over common network and transport protocols, interoperability for the same applications running on different networks would be difficult, if not impossible. Most multimedia applications will continue to build upon enhancements of current Internet protocols, and deploy a wide variety of high-speed Internet networking technologies.

In the specific case of IP, the IETF has developed the notion of an integrated services Internet. This envisages a set of enhancements to IP to allow it to support integrated or multimedia services. These enhancements include traffic management mechanisms that closely match the traffic management mechanisms of telecommunication networks.

Network protocols rely upon a flow specification to characterise the expected traffic patterns for streams of IP packets, which the network can process through packet-level policing, shaping and scheduling mechanisms to deliver a requested QoS. In other words, a flow is a layer-three connection, since it identifies and characterises a stream of packets, even though the protocol remains connectionless.

8.2.2 Elastic and inelastic traffic

There are two main classifications of Internet traffic generated by services and applications:

- *Elastic traffic*: this type of traffic is essentially based on TCP, i.e., it uses TCP as the transport protocol. Elastic traffic is defined as traffic that is capable of adapting its flow rate according to changes in delay and throughput across the network. This capability is built-in to the TCP flow control mechanisms. This type of traffic is also known as opportunistic traffic, i.e., if resources are made available, these applications would try to consume them; on the other hand if the resources are temporarily unavailable they can wait (withholding transmission) without adversely affecting the applications. Examples of elastic traffic include email, file transfers, network news, interactive applications such as remote login (telnet) and web access (HTTP). These applications can cope well with delay and variable throughput in the network. This type of traffic can be further categorised into long-lived and short-lived responsive flows, depending on the length of time the flows are active. FTP is an example of a long-lived responsive flow while HTTP represents a short-lived flow.
- *Inelastic traffic*: this type of traffic is essentially based on UDP, i.e., it uses UDP as the transport protocol. Inelastic traffic is exactly the opposite of elastic traffic – the traffic is incapable of varying its flow rate when faced with changes in delay and throughput across the network. A minimum amount of resources is required to ensure the application works well; otherwise, the applications will not perform adequately. Examples of inelastic traffic include conversational multimedia applications such as voice or video over IP, interactive multimedia applications such as network games or distributed simulations and non-interactive multimedia applications such as distance learning or audio/video broadcasts where a continuous stream of multimedia information is involved. These real-time applications can cope with small delays but cannot tolerate jitter (variations in average delay). This stream traffic is also known as long-lived non-responsive flow.

In terms of applications, at present the Internet carries computer data traffic almost exclusively. Traditionally these have been applications such as file transfer (using the FTP protocol), remote login sessions (telnet) and email (SMTP). However, these applications have been somewhat overshadowed by the World Wide Web (HTTP). Voice over IP and video and audio streaming over IP applications are only emerging and are not yet contributing significantly to the composition of Internet traffic. However, they are expected to be the major bandwidth consumers in the future. While the protocol composition remains roughly the same in proportion, UDP applications are expected to have an increase in the RTP/RTCP portion. This is due to increases in audio/video streaming and online gaming applications.

8.2.3 QoS provision and network performance

As defined in the QoS architecture, best-effort service is the default service that a network gives to a datagram between the source and destination in today's Internet networks. Among other implications, this means that if a datagram changes to a best-effort datagram, all flow controls that apply normally to a best-effort datagram also apply to the datagram.

The controlled load service is intended to support a broad class of applications in the Internet, but is highly sensitive to overloaded conditions. Important members of this class are the 'adaptive real-time applications' currently offered by a number of vendors and researchers. These applications work well on networks with light load conditions, but degrade quickly with overload conditions. A service, which mimics an unloaded network, serves these applications well.

Guaranteed service means that a datagram will arrive within a limited time with limited packet loss ratio, if the flow's traffic stays within its specified traffic parameters. This service is intended for applications requiring a firm guarantee of delay within a certain time limit for the traffic to reach its destination. For example, some audio and video 'playback' applications are intolerant of any datagram arriving after their playback time. Applications that have hard real-time requirements also require guaranteed service.

In playback applications, datagrams often arrive far earlier than the delivery deadline and have to be buffered at the receiving system until it is time for the application to process them.

8.3 Traffic modelling and characterisation

Future network infrastructures will have to handle a huge amount of IP traffic from different types of services, including a significant portion of real-time services. The multi-service characteristics of such a network infrastructure demand a clear requisite: the ability to support different classes of services with different QoS requirements. Moreover, Internet traffic is more variable with time and bandwidth, with respect to traditional traffic in telecommunication networks, and it is not easily predictable. This means that networks have to be flexible enough to react adequately to traffic changes. Besides the requirements of flexibility and multi-service capabilities that lead to different levels of QoS requirements, there is also a need to reduce complexity.

8.3.1 Traffic modelling techniques

Multi-services networks need to support a varied set of applications. These applications contain either one or (more often than one) a combination of the following components: data, audio and video. More widely termed as multimedia applications, these components, together with the applications' requirements, will generate a heterogeneous mixture of traffic with different statistical and temporal characteristics. These applications and services require resources to perform their functions. Of special interest is resource sharing among application, system and network. Traffic engineering is a network function that controls a network's response to traffic demands and other stimuli (such as failures) and encompasses traffic and capacity/resource management. In order for the multi-services networks to efficiently support these applications, while at the same time optimally utilises the networks' resources, traffic engineering mechanisms need to be devised. These mechanisms relate intrinsically to the characteristics of the traffic getting into the network. To devise efficient resource and traffic management schemes requires an understanding of the source traffic characteristics and the development of appropriate traffic models. Hence, source traffic characterisation and modelling is a crucial first step in the overall network design and performance evaluation process. Indeed traffic modelling is identified as one of the key subcomponents of the traffic engineering process model.

8.3.2 Scope of traffic modelling

Traffic characterisation describes what traffic patterns the application/user generates. The goal is to develop an understanding of the nature of the traffic and to devise tractable models that capture the important properties of the data, which can eventually lead to accurate performance prediction. Tractability is an important feature as it infers that the traffic models used in subsequent analysis readily lend themselves to numerical computation, simulation and analytical solutions. They also have wide range of time scales.

Traffic modelling summarises the expected behaviour of an application or an aggregate of applications. Among the primary uses of traffic characteristics are:

- Long-range planning activities (network planning, design and capacity management).
- Performance prediction, real-time traffic control/management and network control.

Traffic models can be utilised in three different applications:

- As a source for generating synthetic traffic to evaluate network protocols and designs. This complements the theoretical part of the analysis, which increases in complexity as networks become complicated.
- As traffic descriptors for a range of traffic and network resource management functions. These include call admission control (CAC), usage parameter control (UPC) and traffic policing. These functions are the key to ensure meeting certain network QoS levels while achieving high multiplexing gains.
- As source models for queuing analysis, they use queuing systems extensively as the primary method for evaluating network performances and as a tool in network design. A reasonably good match to real network traffic will make analytical results more useful in practical situations.

8.3.3 Statistical methods for traffic modelling

The main aim of traffic modelling is to map accurately the statistical characteristics of actual traffic to a stochastic process from which synthetic traffic can be gemerated.

For a given traffic trace (TT), the model finds a stochastic process (SP) defined by a small number of parameters such that:

- TT and SP give the same performance when fed into a single server queue (SSQ) for any buffer size and service rate.
- TT and SP have the same mean and autocorrelation (goodness-of-fit).
- Preferably, SP SSQ is amenable to analysis.

There are many traffic models that have been developed over the years.

8.3.4 Renewal traffic models

A renewal process is defined as a discrete-time stochastic process, $X(t)$, where $X(t)$ are independent identically distributed (iid), non-negative random variables with a general distribution function. Independence here implies that observation at time t does not depend on the past or future observation, i.e., there is no correlation between the present observation and previous observations.

Analysing renewal processes is mathematically simple. However, there is one major shortcoming with this model: the absence of an autocorrelation function. Autocorrelation is a measure of the relationship between two time instances of a stochastic process. It is an important parameter and is used to capture the temporal dependency and burst of the traffic. As mentioned previously, temporal dependencies are important in a multimedia traffic stream while burst traffic expect to dominate broadband networks. Therefore, models, which capture the autocorrelated nature of traffic, are essential for evaluating the performance of these networks.

However, because of its simplicity, the renewal process model is still widely used to model traffic sources. Examples of a renewal process include the Poisson and Bernoulli processes.

8.3.5 Markov models

The Poisson and Bernoulli processes display memory-less properties in the sense that the future does not depend on the past, i.e., the occurrences of new arrivals do not depend on the history of the process. This in turn results in the non-existence of the autocorrelation function since there is no dependency among the random sequence of events.

Markov-based traffic models overcome this shortcoming by introducing dependency into the random sequence. Consequently, autocorrelation is now non-zero and this can capture the characteristics of traffic burst. Markov dependency or a Markov process is defined as a stochastic process $X(t)$ where for any $t_0 < \cdots < t_n < t_{n+1}$ and given the values of $X(t_0), \ldots, X(t_n)$, the distribution of $X(t_{n+1})$ only depends on $X(t_n)$. This implies that the next state in a Markov stochastic process only depends on the current state of the process and not on states assumed previously; this is the minimum possible dependence that can exist between successive states. How the process arrives at the current state is irrelevant.

Another important implication of this Markov property is that the next state only depends on the current state and not on how long the process has already been in that (current) state. This means that the state residence times (also called sojourn times) must be random variables with memory-less distribution. Examples of Markov models include on-off and Markov modulated Poisson process (MMPP).

8.3.6 Fluid traffic models

Fluid traffic models view traffic as a stream of fluid characterised by a flow rate (e.g., bits per second), that traffic volume is better than traffic count in the model. Fluid models are based on the assumption that the number of individual traffic units (packets or cells) generated during the active period is so large that it appears like a continuous flow of fluid. In other words, a single unit of traffic in this case would have little significance and its impact on the overall flow is negligible, i.e., individual units will only add infinitesimal information to the traffic stream.

An important benefit of the fluid traffic model is that it can achieve enormous savings in computing resources when simulating streams of traffic as described above. For example, in an ATM network scenario supporting the transmission of high-quality video, it requires a large number of ATM cells for a compressed video at 30 frames per second. If a model were to distinguish between cells and consider the arrival of each ATM cell as a separate event, processing cell arrivals would quickly consume vast amounts of CPU and memory, even if the simulated time were in the order of a few minutes.

By assuming the incoming fluid flow remains (roughly) constant over much longer periods, a fluid flow simulation performs well. A change in flow rate signals the event that traffic is fluctuating. Because these changes can occur far less frequently than the arrival of individual cells the computing overhead involved is greatly reduced.

8.3.7 Auto-regressive and moving average traffic models

Auto-regressive traffic models define the next random variable in the sequence X_n as an explicit function of previous variables within a time window stretching from present to past. Some of the popular auto-regressive models are:

- Linear auto-regressive processes, AR(p), described as

$$X_n = a_0 + \sum_{r=1}^{p} a_r X_{n-r} + \varepsilon_n, \quad n > 0$$

where X_n are the family of random variables, $a_r (0 \leq r \leq p)$ are real constants and ε_n are zero-mean, uncorrelated random variables also called white noise, which are independent of X_n.

- Moving average processes, MA(q), described as

$$X_n = \sum_{r=0}^{q} b_r \varepsilon_{n-r}, \quad n > 0$$

- Auto-regressive moving average processes, ARMA (p, q), described as

$$X_n = a_0 + \sum_{r=1}^{p} a_r X_{n-r} + \sum_{r=0}^{q} b_r \varepsilon_{n-r}$$

8.3.8 Self-similar traffic models

The development of these models were based on the observation that Internet traffic dynamics resulting from interactions among users, applications and protocols, is best represented by the notion of 'fractals' a well-established theories with wide applications in physics, biology and image processing. Therefore, it is natural to apply traffic models that are inherently fractal for characterisation of Internet traffic dynamics and generating synthesised traffic in a computationally efficient manner.

Wavelet modelling offers a powerful and flexible technique for mathematically representing network traffic at multiple time scales. A wavelet is a mathematical function having principles similar to that of Fourier analysis; it is widely used in digital signal processing and image compression techniques.

8.4 The nature of internet traffic

Internet traffic is due to a very large pool of uncoordinated, i.e. independent users accessing and using the various applications. Each Internet communication consists of a transfer of information from one computer to another for e.g. downloading web pages or sending/receiving emails. Packets containing bits of information transmitted over the Internet are the result of simultaneous active communications between two or more computers (usually termed as hosts) on the Internet.

8.4.1 The world wide web (WWW)

A web page typically consists of multiple elements or objects when downloaded. These objects are loaded using separate HTTP GET requests, serialised in one or more parallel TCP connections to the corresponding server(s). In practice, web access is request–response oriented with bursts of numerous requests and small, unidirectional responses. Retrieval of a complete web page requires separate requests for text and each embedded image, thus making traffic inherently burst. Figure 8.1 illustrates a typical message sequence in a web surf session.

The characteristics of web traffic have been studied over the years to understand the nature of the traffic. One of the key findings is that web traffic comes in bursts, rather than in steady flows, and the same patterns of bursts repeat themselves regardless of whether the time interval studied is a few seconds long or a millionth of a second. This particular type

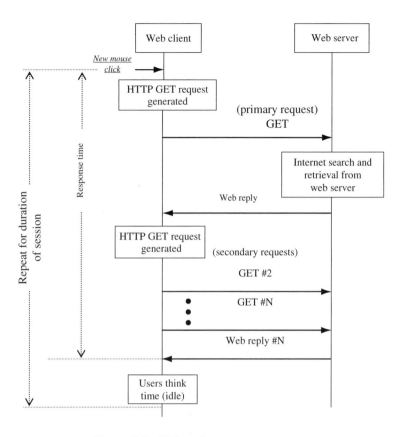

Figure 8.1 Web surfing message sequence

of traffic is called the self-similar or fractal or scale-invariant traffic. Fractals are objects whose appearances are unchanged at different time scales. Essentially, a self-similar process behaves in a similar way (looks statistically the same) over all time scales.

Analyses of actual web traffic traces can help to understand the causes of this phenomenon. Parameters of traffic traces in statistics include size of HTTP files, number of files per web page and user browsing behaviour (user think times and successive document retrievals) and the distribution properties of web requests, file sizes and file popularity. Studies indicated that the self-similarity phenomenon is highly variable depending on measurement. It was found that the best distributional model for such highly variable datasets as file sizes and requests inter-arrivals is one with a heavy tail. This self-similarity phenomenon is due to the superposition of many on/off sources, each of which exhibits the infinite variance syndrome.

Processes with heavy-tailed sojourn-time distributions have long-term (slowly decaying) correlations, also known as long range dependence (LRD). The following formula shows the autocorrelation function of such processes:

$$r(k) \approx \frac{1}{k^a} \text{ when } k \to \infty \text{ and } 0 < a < 1 \tag{8.1}$$

The autocorrelation function thus decays hyperbolically, which is much slower than exponential decay. In addition, the sum of the autocorrelation values approaches infinity (the autocorrelation function is non-summable) since $a < 1$. One of the consequences of the non-degenerative correlation is the 'infinite' influence of LRD on the data. Aggregation of LRD sources produces a traffic stream with self-similar characteristics as indicated by actual traffic traces.

8.4.2 Pareto distribution model for self-similar traffic

One of the classes of distributions that are heavy-tailed is the Pareto distribution and its probability distribution function (pdf) is defined as

$$p(x) = \frac{\alpha\beta^\alpha}{x^{\alpha+1}} \text{ for } \alpha, \beta > 0, x > \beta \tag{8.2}$$

Its cumulative distribution function (cdf) is given by

$$P[X \le x] = 1 - \left(\frac{\beta}{x}\right)^\alpha \text{ for } \alpha, \beta > 0, x > \beta \tag{8.3}$$

The mean and variance of the Pareto distribution are given respectively by

$$\mu = \frac{\alpha\beta}{\alpha - 1} \tag{8.4}$$

$$\sigma^2 = \frac{\alpha\beta^2}{(\alpha - 1)^2(\alpha - 2)} \tag{8.5}$$

α is the shape parameter and β the location parameter, hence $\alpha > 2$ for this distribution to have a finite mean and variance. However from Equation (8.2) $0 < \alpha < 2$ for the heavy tail definition to hold, therefore Pareto distribution is a distribution with an infinite variance. A random variable whose distribution has an infinite variance implies that the variable can take on extremely large values with non-negligible probability.

8.4.3 Fractional Brownian motion (FBM) process

There have been advances in the development of reliable analytical models representing self-similar traffic. The fractional Brownian motion (FBM) process can be used as the basis of a workload model for generating synthesised self-similar traffic, resulting in a simple but useful relationship between the number of customers in the system, q and the system utilisation, ρ. Assuming an infinite buffer with a constant service time, this relationship is given as

$$q = \frac{\rho^{\frac{1}{2(1-H)}}}{(1-\rho)^{\frac{H}{1-H}}} \tag{8.6}$$

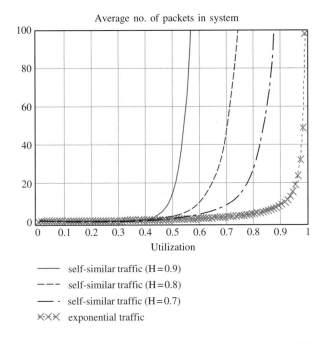

Figure 8.2 Comparison between self-similar traffic and exponential traffic

where H is the Hurst parameter $(0.5 < H < 1)$, a parameter that is often used as a measure of the degree of self-similarity in a time series. Note that when $H = 0.5$ the above equation reduces to the classical result, the M/M/1 relationship (a queuing system with exponential inter-arrival time and exponential service time). Hence, a value of 0.5 represents a memoryless process whereas a value of one corresponds to a process, which is the same in all respects, at whatever time scale.

Using the above relationship, we plotted the distribution of the average number of packets in the system as a function of the system utilisation for a range of H values and compared this with the exponential traffic (see Figure 8.2). We can see that the traces showed the same trend, with a characteristic 'knee' beyond which the number of packets increases rapidly. We can also see that as the H parameter value increases, i.e. the traffic becomes more self-similar, it is difficult to achieve high utilisation factors. In order to operate at high system utilisation, it requires considerable buffer provisions to avoid overflow. That is to say if we were to design the system according to what is predicted by the exponential traffic, we would not be able to operate at high utilisation when subjected to self-similar traffic because the buffer would very quickly overflow.

8.4.4 Consideration of user behaviour in traffic modelling

Traffic sources are random or stochastic in nature and the only tool to describe it is in statistical terms. Numerous models have been developed to capture and represent the randomness of this behaviour in the form of tractable mathematical equations.

Among the traffic, characteristics of interest include arrival rate, inter-arrival time, packet sizes, burst, duration of connection and distribution of arrival times between application invocations. Another important characteristic is the correlation between successive arrivals or between arrivals from different sources. Correlation functions are important measurements used to describe temporal dependencies between sources and bursts of traffic. A temporal or timing relation between sources is especially important in multimedia traffic.

The most widely used assumption in modelling these characteristics has considered them as independent identically distributed (iid) random arrival events. It describes the joint distribution of two or more random variables; in such cases, there is no correlation between the variables. This implies that users are independent of each other; the generation of traffic from one user does not affect the behaviour of another user. This property can simplify the mathematical analysis and gives rise to a unique formula representing certain characteristics of interest. While this assumption has been useful, it also gives rise to an independent, uncorrelated arrival process. In real-life scenarios, traffic often has complex correlation structures especially with video applications.

Several modelling approaches have attempted to capture these correlation structures. There are two modelling approaches – auto-regressive model and Markov-modulated fluid model – which capture the effects of coded video within a scene. We can also use an augmented auto-regressive model to capture the effects of scene changes or consider multimedia source as a superposition of on-off processes to model the individual components of the multimedia source (voice/audio, video and data).

User behaviour is another important factor that can have an effect on the characteristics of traffic. This is even more so with the explosive growth of the Internet and the corresponding increase in Internet-related traffic. Models, which capture this behaviour (also called behavioural modelling), would be useful to model both packet generation and user interaction with applications and services by representing the user behaviour as a profile. This profile defines a hierarchy of independent processes and different types of stochastic process for modelling these processes. Another related characteristic with regards to Internet or web traffic that is being researched currently is the structure of the web server itself as this has a bearing on the web page response times (document transfer times), which in turn affects the user session. The development in traffic modelling often results in development of specialised software to generate workload for stress-testing web servers.

8.4.5 Voice traffic modelling

Here we consider current and future multi-service packet-switch networks. Analogue speech is first digitised into pulse code modulation (PCM) signals by a speech/voice codec (coder-decoder). The PCM samples pass on to a compression algorithm, which compresses the voice into packet format prior to transmission on this packet-switched network. At the destination end, the receiver performs the same functions in reverse order. Figure 8.3 shows this flow of end-to-end packet voice. Voice applications that utilise IP-based packet networks are commonly referred to as Internet telephony or voice over IP (VoIP).

The most distinctive feature of speech signals is that in conversational speech there are alternating periods of signal (speech) and no signal (silence). Human speech consists of an alternating sequence of active interval (during which time one is talking) followed by silence or inactive interval (during which one pauses or listens to the other party). Since the encoded

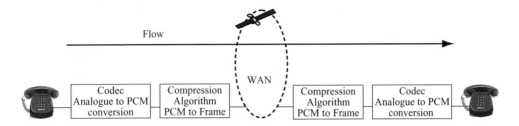

Figure 8.3 Packet voice end-to-end flow

bit rate for speech signals is at most 64 kbit/s, it is acceptable to treat the maximum rate during intervals of speech as 64 kbit/s. However, there are speech-coding techniques, which result in 32, 16 or 8 kbit/s rates, in which case the maximum rate during intervals of speech can assume the corresponding coding rates.

As an example, the G.729 coder-decoder (codec) has a default payload size of two voice samples of 10 ms each sampled at 8 kHz rate. With a coding rate of 8 kbit/s, this results in a payload size of 20 bytes. This payload is then packetised, in the case of VoIP, into IP packets consisting of real-time transport protocol (RTP)/UDP/IP and multi-link PPP (MLPPP) headers. The RTP is a media packet protocol for transmitting real-time media data. It provides mechanisms for sending and receiving applications to support streaming data (facilitates the delivery and synchronisation of media data). The MLPPP is an extension of PPP that allows the combination of multiple PPP links into one logical data pipe. Note that the addition of this header is dependent on the link layer; in this case, it is the PPP link.

Without RTP header compression, the RTP/UDP/IP overhead amounts to 40 bytes (this reduces to 2 bytes with compression, offering significant bandwidth saving, while the MLPPP header is 6 bytes. The resulting voice packet size is then 66 bytes with RTP or 28 bytes with compressed RTP (cRTP). Table 8.1 shows the voice payload and packet sizes for the different speech codecs.

An important feature of the above codecs is the voice activity detection (VAD) scheme. When voice is packetised both speech and silence packets are packetised. Using VAD, packets of silence can be suppressed to allow data traffic to interleave with packetised voice traffic to allow for more efficient utilisation of the finite network bandwidth. It is estimated that VAD can save a further 30–40% of bandwidth.

VoIP is a real-time service, i.e. data representing the actual conversation must be processed as it is created. This processing affects the ability to carry out conversation over

Table 8.1 Parameters for G.711, G.729, G.723.1 and G.726 codecs

Codec	Bit rate (kbit/s)	Frame size (ms)	Voice payload (bytes)	Voice packet (bytes)	
				With cRTP	Without cRTP
G.711	64	10	160	168	208
G.729 Annex A	8	10	20	28	66
G.723.1 (MP-MLQ)	5.3	30	20	28	66
G.723.1 (CS-ACELP)	6.4	30	24	32	70
G.726	32	5	80	88	146

Table 8.2 Network delay specifications for voice applications
(ITU-T, G114)

Range (ms)	Description
0–150	Acceptable for most services and applications by users
150–400	Acceptable provided that administrators are aware of the transmission time and its impact on the transmission quality of user applications
>400	Unacceptable for general network planning purposes, however, only some exceptional cases exceed this limit

the communications channel (in this case the Internet). Excessive delays will mean that this ability is severely restricted. Variations in this delay (jitter) can possibly insert pauses or even break up words making the voice communication unintelligible. This is why most packetised voice applications use UDP to avoid recovering any packet loss or error.

The ITU-T considers network delay for voice applications in Recommendation G.114. This recommendation defines three bands of one-way delay as shown in Table 8.2.

8.4.6 On-off model for voice traffic

It has been widely accepted that modelling packet voice can be conveniently based on mimicking the characteristics of conversation – the alternating active and silent periods. A two-phase on-off process can represent a single packetised voice source. Measurements indicate that the average active interval is 0.352 s in length while the average silent interval is 0.650 s. An important characteristic of a voice source to capture is the distribution of these intervals. A reasonable good approximation for the distribution of the active interval is an exponential distribution; however, this distribution does not represent the silent interval well. Nevertheless, it often assumes that both these intervals are exponentially distributed when modelling voice sources. The duration of voice calls (call holding time) and inter-arrival time between the calls can be characterised using telephony traffic models.

During the active (on) interval, voice generates fixed size packets with a fixed inter-packet spacing. This is the nature of voice encoders with fixed bit rate and fixed packetisation delay. This packet generation process follows a Poisson process with exponentially distributed inter-arrival times of mean T second or packet per second (pps) $1/T$. As mentioned above, both the on and off intervals are exponentially distributed, giving rise to a two-state MMPP model. No packets are generated during the silent (off) interval. Figure 8.4 represents a single voice source.

The mean on period is $1/\alpha$ while the mean off period is $1/\lambda$. The mean packet inter-arrival time is T s. A superposition of N such voice sources results in the following N-state birth–death model, Figure 8.5, where a state represents the number of sources in the on state.

This approach can model different voice codecs, with varying mean opinion score (MOS). MOS is a system of grading the voice quality of telephone connections. A wide range of listeners

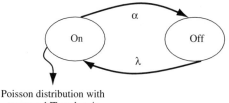

Figure 8.4 A single voice source, represented by a two-state MMPP

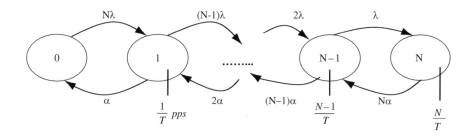

Figure 8.5 Superposition of N voice sources with exponentially distributed inter-arrivals

judges the quality of a voice sample on a scale of one (bad) to five (excellent). The scores are averaged to provide the MOS for the codec. The respective scores are 4.1 (G.711), 3.92 (G.729) and 3.8 (G.726). The parameters for this model are given in Table 8.2 with the additional parameter representing packet inter-arrival time calculated using the following formula:

$$Inter\_arrival\_time = \frac{1}{average\_traffic\_sent\_(pps)} \qquad (8.7)$$

where

$$average\_traffic\_sent = \frac{codec\_bit\_rate}{payload\_size\_(bits)} \qquad (8.8)$$

The mean off interval is typically 650 ms while the mean on interval is 350 ms.

8.4.7 Video traffic modelling

An emerging service of future multi-service networks is packet video communication. Packet video communication refers to the transmission of digitised and packetised video signals in real time. The recent development in video compression standards, such as ITU-T H.261, ITU-T H.263, ISO MPEG-1, MPEG-2 and MPEG-4 [ISO99], has made it feasible to transport video over computer communication networks. Video images are represented by a series of frames in which the motion of the scene is reflected in small changes in sequentially displayed frames. Frames are displayed at the terminal at some constant rate (e.g. 30 frames/s) enabling the human eye to integrate the differences within the frame into a moving scene.

In terms of the amount of bandwidth consumed, video streaming is high on the list. Uncompressed, a one-second worth of video footage with a 300×200 pixels resolution at a playback rate of 30 frames per second would require 1.8 byte/s. Apart from the high throughput requirements, video applications also put a stringent requirement in terms of loss and delay.

There are several factors affecting the nature of video traffic. Among these are compression techniques, coding time (on- or off-line), adaptiveness of the video application, supported level of interactivity and the target quality (constant or variable). The output bit rate of the video encoder can either be controlled to produce a constant bit-rate stream which can significantly vary the quality of the video (CBR encoding), or left uncontrolled to produce a more variable bit-rate stream for a more fixed quality video (VBR encoding). Variable bit-rate encoded video is expected to become a significant source of network traffic because of its advantages in statistical multiplexing gains and consistent video quality.

Statistical properties of a video stream are quite different from that of voice or data. An important property of video is the correlation structure between successive frames. Depending on the type of video codecs, video images exhibit the following correlation components:

- Line correlation is defined as the level of correlation between data at one part of the image with data at the same part of the next line; also called spatial correlation.
- Frame correlation is defined as the level of correlation between data at one part of the image with data at the same part of the next image; also called temporal correlation.
- Scene correlation is defined as the level of correlation between sequences of scenes.

Because of this correlation structure, it is no longer sufficient to capture the burst of video sources. Several other measurements are required to characterise video sources as accurately as possible. These measurements include:

- Autocorrelation function: measures the temporal variations.
- Coefficient of variation: measures the multiplexing characteristics when variable rate signals are statistically multiplexed.
- Bit-rate distribution: indicates together with the average bit rate and the variance, an approximate requirement for the capacity.

As mentioned previously, VBR encoded video source is expected to be the dominant video traffic source in the Internet. There are several statistical VBR source models. The models are grouped into four categories – auto-regressive (AR)/Markov-based models, transform expand sample (TES), self-similar and analytical/IID. These models were developed based on several attributes of the actual video source. For instance, a video conferencing session, which is based on the H.261 standards, would have very little scene changes and it is recommended to use the dynamic AR (DAR) model. To model numerous scene changes (as in MPEG-coded movie sequences), Markov-based models or self-similar models can be used. The choice of which one to use is based on the number of parameters needed by the model and the computational complexity involved. Self-similar models only require a single parameter (Hurst or H parameter) but their computational complexity in generating samples is high (because each sample is calculated from all previous samples). Markov chain models on the other hand, require many parameters (in the form of transitional probabilities to model the scene changes), which again increase the computational complexity because it requires many calculations to generate a sample.

8.4.8 Multi-layer modelling for internet WWW traffic

The Internet operations consist of a chain of interactions between the users, applications, protocols and the network. This structured mechanism can be attributed to the layered architecture employed in the Internet – a layering methodology was used in designing the Internet protocol stack. Hence, it is only natural to try to model Internet traffic by taking into account the different effects each layer of the protocol stack has on the resulting traffic.

The multi-layer modelling approach attempts to replicate the packet generation mechanism as activated by the human users of the Internet and the Internet applications themselves. In a multi-layer approach, packets are generated in a hierarchical process. It starts with a human user arriving at a terminal and starting one or more Internet applications. This action of invoking an application will start the chain of a succession of interactions between the application and the underlying protocols on the source terminal and the corresponding protocols and application on the destination terminal, culminating in the generation of packets to be transported over the network. These interactions can generally be seen as 'sessions'; the definition of a session is dependent on the application generating it, as we will see later when applying this method in modelling the WWW application. An application generates at least one, but usually more, sessions. Each session comprises one or more 'flows'; each flow in turn comprises packets. Therefore, there are three layers or levels encountered in this multi-layer modelling approach – session, flow and packet levels.

Take a scenario where a user arrives at a terminal and starts a WWW application by launching a web browser. The user then clicks on a web link (or types in the web address) to access the web sites of interest. This action generates what we call HTTP sessions. The session is defined as the downloading of web pages from the same web server over a limited period; this does not discount the fact that other definitions of a session are also possible. The sessions in turn generate flows. Each flow is a succession of packets carrying the information pertaining to a particular web page and packets are generated within flows. This hierarchical process is depicted in Figure 8.6.

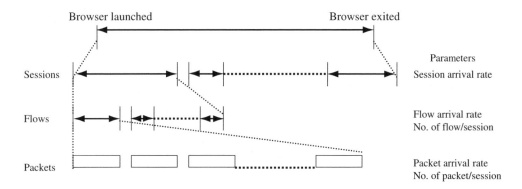

Figure 8.6 Multi-layer modelling

Depicted in the diagram are the suggested parameters for this model. More complex models attempting to capture the self-similarity of web traffic might include the use of heavy-tailed distributions to model any of the said parameters. Additional parameters such as user think time and packet sizes are also modelled by heavy-tailed distributions. While this type of model might be more accurate in capturing the characteristics of web traffic, it comes with the added parameters and complexity.

8.5 Traffic engineering

A dilemma emerges for carriers and network operators: the cost to upgrade the infrastructure as it is nowadays for fixed and mobile telephone networks, is too high to be supported by revenues coming from Internet services. Actually, revenues coming from voice-based services are quite high with respect to the ones derived by current Internet services. Therefore, to obtain cost effectiveness it is necessary to design networks that make an effective use of bandwidth or, in a broader sense, of network resources.

Traffic engineering (TE) is a solution that enables the fulfilment of all those requirements, since it allows network resources to be used when necessary, where necessary and for the desired amount of time. TE can be regarded as the ability of the network to control traffic flows dynamically in order to prevent congestion, to optimise the availability of resources, to choose routes for traffic flows while taking into account traffic loads and network state, to move traffic flows towards less congested paths, to react to traffic changes or failures timely.

The Internet has seen such a tremendous growth in the past few years. This growth has correspondingly increased the requirements for network reliability, efficiency and service quality. In order for the Internet service providers to meet these requirements, they need to examine every aspect of their operational environment critically, assessing the opportunities to scale their networks and optimise performance. However, this is not a trivial task. The main problem is with the simple building block on which the Internet was built – namely IP routing based on the destination address and simple metrics like hop count or link cost. While this simplicity allows IP routing to scale to very large networks, it does not always make good use of network resources. Traffic engineering (TE) has thus emerged as a major consideration in the design and operation of large public Internet backbone networks. While its beginnings can be traced back to the development of the public switched telephone networks (PSTN), TE is fast finding a more crucial role to play in the design and operation of the Internet.

8.5.1 Traffic engineering principles

Traffic engineering is 'concerned with the performance optimisation of networks'. It seeks to address the problem of efficient allocation of network resources to meet user constraints and to maximise service provider benefit. The main goal of TE is to balance service and cost. The most important task is to calculate the right amount of resources; too much and the cost will be excessive, too little will result in loss of business or lower productivity. As this service/cost balance is sensitive to the changes in business conditions, TE is thus a continuous process to maintain an optimum balance.

TE is a framework of processes whereby a network's response to traffic demand (in terms of user constraints such as delay, throughput and reliability) and other stimuli such as failure can be efficiently controlled. Its main objective is to ensure the network is able to support as much traffic as possible at their required level of quality and to do so by optimally utilising its (the network's) shared resources while minimising the costs associated with providing the service. To do this requires efficient control and management of the traffic. This framework encompasses:

- traffic management through control of routing functions and QoS management;
- capacity management through network control;
- network planning.

Traffic management ensures that network performance is maximised under all conditions including load shifts and failures (both node and link failures). Capacity management ensures that the network is designed and provisioned to meet performance objectives for network demands at minimum cost. Network planning ensures that the node and transport capacity is planned and deployed in advance of forecasted traffic growth. These functions form an interacting feedback loop around the network as shown in Figure 8.7.

The network (or system) shown in the figure is driven by a noisy traffic load (or signal) comprising predictable average demand components added to unknown forecast errors and load variation components. The load variation components have different time constants ranging from instantaneous variations, hour-to-hour variations, day-to-day variations and week-to-week or seasonal variations. Accordingly, the time constants of the feedback controls are matched to the load variations and function to regulate the service provided by the network through routing and capacity adjustments. Routing control typically applies on minutes, days or possibly real-time time scales while capacity and topology changes are much longer term (months to a year).

Advancement in optical switching and transmission systems enables ever-increasing amounts of available bandwidth. The effect is that the marginal cost (i.e. the cost associated with producing one additional unit of output) of bandwidth is rapidly being reduced: bandwidth is getting cheaper. The widespread deployment of such technologies is accelerating and network providers are now able to sell high-bandwidth transnational and international

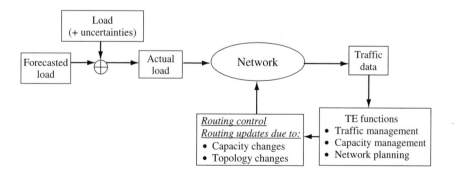

Figure 8.7 The traffic engineering process model

connectivity simply by overprovisioning their networks. Logically, it would seem that in the face of such developments and the abundance of available bandwidth, the need for TE would be invalidated. On the contrary, TE still maintains its importance due principally to the fact that both the number of users and their expectations are exponentially increasing in parallel to the exponential increase in available bandwidth. A corollary of Moore's law says, 'As you increase the capacity of any system to accommodate user demand, user demand will increase to consume system capacity'. Companies that have invested in such overprovisioned networks will want to recoup their investments. Service differentiation charging and usage-proportional pricing are mechanisms widely accepted for doing so. To implement these mechanisms, simple and cost-effective mechanisms for monitoring usage and ensuring customers are receiving what they are requesting are required to make usage-proportional pricing practical. Another important function of TE is to map traffic onto the physical infrastructure to utilise resources optimally and to achieve good network performance. Hence, TE still performs a useful function for both network operators and customers.

8.5.2 Internet traffic engineering

Internet TE is defined as that aspect of Internet network engineering dealing with the issue of performance evaluation and performance optimisation of operational IP networks. Internet TE encompasses the application of technology and scientific principles to the measurement, characterisation, modelling and control of Internet traffic. One of the main goals of Internet TE is to enhance the performance of an operational network, both in terms of traffic-handling capability and resource utilisation. Traffic-handling capability implies that IP traffic is transported through the network in the most efficient, reliable and expeditious manner possible. Network resources should be utilised efficiently and optimally while meeting the performance objectives (delay, delay variation, packet loss and throughput) of the traffic.

There are several functions contributing directly to this goal. One of them is the control and optimisation of the routing function, to steer traffic through the network in the most effective way. Another important function is to facilitate reliable network operations. Mechanisms should be provided that enhance network integrity and by embracing policies emphasising network survivability. This results in a minimisation of the vulnerability of the network to service outages arising from errors, faults and failures occurring within the infrastructure.

Effective TE is difficult to achieve in public IP networks due to the limited functional capabilities of conventional IP technologies. One of the major problems lies in mapping traffic flows onto the physical topology. In the Internet, mapping of flows onto a physical topology was heavily influenced by the routing protocols used. Traffic flows simply followed the shortest path calculated by interior gateway protocols (IGP) used within autonomous systems (AS) such as open shortest path first (OSPF) or intermediate system – intermediate system (IS-IS) and exterior gateway protocols (EGP) used to interconnect ASs such as border gateway protocol 4 (BGP-4). These protocols are topology-driven and employ per-packet control. Each router makes independent routing decisions based on the information in the packet headers. By matching this information to a corresponding entry of a local instantiation of a synchronised routing area link state database, the next hop or route for the packet is then determined. This determination is based on shortest path computations (often equated to lowest cost) using simple additive link metrics.

While this approach is highly distributed and scalable, there is a major flaw – it does not consider the characteristics of the offered traffic and network capacity constraints when determining the routes. The routing algorithm tends to route traffic onto the same links and interfaces, significantly contributing to congestion and unbalanced networks. This results in parts of the network becoming over-utilised while other resources along alternate paths remain under-utilised. This condition is commonly referred to as hyper aggregation. While it is possible to adjust the value of the metrics used in calculating the IGP routes, it soon became too complicated as the Internet core grows. Continuously adjusting the metrics also adds instability to the network. Hence, congested parts are often resolved by adding more bandwidth (overprovisioning), which is not treating the actual symptom of the problem in the first place resulting in poor resource allocation or traffic mapping.

The requirements for Internet TE is not that much different than that of telephony networks – to have a precise control over the routing function in order to achieve specific performance objectives both in terms of traffic-related performance and resource-related performance (resource optimisation). However, the environment in which Internet TE is applied is much more challenging due to the nature of the traffic and the operating environment of the Internet itself. Traffic on the Internet is becoming more multi-class (compared to fixed 64 kbit/s voice in telephony networks) with different service requirements but contending for the same network resources. In this environment, TE needs to establish resource-sharing parameters to give preferential treatment to some service classes in accordance with a utility model. The characteristics of the traffic are also proving to be a challenge – it exhibits very dynamic behaviour, which is still to be understood and tends to be highly asymmetric. The operating environment of the Internet is also an issue. Resources are augmented constantly and they fail on a regular basis. Routing of traffic, especially when traversing autonomous system boundaries, makes it difficult to correlate network topology with the traffic flow. This makes it difficult to estimate the traffic matrix, the basic dataset needed for TE.

An initial attempt at circumventing some of the limitations of IP with respect to TE was the introduction of a secondary technology with virtual circuits and traffic management capabilities (such as ATM) into the IP infrastructure. This is the overlay approach that it consists of ATM switches at the core of the network surrounded by IP routers at the edges. The routers are logically interconnected using ATM PVC, usually in a fully meshed configuration. This approach allows virtual topologies to be defined and superimposed onto the physical network topology. By collecting statistics on the PVC, a rudimentary traffic matrix can be built. Overloaded links can be relieved by redirecting traffic to under-utilised links.

ATM was used mainly because of its superior switching performance compared to IP routing at that time (there are currently IP routers that are as fast if not faster than an ATM switch). ATM also afforded QoS and TE capabilities. However, there are fundamental drawbacks to this approach. Firstly, two networks of dissimilar technologies need to be built and managed, adding to the increased complexity of network architecture and design. Reliability concerns also increase because the number of network elements existing in a routed path increases. Scalability is another issue especially in a fully meshed configuration whereby the addition of another edge router would increase the number of PVC required by $(n(n-1))/2$, where n is the number of nodes (the 'n-squared' problem). There is also the possibility of IP routing instability caused by multiple PVC failures following single link impairment in the ATM core. Concerning ATM itself, segmentation and reassembly (SAR) is difficult to perform at high speeds. SAR is required because of the difference in packet formats between IP and ATM – ATM is cell-based with a fixed size of 53 bytes. IP packets would need to be segmented into ATM cells at the ingress of an ATM network. At the egress, the cells would need to be reassembled into packets. Because of cell interleave, SAR must perform queuing and scheduling for a large number of VCs. Implementing this at STM-32 (10 Gbit/s) or higher speed is a very difficult task. Finally, the well-known problem of ATM cell tax – the overhead penalty with the use of ATM, which is approximately 20% of the link bandwidth (e.g. 498 Mbit/s is wasted on ATM cell overhead on an STM-16 or 2.4 Gbit/s link,). Hence, there is a need to move away from the overlay model to a more integrated solution. This was one of the motivations for the development of MPLS.

8.6 Multi-protocol label switching (MPLS)

To improve on the best-effort service provided by the IP network layer protocol, new mechanisms such as differentiated services (Diffserv) and integrated services (Intserv), have been developed to support QoS. In the Diffserv architecture, services are given different priorities and resource allocations to support various types of QoS. In the Intserv architecture, resources have to be reserved for individual services. However, resource reservation for individual services does not scale well in large networks, since a large number of services have to be supported, each maintaining its own state information in the network's routers. Flow-based techniques such as multi-protocol label switching (MPLS) have also been developed to combine layer 2 and layer 3 functions to support QoS requirements.

MPLS introduces a new connection-oriented paradigm, based on fixed-length labels. This fixed-length label-switching concept is similar but not the same as that utilised by ATM. Among the key motivation for its development was to provide a mechanism for the seamless integration of IP and ATM. As discussed in the previous chapter, the occurrence of IP and ATM co-existence is something that is unavoidable in the pursuit for end-to-end QoS guarantees. However, the architectural differences between the two technologies prove to be a stumbling block for their smooth interoperation. Overlay models have been proposed as solutions but they do not provide the single operating paradigm, which would simplify network management and improve operational efficiency. MPLS is a peer model technology. Compared to the overlay model, a peer model integrates layer 2 switching with layer 3 routing, yielding a single network infrastructure. Network nodes would typically have integrated routing and switching functions. This model also allows IP routing protocols to set up ATM connections and do not require address resolution protocols. While MPLS has successfully merged the benefits of both IP and ATM, another application area in which MPLS is fast establishing its usefulness is traffic engineering (TE). This also addresses other major network evolution problems – throughput and scalability.

8.6.1 MPLS forwarding paradigm

MPLS is a technology that combines layer 2 switching technologies with layer 3 routing technologies. The primary objective of this new technology is to create a flexible networking fabric that provides increased performance and scalability. This includes TE capabilities. MPLS is designed to work with a variety of transport mechanisms; however, initial deployment will focus on leveraging ATM and frame relay, which are already deployed in large-scale providers' networks.

MPLS was initially designed in response to various inter-related problems with the current IP infrastructure. These problems include scalability of IP networks to meet growing demands, enabling differentiated levels of IP services to be provisioned, merging disparate traffic types into a single network and improving operational efficiency in the face of tough competition. Network equipment manufacturers were among the first to recognise these problems and worked individually on their own proprietary solutions including tag switching, IP switching, aggregate route-based IP switching (ARIS) and cell switch router (CSR). MPLS draws on these implementations in an effort to produce a widely applicable standard.

Because the concepts of forwarding, switching and routing are fundamental in MPLS, a concise definition of each one of them is given below:

- Forwarding is the process of receiving a packet on an input port and sending it out on an output port.
- Switching is forwarding process following the choosen path based information or knowledge of current network resources and loading conditions. Switching operates on layer 2 header information.
- Routing is the process of setting routes to understand the next hop a packet should take towards its destination within and between networks. It operates on layer 3 header information.

The conventional IP forwarding mechanism (layer 3 routing) is based on the source–destination address pair gleaned from a packet's header as the packet enters an IP network via a router. The router analyses this information and runs a routing algorithm. The router will then choose the next hop for the packet based on the results of the algorithm calculations (which are usually based on the shortest path to the next router). More importantly, this full packet header analysis must be performed on a hop-by-hop basis, i.e. at each router traversed by the packet. Clearly, the IP packet forwarding paradigm is closely coupled to the processor-intensive routing procedure.

While the efficiency and simplicity of IP routing is widely acknowledged, there are a number of issues brought about by large routed networks. One of the main issues is the use of software components to realise the routing function. This adds latency to the packet. Higher speed, hardware-based routers are being designed and deployed, but these come at a cost, which could easily escalate for large service providers' or enterprise networks. There is also difficulty in predicting the performance of a large meshed network based on traditional routing concepts.

Layer 2 switching technologies such as ATM and frame relay utilise a different forwarding mechanism, which is essentially based on a label-swapping algorithm. This is a much simpler mechanism and can readily be implemented in hardware, making this approach much faster and yielding a better price/performance advantage when compared to IP routing. ATM is also a connection-oriented technology, between any two points, traffic flows along a predetermined path are established prior to the traffic being submitted to the network. Connection-oriented technology makes a network more predictable and manageable.

8.6.2 MPLS basic operation

MPLS tries to solve the problem of integrating the best features of layer 2 switching and layer 3 routing by defining a new operating methodology for the network. MPLS separates packet forwarding from routing, i.e. separating the data-forwarding plane from the control plane. While the control plane still relies heavily on the underlying IP infrastructure to disseminate routing updates, MPLS effectively creates a tunnel underneath the control plane using packet tags called labels. The concept of a tunnel is the key because it means the forwarding process is no more IP-based and classification at the entry point of an MPLS network is not relegated to IP-only information. The functional components of this solution are shown in Figure 8.8, which do not differ much from the traditional IP router architecture.

The key concept of MPLS is to identify and mark IP packets with labels. A label is a short, fixed-length, unstructured identifier that can be used to assist in the forwarding process. Labels are analogous to the VPI/VCI used in an ATM network. Labels are normally local to a single data link, between adjacent routers and have no global significance (as would an IP address). A modified router or switch will then use the label to forward/switch the packets through the network. This modified switch/router termed label switching router (LSR) is a key component within an MPLS network. LSR is capable of understanding and participating in both IP routing and layer 2 switching. By combining these technologies into a single integrated operating environment, MPLS avoids the problem associated with maintaining two distinct operating paradigms.

Label switching utilised in MPLS is based on the so-called MPLS shim header inserted between the layer 2 header and the IP header. The structure of this MPLS shim header is shown in Figure 8.9. Note that there can be several shim headers inserted between the layer 2 and IP headers. This multiple label insertion is called label stacking, allowing MPLS to utilise a network hierarchy, provide virtual private network (VPN) services (via tunnelling) and support multiple protocols [RFC3032].

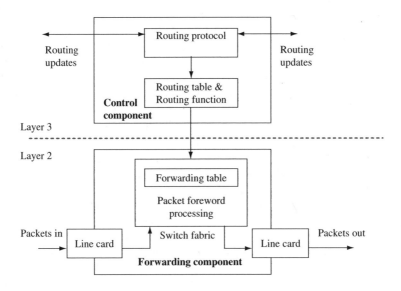

Figure 8.8 Functional components of MPLS

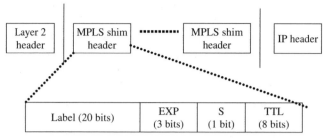

EXP: Experimental functions
S: Level of stack indicator, 1 indicates the bottom of the stack
TTL: Time to live

Figure 8.9   MPLS shim header structure

The MPLS forwarding mechanism differs significantly from conventional hop-by-hop routing. The LSR participates in IP routing to understand the network topology as seen from the layer 3 perspective. This routing knowledge is then applied, together with the results of analysing the IP header, to assign labels to packets entering the network. Viewed on an end-to-end basis, these labels combine to define paths called label switched paths (LSP). LSP are similar to VCs utilised by switching technologies. This similarity is reflected in the benefits afforded in terms of network predictability and manageability. LSP also enable a layer 2 forwarding mechanism (label swapping) to be utilised. As mentioned earlier, label swapping is readily implemented in hardware, allowing it to operate at typically higher speeds than routing. To control the path of LSP effectively, each LSP can be assigned one or more attributes (see Table 8.3). These attributes will be considered in computing the path for the LSP. There are two ways to set up an LSP – control-driven (i.e. hop-by-hop) and explicitly

Table 8.3   LSP attributes

Attribute name	Meaning of attribute
Bandwidth	The minimum requirement on the reservable bandwidth for the LSP to be set up along that path
Path attribute	An attribute that decides whether the path for the LSP should be manually specified or dynamically computed by constraint-based routing
Setup priority	The attribute that decides which LSP will get the resource when multiple LSPs compete for it
Holding priority	The attribute that decides whether an established LSP should be pre-empted by a new LSP
Affinity	An administratively specified property of an LSP to achieve some desired LSP placement
Adaptability	Whether to switch the LSP to a more optimal path when one becomes available
Resilience	The attribute that decides to re-route the LSP when the current path fails

routed LSP (ER-LSP). Since the overhead of manually configuring LSP is very high, there is a need on service providers' behalf to automate the process by using signalling protocols. These signalling protocols distribute labels and establish the LSP forwarding state in the network nodes. A label distribution protocol (LDP) is used to set up a control-driven LSP while RSVP-TE and CR-LDP are the two signalling protocols used for setting up ER-LSP.

The label swapping algorithm is a more efficient form of packet forwarding, compared to the longest address match-forwarding algorithm used in conventional layer 3 routing. The label-swapping algorithm requires packet classification at the point of entry into the network from the ingress label edge router (LER) to assign an initial label to each packet. Labels are bound to forwarding equivalent classes (FEC). An FEC is defined as a group of packets that can be treated in an equivalent manner for purposes of forwarding (share the same requirements for their transport). The definition of FEC can be quite general. FEC can relate to service requirements for a given set of packets or simply on source and destination address prefixes. All packets in such a group get the same treatment en route to the destination. In a conventional packet forwarding mechanism, FEC represent groups of packets with the same destination address; then the FEC should have their respective next hops. However, it is the intermediate nodes processing the FEC grouping and mapping. As opposed to conventional IP forwarding, in MPLS, it is the edge-to-edge router assigning packets to a particular FEC when the packet enters the network. Each LSR then builds a table to specify how to forward packets. This forwarding table, called a label information base (LIB), comprises FEC-to-label bindings.

In the core of the network, LSR ignore the header of network layer packets and simply forward the packet using the label with the label-swapping algorithm. When a labelled packet arrives at a switch, the forwarding component uses the pairing, {input port number/incoming interface, incoming label value}, to perform an exact match search of its forwarding table. When a match is found, the forwarding component retrieves the pairing, {output port number/outgoing interface, outgoing label value}, and the next-hop address from the forwarding table. The forwarding component then replaces the incoming label with the outgoing label and directs the packet to the outbound interface for transmission to the next hop in the LSP. When the labelled packet arrives at the egress LER (point of exit from the network), the forwarding component searches its forwarding table. If the next hop is not a label switch, the egress LSR discards (pop-off) the label and forwards the packet using conventional longest match IP forwarding. Figure 8.10 shows the label swapping process.

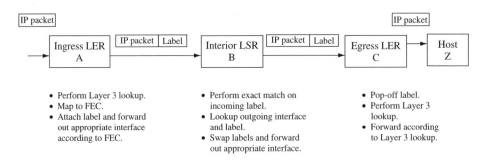

Figure 8.10 Label swapping and forwarding process

LSP can also allow minimising the number of hops, to meet certain bandwidth requirements, to support precise performance requirements, to bypass potential points of congestion, to direct traffic away from the default path, or simply to force traffic across certain links or nodes in the network. Label swapping gives a huge flexibility in the way that it assigns packets to FEC. This is because the label swapping forwarding algorithm is able to take any type of user traffic, associate it with an FEC, and map the FEC to an LSP that has been specifically designed to satisfy the FEC requirements; therefore allowing a high level of control in the network. These are the features, which lend credibility to MPLS to support traffic engineering (TE). We will discuss further the application of MPLS in TE in a later section.

8.6.3 MPLS and Diffserv interworking

The introduction of a QoS enabled protocol into a network supporting various other QoS protocols would undoubtedly lead to the requirement for these protocols to interwork with each other in a seamless fashion. This requirement is essential to the QoS guarantees to the packets traversing the network. It is an important issue of interworking MPLS with Diffserv and ATM.

The combination of MPLS and Diffserv provides a scheme, which is mutually beneficial for both. Path-oriented MPLS can provide Diffserv with a potentially faster and more predictable path protection and restoration capabilities in the face of topology changes, as compared to conventional hop-by-hop routed IP networks. Diffserv, on the other hand, can act as QoS architecture for MPLS. Combined, MPLS and Diffserv can provide the flexibility to provide different treatments to certain QoS classes requiring path protection.

IETF3270 specifies a solution for supporting Diffserv behaviour aggregates (BA) and their corresponding per hop behaviours (PHB) over an MPLS network. The key issue for supporting Diffserv over MPLS is how to map Diffserv to MPLS. This is because LSR cannot see an IP packet's header and the associated DSCP values, which links the packet to its BA and consequently to its PHB, as PHB determines the scheduling treatment and, in some cases, the drop probability of a packet. LSR only look for labels, read their contents and decide the next hop. For an MPLS domain to handle a Diffserv packet appropriately, the labels must contain some information regarding the treatment on the packet.

The solution to this problem is to map the six-bit DSCP values to the three-bit EXP field of the MPLS shim header. This solution relies on the combined use of two types of LSP:

- A LSP that can transport multiple ordered aggregates, so that the EXP field of the MPLS shim header conveys to the LSR with the PHB applied to the packet (covering both information about the packet's scheduling treatment and its drop precedence). An ordered aggregate (OA) is a set of BAs sharing an ordering constraint. Such an LSP refers to as EXP-Inferred-PSC-LSP (E-LSP), when defining PSC as a PHB scheduling class. The set of one or more PHB applies to the BAs belonging to a given OA. With this method, it can map up to eight DSCPs to a single E-LSP.
- A LSP that can transport only a single ordered aggregate, so that the LSR exclusively infer the packet scheduling treatment exclusively from the packet label value. The packet drop precedence is conveyed in the EXP field of the MPLS shim header or in the encapsulating link layer specific selective drop mechanism, where in such cases the MPLS shim header

is not used (e.g. MPLS over ATM). Such LSP refer to label-only-inferred-PSC-LSP (L-LSP). With this method, an individual L-LSP has a dedicated Diffserv code point.

8.6.4 MPLS and ATM interworking

MPLS and ATM can interwork at network edges to support and bring multiple services into the network core of an MPLS domain. In this instance, ATM connections need to be transparent across the MPLS domain over MPLS LSP. Transparency in this context means that ATM-based services should be carried over the domain unaffected.

There are several requirements that need to be addressed concerning MPLS and ATM interworking. Some of these requirements are:

- The ability to multiplex multiple ATM connections (VPC and/or VCC) into an MPLS LSP.
- Support for the traffic contracts and QoS commitments made to the ATM connections.
- The ability to carry all the AAL types transparently.
- Transport of RM cells and CLP information from the ATM cell header.

Transport of ATM traffic over the MPLS uses only the two-level LSP stack. The two-level stack specifies two types of LSP. A transport LSP (T-LSP) transports traffic between two ATM-MPLS interworking devices located at the boundaries of the ATM-MPLS networks. This traffic can consist of a number of ATM connections, each associated with an ATM service category. The outer label of the stack (known as a transport label) defines a T-LSP, i.e. the S field of the shim header is set to 0 to indicate it is not the bottom of the stack. The second type of LSP is an interworking LSP (I-LSP), nested within the T-LSP (identified by an interworking label), which carries traffic associated with a particular ATM connection, i.e. one I-LSP is used for an ATM connection. I-LSP also provides support for VP/VC switching functions. One T-LSP may carry more than one I-LSP. Because an ATM connection is bi-directional while an LSP is unidirectional, two different I-LSPs, one for each direction of the ATM connection, are required to support a single ATM connection. Figure 8.11 shows the relationship between T-LSP, I-LSP and ATM connections. The interworking unit (IWU) encapsulates ATM cells in the ATM-to-MPLS direction, into a MPLS frame. For the MPLS-to-ATM direction, the IWU reconstructs the ATM cells.

With regarding to support of ATM traffic contracts and QoS commitments to ATM connections, the mapping of ATM connections to I-LSP and subsequently to T-LSP must take into consideration the TE properties of the LSP. There are two methods to implement this.

Firstly, a single T-LSP can multiplex all the I-LSP associated to several ATM connections with different service categories. This type of LSP is termed class multiplexed LSP. It groups the ATM service categories into groups and maps each group into a single LSP. As an example for the second scenario, it groups the categories initially into real-time traffic (CBR and rt-VBR) and non-real-time traffic (nrt-VBR, ABR, UBR). It transports the real-time traffic over one T-LSP while the non-real-time traffic over another T-LSP. It can implement class multiplexed LSP by using either L-LSP or E-LSP. Class multiplexed L-LSP must meet the most stringent QoS requirements of the ATM connections transported by the LSP. This is because L-LSP treats every packet going through it the same. Class multiplexed E-LSP, on

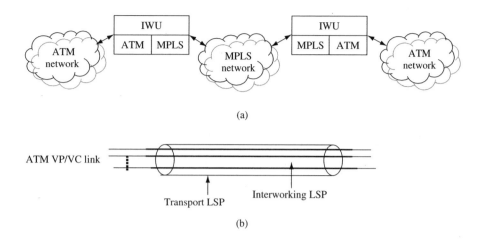

(a)

(b)

Figure 8.11 ATM-MPLS networks interworking. (a) ATM-MPLS network interworking architecure. (b) the relationship between transport LSP, interworking LSP and ATM link

the other hand, identifies the scheduling and dropping treatments applied to a packet based on the value of the EXP field inside the T-LSP label. Each LSR can then apply different scheduling treatments for each packet transported over the LSP. This method also requires a mapping between ATM service categories and the EXP bits.

Secondly, an individual T-LSP is allocated to each ATM service class. This LSP is termed class based LSP. There can be more than one connection per ATM service class. In this case, the MPLS domain would search for a path that meets the requirement of one of the connections.

8.6.5 MPLS with traffic engineering (MPLS-TE)

An MPLS domain still requires IGP such as OSPF and IS-IS to calculate routes through the domain. Once it has computed a route, it uses signalling protocols to establish LSP along the route. Traffic that satisfies a given FEC associated with a particular LSP is then sent down the LSP.

The basic problem addressed by TE is the mapping of traffic onto routes to achieve the performance objectives of the traffic while optimising the resources at the same time. Conventional IGP such as open shortest path first (OSPF), makes use of pure destination address-based forwarding. It selects routes based on simply the least cost metric (or shortest path). Traffic from different routers therefore converge on this particular path, leaving the other paths under-utilised. If the selected path becomes congested, there is no procedure to off-load some of the traffic onto the alternative path.

For TE purposes, the LSR should build a TE database within the MPLS domain. This database holds additional information regarding the state of a particular link. Additional link attributes may include maximum link bandwidth, maximum reserverable bandwidth, current bandwidth utilisation, current bandwidth reservation and link affinity or colour (an administratively specified property of the link). These additional attributes are carried by TE extensions of existing IGP – OSPF-TE and IS-IS TE. This enhanced database will then be used by the signalling protocols to establish ER-LSP.

The IETF has specified LDP as the signalling protocol for setting up LSP. LDP is usually used for hop-by-hop LSP set up, whereby each LSR determines the next interface to route the LSP based on its layer 3 routing topology database. This means that hop-by-hop LSP follow the path that normal layer 3 routed packets will take. There are two signalling protocols: RSVP-TE (RSVP with TE extension) and CR-LDP (constraint-based routing LDP) control the LSP for TE applications. These protocols are used to establish traffic-engineered ER-LSP. An explicit route specifies all the routers across the network with a precise sequence of steps from ingress to egress. Packets must follow this route strictly. Explicit routing is useful to force an LSP down a path that is different from the one offered by the routing protocol. Explicit routing can also be used to distribute traffic in a busy network, to route around failed or congestion hot spots, or to provide pre-allocated back-up LSP to protect against network failures.

8.7 Internet protocol version 6 (IPv6)

Recently, there has been increasing interest in research, development and deployment in IPv6. The protocol itself it very easy to understands. Like any new protocols and networks, it faces a great challenge in compatibility with the existing operational networks, balancing economic cost and benefit of the evolution towards IPv6, and smooth change over from IPv4 to IPv6. It is also a great leap. However, most of these are out of the scope of this book. Here we only discuss the basics of IPv6 and issues on IPv6 networking over satellites.

8.7.1 Basics of internet protocol version 6 (IPv6)

The IP version 6 (IPv6), which the IETF have developed as a replacement for the current IPv4 protocol, incorporates support for a flow label within the packet header, which the network can use to identify flows, much as VPI/VCI are used to identify streams of ATM cells. RSVP helps to associate with each flow a flow specification (flow spec) that characterises the traffic parameters of the flow, much as the ATM traffic contract is associated with an ATM connection.

IPv6 can support integrated services with QoS with such mechanisms and the definition of protocols like RSVP. It extends the IPv4 protocol to address the problems of the current Internet to:

- support more host addresses;
- reduce the size of the routing table;
- simplify the protocol to allow routers to process packets faster;
- have better security (authentication and privacy);
- provide QoS to different types of services including real-time data;
- aid multicasting (allow scopes);
- allow mobility (roam without changing address);
- allow the protocol to evolve;
- permit coexisting of old and new protocols.

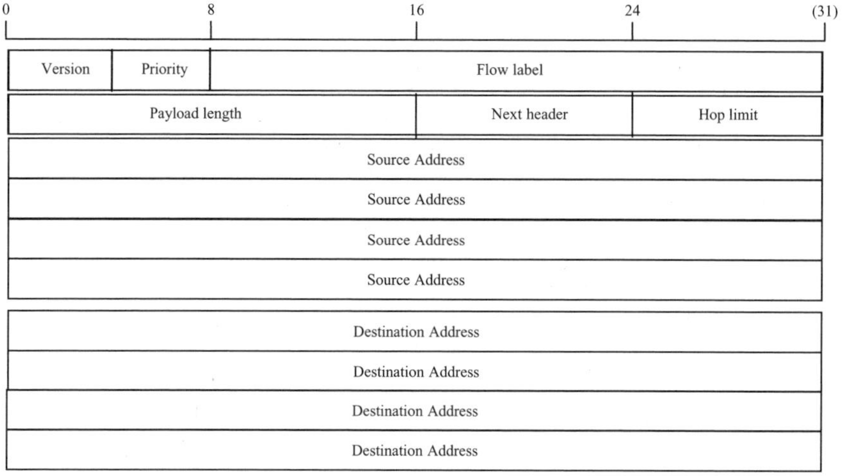

Figure 8.12   IPv6 packet header format

Compared to IPv4, IPv6 has made significant changes to the IPv4 packet format in order to achieve the objectives of the next generation Internet with the network layer functions. Figure 8.12 shows the IPv6 packet header format. The functions of its fields is summarised as the following:

- The version field has the same function as IPv4. It is 6 for IPv6 and 4 for IPv4.
- The priority field identifies packets with different real-time delivery requirements.
- The flow label field is used to allow source and destination to set up a pseudo-connection with particular properties and requirements.
- The payload field is the number of bytes following the 40-byte header, instead of total length in IPv4.
- The next header field tells which transport handler to pass the packet to, like the protocol field in the IPv4.
- The hop limit field is a counter used to limit packet lifetime to prevent the packet staying in the network forever, like the time to live field in IPv4.
- The source and destination addresses indicate the network number and host number, four times larger than IPv4
- There are also extension headers like the options in IPv4. Table 8.4 shows the IPv6 extension header.

Each extension header consists of *next header* field, and fields of *type*, *length* and *value*. In IPv6, the optional features become mandatory features: security, mobility, multicast and transitions. IPv6 tries to achieve an efficient and extensible IP datagram in that:

- the IP header contains less fields that enable efficient routing and performance;
- extensibility of header offers better options;
- the flow label gives efficient processing of IP datagram.

Table 8.4   IPv6 extension headers

Extension header	Description
Hop-by-hop options	Miscellaneous information for routers
Destination options	Additional information for the destination
Routing	Loose list of routers to visit
Fragmentation	Management of datagram fragments
Authentication	Verification of the sender's identity
Encrypted security payload	Information about the encrypted contents

8.7.2 IPv6 addressing

IPv6 has introduced a large addressing space to address the shortage of IPv4 addresses. It uses 128 bits for addresses, four times the 32 bits of the current IPv4 address. It allows about 3.4×10^{38} possible addressable nodes, equivalent to 1030 addresses per person on the planet. Therefore, we should never exhaust IPv6 addresses in the future Internet.

In IPv6, there is no hidden network and host. All hosts can be servers and are reachable from outside. This is called global reachability. It supports end-to-end security, flexible addressing and multiple levels of hierarchy in the address space.

It allows autoconfiguration, link-address encapsulation, plug & play, aggregation, multi-homing and renumbering.

The address format is x:x:x:x:x:x:x:x, where x is a 16-bit hexadecimal field. For example, herewith is an IPv6 address:

2001:FFFF:1234:0000:0000:C1C0:ABCD:8760

It is case sensitive and is different from the following address:

2001:FFFF:1234:0000:0000:**c1c0:abcd**:8760

Leading zeros in a field are optional:

2001:0:1234:0:0:C1C0:ABCD:8760

Successive fields of 0 can be written as ': :'. For example:

2001:0:1234::C1C0:FFCD:8760

We can also rewrite the following addresses:

FF02:0:0:0:0:0:0:1 into FF02::1

0:0:0:0:0:0:0:1 into ::1, and

0:0:0:0:0:0:0:0 into ::

But we can only use ': :' once in an address. An address like this is not valid:

$$2001::1234::C1C0:FFCD:8760$$

IPv6 addresses are also different in a URL. It only allows fully qualified domain names (FQDN). An IPv6 address is enclosed in brackets such as http://[2001:1:4F3A::20F6:AE14]:8080/index.html. Therefore, URL parsers have to be modified, and it could be a barrier for users.

IPv6 address architecture defines different types of address: unicast, multicast and anycast. There are also unspecified and loop back addresses. Unspecified addresses can be used as a placeholder when no address is available, such as in an initial DHCP request and duplicate address detection (DAD). Loop back addresses identify the node itself as the local host using 127.0.0.1 in IPv4 and 0:0:0:0:0:0:0:1 or simply ::1 in IPv6. It can be used for testing IPv6 stack availability, for example, ping6 ::1.

The scope of IPv6 addresses allows link-local and site-local. It allows aggregatable global addresses including multicast and anycast, but there is no broadcast address in IPv6.

The link-local scoped address is new in IPv6: 'scope = local link' (i.e. WLAN, subnet). It can only be used between nodes of the same link, but cannot be routed. It allows autoconfiguration on each interface using a prefix plus interface identifier (based on MAC address) in the format of 'FE80:0:0:0:<interface identifier>'. It gives every node an IPv6 address for start-up communications.

The site-local scoped address has 'scope = site (a network of links)'. It can only be used between nodes of the same site, but cannot be routed outside the site, and is very similar to IPv4 private addresses. There is no default configuration mechanism to assign it. It has the format of 'FEC0:0:0:<subnet id>:<interface id>' where the <subnet id> has 16 bits capable of addressing 64 k subnets. It can be used to number a site before connecting to the Internet or for private addresses (e.g. local printers).

The aggregatable global address is for generic use and allows globally reach. The address is allocated by IANA (Internet assigned number authority) with a hierarchy of tier-1 providers as top-level aggregator (TLA), intermediate providers as next-level aggregator (NLA), and finally sites and subnets at the bottom, as shown in Figure 8.13.

IPv6 support multicast, i.e. one-to-many communications. Multicast is used instead, mostly on local links. The scope of the addresses can be node, link, site, organisation and global. Unlike IPv4, it does not use time to live (TTL). IPv6 multicast addresses have a format of 'FF<flags><scope>::<multicast group>'. Any IPv6 node should recognise the following addresses as identifying itself (see Table 8.5):

- link-local address for each interface;
- assigned (manually or automatically) unicast/anycast addresses;

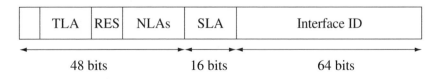

Figure 8.13 Structure of the aggregatable global address

Table 8.5 Some reserved multicast addresses

Address	Scope	Use
FF01::1	Interface-local	All nodes
FF02::1	Link-local	All nodes
FF01::2	Interface-local	All routers
FF02::2	Link-local	All routers
FF05::2	Site-local	All routers
FF02::1:FFXX:XXXX	Link-local	Solicited nodes

- loop back address;
- all-nodes multicast address;
- solicited-node multicast address for each of its assigned unicast and anycast address;
- multicast address of all other groups to which the host belongs.

The anycast address is one-to-nearest, which is great for discovery functions. Anycast addresses are indistinguishable from unicast addresses, as they are allocated from the unicast address space. Some anycast addresses are reserved for specific uses, for example, router-subnet, mobile IPv6 home-agent discovery and DNS discovery. Table 8.6 shows the IPv6 address architecture.

Table 8.6 IPv6 addressing architecture

Prefix	Hex	Size	Allocation
0000 0000	0000-00FF	1/256	Reserved
0000 0001	0100-01FF	1/256	Unassigned
0000 001	0200-03FF	1/128	NSAP
0000 010	0400-05FF	1/128	Unassigned
0000 011	0600-07FF	1/128	Unassigned
0000 1	0800-0FFF	1/32	Unassigned
0001	1000-1FFF	1/16	Unassigned
001	2000-3FFF	1/8	Aggregatable: IANA to registry
010, 011, 100, 101, 110	4000-CFFF	5/8	Unassigned
1110	D000-EFFF	1/16	Unassigned
1111 0	F000-F7FF	1/32	Unassigned
1111 10	F800-FBFF	1/64	Unassigned
1111 110	FC00-FDFF	1/128	Unassigned
1111 1110 0	FE00-FE7F	1/512	Unassigned
1111 1110 10	F800-FEBF	1/1024	Link-local
1111 1110 11	FEC0-FEFF	1/1024	Site-local
1111 1111	FF00-FFFF	1/256	Multicast

When a node has many IPv6 addresses, to select which one to use for the source and destination addresses for a given communication, one should address the following issues:

- scoped addresses are unreachable depending on the destination;
- preferred vs. deprecated addresses;
- IPv4 or IPv6 when DNS returns both;
- IPv4 local scope (169.254/16) and IPv6 global scope;
- IPv6 local scope and IPv4 global scope;
- mobile IP addresses, temporary addresses, scope addresses, etc.

8.7.3 IPv6 networks over satellites

We have learnt through the book to treat the satellite networks as generic networks with different characteristics and IP networks interworking with other different networking technologies. Therefore, all the concepts, principles and techniques can be applied to IPv6 over satellites. Though IP has been designed for internetworking purposes, the implementation and deployment of any new version or new type of protocol always face some problems. These also have potential impacts on all the layers of protocols including trade-offs between processing power, buffer space, bandwidth, complexity, implementation costs and human factors. To be concise, we will only summarise the issues and scenarios on internetworking between IPv4 and IPv6 as the following:

- **Satellite network is IPv6 enabled**: this raises issues on user terminals and terrestrial IP networks. We can imagine that it is not practical to upgrade them all at the same time. Hence, one of the great challenges is how to evolve from current IP networking over satellite towards the next generation network over satellites. Tunnelling from IPv4 to IPv6 or from IPv6 to IPv4 is inevitable, hence generating great overheads. Even if all networks are IPv6 enabled, there is still a bandwidth efficiency problem due to the large overhead of IPv6.
- **Satellite network is IPv4 enabled**: this faces similar problems to the previous scenario, however, satellite networks may be forced to evolve to IPv6 if all terrestrial networks and terminals start to run IPv6. In terrestrial networks when bandwidth is plentiful, we can afford to delay the evolution. In satellite networks, such a strategy may not be practical. Hence, timing, stable IPv6 technologies and evolution strategies all play an important role.

8.7.4 IPv6 transitions

The transition of IPv6 towards next-generation networks is a very important aspect. Many new technologies failed to succeed because of the lack of transition scenarios and tools. IPv6 was designed with transition in mind from the beginning. For end systems, it uses a dual stack approach as show in Figure 8.14; and for network integration, it uses tunnels (some sort of translation from IPv6-only networks to IPv4-only networks).

Figure 8.14 illustrates a node that has both IPv4 and IPv6 stacks and addresses. The IPv6-enabled application requests both IPv4 and IPv6 destination addresses. The DNS resolver returns IPv6, IPv4 or both addresses to the application. IPv6/IPv4 applications choose the address and then can communicate with IPv4 nodes via IPv4 or with IPv6 nodes via IPv6.

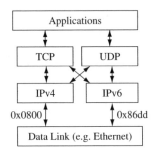

Figure 8.14   Illustration of dual stack host

8.7.5 IPv6 tunnelling through satellite networks

Tunnelling IPv6 in IPv4 is a technique use to encapsulate IPv6 packets into IPv4 packets with protocol field 41 of the IP packet header (see Figure 8.15). Many topologies are possible including router to router, host to router, and host to host. The tunnel endpoints take care of the encapsulation. This process is 'transparent' to the intermediate nodes. Tunnelling is one of the most vital transition mechanisms.

In the tunnelling technique, the tunnel endpoints are explicitly configured and they must be dual stack nodes. If the IPv4 address is the endpoint for the tunnel, it requires reachable IPv4 addresses. Tunnel configuration implies manual configuration of the source and destination IPv4 addresses and the source and destination IPv6 addresses. Tunnel configuration cases can be between two hosts, one host and one router as shown in Figure 8.16, or two routers of two IPv6 networks as shown in Figure 8.17.

8.7.6 The 6to4 translation via satellite networks

The 6to4 translation is a technique used to interconnect isolated IPv6 domains over an IPv4 network with automatic establishment of a tunnel. It avoids the explicit tunnels used in the tunnelling technique by embedding the IPv4 destination address in the IPv6 address. It uses the reserved prefix '2002::/16' (2002::/16 ≡ 6to4). It gives the full 48 bits of the address to a site based on its external IPv4 address. The IPv4 external address is embedded: 2002:<ipv4 ext address>::/48 with the format, '2002:<ipv4add>:<subnet>::/64'. Figures 8.18 and 8.19 show the tunnelling techniques.

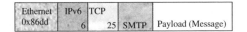

Original IPv6 packet

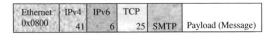

Encapsulated IPv6 packet

Figure 8.15   Encapsulation of IPv6 packet into IPv4 packet

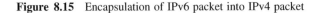

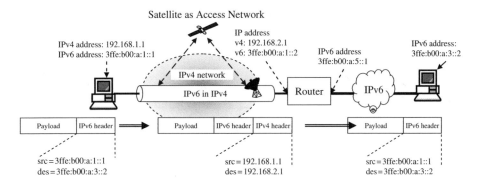

Figure 8.16 Host to router tunnelling through satellite access network

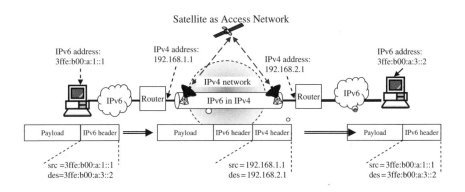

Figure 8.17 Router to router tunnelling through satellite core network

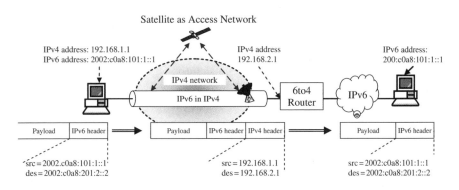

Figure 8.18 The 6to4 translation via satellite access network

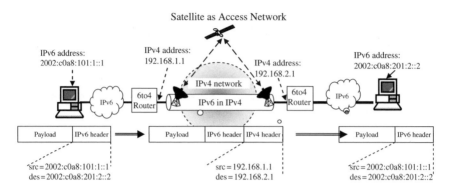

Figure 8.19   The 6to4 translation via satellite core network

To support 6to4, the egress router implementing 6to4 must have a reachable external IPv4 address. It is a dual-stack node. It is often configured using a loop back address. Individual nodes do not need to support 6to4. The prefix 2002 may be received from router advertisements. It does not need to be dual stack.

8.7.7 Issues with 6to4

IPv4 external address space is much smaller than IPv6 address space. If the egress router changes its IPv4 address, then it means that the full IPv6 internal network needs to be renumbered. There is only one entry point available. It is difficult to have multiple network entry points to include redundancy.

Concerning application aspects of IPv6 transitions, there also other problems with IPv6 at the application layer: the support of IPv6 in the operating systems (OS) and applications is unrelated; dual stack does not mean having both IPv4 and IPv6 applications; DNS does not indicate which IP version to be used; and it is difficult to support many versions of applications.

Therefore, the application transitions of different cases can be summarised as the following (also see Figure 8.20):

- For IPv4 applications in a dual-stack node, the first priority is to port applications to IPv6.
- For IPv6 applications in a dual-stack node, use IPv4-mapped IPv6 address ‘::FFFF:x.y.z.w’ to make IPv4 applications work in IPv6 dual stack.
- For IPv4/IPv6 applications in a dual-stack node, it should have a protocol-independent API.
- For IPv4/IPv6 applications in an IPv4-only node, it should be dealt with on a case-by-case basis, depending on applications/OS support.

8.7.8 Future development of satellite networking

It is difficult to predict the future, sometime impossible, but it is not too difficult to predict the trends towards future development if we have enough past and current knowledge. In addition to integrating satellites into the global Internet infrastructure, one of the major tasks is to create new services and applications to meet the needs of people. Figure 8.21 illustrates an abstract vision of future satellite networking.

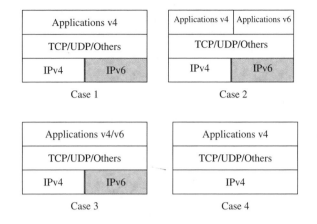

Figure 8.20 IPv6 application transitions

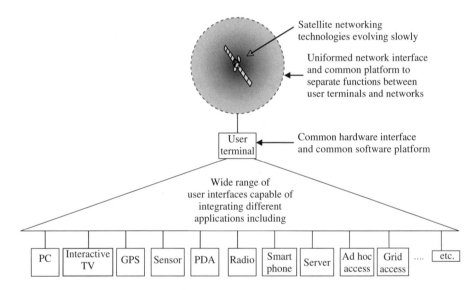

Figure 8.21 An illustration of future development of satellite networking

The main difficulties are due to evolution, integration and convergence:

- It becomes difficult to separate satellite networking concepts from others.
- It will not be easy to tell the differences between protocols and satellite-friendly protocols due to network convergence (see Figure 8.22), except in the physical and link layers.

The trends are due to the following reasons:

- The services and applications will converge to common applications for both satellite networking terminals and terrestrial mobile networking terminals. Even satellite-specific services such as global positioning systems (GPS) have been integrated with the new generation of 2.5G and 3G mobile terminals (see Figures 8.21 and 8.22).

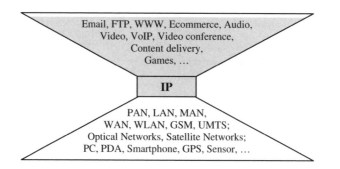

Figure 8.22 Protocol convergence

- The hardware platforms and networking technologies will be well developed, powerful and standardised. This will allow quick and economic development of specialised user terminals.
- We will see significant development in system software, and face the challenge of managing complexity of large software.

In the last 25 years, satellite capacities have increased tremendously due to technology development. The weight of satellites has increased from 50 kg to 3000 kg, and power from 40 W to 1000 W. Weight and power will increase to 10 000 kg and 20 000 W in the near future. Satellite earth terminals have decreased from 20–30 m to 0.5–1.5 m. Handheld terminals have also been introduced. Such trends will continue but perhaps in different ways, such as constellations and clusters of satellites. User terminals can also function as interworking devices to private networks or a hub of sensor networks.

From a satellite networking point of view, we will see end systems such as servers providing information services directly from onboard satellites with multimedia terminals on board satellites to watch and safeguard our planet, and routers on board as network nodes to extend our Internet into space. Satellites are mysterious stars. We create them and know them better than any other stars. The capability of satellite technologies and human creativity will exceed our current imaginations. Thank you for reading through this book and please feel free to contact me should you need any help on teaching satellite networking based on this textbook.

Further reading

[1] Awduche, D., Berger, L., Gan, D., Li, T., Srinivasan, V. and Swallow, G., RSVP-TE: extensions to RSVP for LSP tunnels, IETF RFC 3209, December 2001.

[2] Awduche, D., Chiu, A., Elwalid, A., Widjaja, I. and Xiao. X., Overview and principles of Internet traffic engineering, IETF RFC 3272 (informational), May 2002.

[3] Blake, S., Black, D., Carlson, M., Davies, E., Wang, Z. and Weiss, W., An architecture for differentiated services, IETF RFC 2475 (informational), December 1998.

[4] Braden, R., Clark, D. and Shenker, S., Integrated services in the Internet architecture: an overview, IETF RFC 1633 (informational), June 1994.

[5] Braden, R., Zhang, L., Berson, S., Herzog, S. and Jamin, S., Resource ReSerVation Protocol (RSVP) – Version 1 Functional Specification, IETF RFC 2205 (standard track), September 1997.

[6] Bradner, S. and Mankin, A., The recommendation for the IP next generation protocol, IETF RFC 1752 (standard track), January 1995.

 [7] Davie, B., Lawrence, J., McCloughrie, K., Rosen, E., Swallow, G., Rekhter, Y. and Doolan, P., MPLS using
     LDP and ATM VC switching, IETF RFC 3035 (standard track), January 2001.
 [8] Davie, B., Charny, A., Bennet, J.C.R., Benson, K., Le Boudec, J-Y., Courtney, W., Davari, S., Firoiu, V. and
     Stiliadis, D., An expedited forwarding PHB (per hop behaviour), IETF RFC 3246 (proposed standard), March
     2002.
 [9] Deering, S. and Hinden, R., Internet protocol, version 6 (IPv6) specification, IETF RFC 2460 (standard track),
     December 1998.
[10] Faucheur, F. Le, Davie, B., Davari, S., Vaananen, P., Krishnan, R., Cheval, P. and Heinanen, J., Multi-protocol
     label switching (MPLS) support of differentiated services, IETF RFC 3270 (standard track), May 2002.
[11] Gilligan, R. and Nordmark, E., Transition mechanisms for IPv6 hosts and routers, IETF RFC 2893 (standard
     track), August 2000.
[12] Heinanen, J., Baker, F., Weiss, W. and Wroclawski, J., Assured forwarding PHB group, RFC 2597 (standard
     track), June 1999.
[13] ISO/IEC 11172, Coding of moving pictures and associated audio for digital storage media at up to about
     1.5 Mbit/s, 1993.
[14] ISO/IEC 13818, Generic coding of moving pictures and associated audio information, 1996.
[15] ISO/IEC 14496, Coding of audio-visual objects, 1999.
[16] ITU-T G.723.1, Speech coders: Dual rate speech coder for multimedia communications transmitting at 5.3
     and 6.3 kbit/s, 1996.
[17] ITU-T G.729, Coding of speech at 8 kbit/s using conjugate-structure algebraic-code-excited linear-prediction
     (CS-ACELP), 1996.
[18] ITU-T E.800, Terms and definitions related to quality of service and network performance including depend-
     ability, 1994.
[19] ITU-T H.261, Video codec for audiovisual services at px64 kbit/s, 1993.
[20] ITU-T H.263, Video coding for low bit rate communication, 1998.
[21] Marot, M., Contributions to the study of traffic in networks, PhD thesis, INT and University of Paris VI,
     France, 2001.
[22] Nichols, K., Blake, S., Baker, F. and Black, D., Definition of the differentiated services field (DS field) in the
     IPv4 and IPv6 headers, IETF RFC 2474 (standard track), December 1998.
[23] RFC2375, IPv6 Multicast address assignments, R. Hinden, July 1998.
[24] RFC2529, Transmission of IPv6 over IPv4 domains without explicit tunnels, B. Carpenter, C. Jung, IETF,
     March 1999.
[25] RFC2766, Network address translation – protocol translation (NAT-PT), G. Tsirtsis, P. Srisuresh, IETF,
     February 2000.
[26] RFC2767, Dual stack hosts using the 'bump-in-the-stack' technique (BIS), K. Tsuchiya, H. Higuchi,
     Y. Atarashi, IETF, 2000-02-01,
[27] RFC2893, Transition mechanisms for IPv6 hosts and routers, R. Gilligan, E.Nordmark, 2000-08-01
[28] Rosen, E., Viswanathan, A. and Callon, R., Multiprotocol label switching architecture, IETF RFC 3031
     (standard track), January 2001.
[29] Salleh, M., New generation IP quality of service over broadband networks, PhD thesis, University of Surrey,
     UK, 2004.
[30] Sánchez, A., Contribution to the study of QoS for real-time multimedia services over IP networks, PhD thesis,
     University of Valladolid, Spain, 2003.
[31] Shenker, S., Partridge, C. and Guerin, R., Specification of guaranteed quality of service, IETF RFC 2212
     (standard track), September 1997.

Exercises

1. Understand the concepts of new services and applications in future networks and
   terminals.
2. Understand the basic principles and techniques for traffic modelling and traffic
   characterisation.

Exercises (*continued*)

3. Describe the concepts of traffic engineering in general and Internet traffic engineering in particular.
4. Explain the principles of MPLS, and interworking with different technologies and traffic engineering concepts.
5. Understand IPv6 and its main differences from IPv4.
6. Explain different techniques for IPv6 over satellites such as IPv6 tunnelling thorough satellite networks and 6to4 translation through satellite networks.
7. Discuss the new development of IPv6 over satellites and future development of satellite networking.

Index

Satellite Networking: Principles and Protocols   Zhili Sun
© 2005 John Wiley & Sons, Ltd